Growth of the Gastrointestinal Tract: Gastrointestinal Hormones and Growth Factors

Editors

Jean Morisset, Ph.D.
Professor
Department of Biology
Faculty of Sciences
University of Sherbrooke
Sherbrooke, Quebec, Canada

Travis E. Solomon, M.D., Ph.D.
Professor
Kansas University Medical Center
Kansas City, Kansas
Associate Chief of Staff for Research
Medical Research Service
VA Medical Center
Kansas City, Missouri

CRC Press
Boca Raton Ann Arbor Boston

This book represents information obtained from authentic and highly regarded sources. Reprinted material is quoted with permission, and sources are indicated. A wide variety of references are listed. Every reasonable effort has been made to give reliable data and information, but the author and the publisher cannot assume responsibility for the validity of all materials or for the consequences of their use.

Direct all inquiries to CRC Press, Inc., 2000 Corporate Blvd., N.W., Boca Raton, Florida, 33431.

International Standard Book Number 0-8493-4617-7

Printed in the United States

PREFACE

Cell proliferation and differentiation are key elements in the ability of the gastrointestinal tract to respond to physiologic and disease-induced demands on digestive and absorptive functions. To meet the changing needs for efficient assimilation of nutrients, the gastrointestinal tract has the capability of replacing many of its secretory and absorptive cells and altering the relative proportions of digestive enzymes during postnatal development, pregnancy, lactation, and intake of diets with varying macronutrient composition. Many diseases may also affect these responses or may be due in part to disordered patterns of growth and differentiation. In the past 20 years, much work has been done to increase our understanding of the factors which regulate the growth responses of the gastrointestinal tract. The patterns of cellular proliferation and differentiation in response to many physiologic stimuli have been characterized, and some knowledge about intracellular mechanisms which mediate growth responses in the gastrointestinal tract is beginning to accumulate. Considerable information about the effects of gastrointestinal hormones on growth and differentiation of the many specialized cell types in the gastrointestinal tract is now available.

This book will provide an up-to-date summary of this large body of data for investigators working in the area, as well as for gastroenterologists and basic scientists who are interested in many topics related to gastrointestinal function. For the stomach, small intestine, colon, and pancreas, regulation of growth and differentiation is reviewed by experts in developmental and adult physiology and in pathophysiology of diseases involving each organ. The goal is to provide an in-depth yet readable collection of reviews which will have broad interest to many scientists and physicians.

EDITORS

Jean Morisset, Ph.D., is Professor of Endocrinology in the Department of Biology at the University of Sherbrooke in Quebec, Canada. He is also Director of the Centre de Recherche sur les Mécanismes de Sécrétion of the University of Sherbrooke.

Dr. Morisset received his B.Sc. degree from the University of Sherbrooke in 1965. He obtained his Ph.D. degree in 1968 at the same university. After doing post-doctoral work at the Medical College of Georgia, Augusta, from 1968 to 1970, he was appointed Associate Professor of Biology at the University of Sherbrooke in 1970. He became an Associate Professor in 1974 and Professor in 1979.

Dr. Morisset is a member of the American Physiological Society, American Gastroenterological Association, Canadian Gastroenterological Association, American Pancreatic Association, Society for Experimental Biology and Medicine, and the European Pancreatic Club. He is also an Associate Editor for the *Canadian Journal of Physiology and Pharmacology* and the *International Journal of Pancreatology*.

Dr. Morisset has been the recipient of research grants from the Natural Science and Engineering Research Council and Medical Research Council of Canada, from the Ministere de l'Education du Quebec, and the Canadian Cystic Fibrosis Foundation. His current research interests are in hormonal control of growth and maturation of gastrointestinal tract organs and in control of pancreatic enzyme secretion. Dr. Morisset is the author of more than 100 papers on pancreatic physiology.

Travis E. Solomon, M.D., Ph.D., is Professor of Medicine and Physiology at the Kansas University Medical Center in Kansas City, Kansas and Associate Chief of Staff for Research at the Kansas City VA Medical Center in Kansas City, Missouri.

Dr. Solomon received his Ph.D. and M.D. from the University of Texas Medical School at Houston in 1973 and 1974. After an internship in Medicine at the University of Missouri, he did post-doctoral work at Yale University and UCLA. In 1977 he was appointed Assistant Professor of Medicine at UCLA. He moved to the University of Missouri in 1981 to become Associate Professor of Medicine and Physiology and Associate Chief of Staff for Research at the Truman VA Hospital. He became Vice-Chairperson for Research in the Department of Medicine in 1985 and Professor of Medicine and Physiology in 1986. In 1987 he moved to his present positions.

Dr. Solomon is a member of the American Physiological Society, the American Gastroenterological Association, the American Pancreatic Association, and many other groups. He has served on the editorial board of the *American Journal of Physiology, Pancreas,* the *Scandinavian Journal of Gastroenterology,* and other journals.

Dr. Solomon has received grants from the National Institutes of Health and the Veterans Administration Research Service. His major research interests are regulation of normal and abnormal growth of gastrointestinal tissues, pancreatic secretory function, and the physiology and pathophysiology of gastrointestinal hormones.

CONTRIBUTORS

Raymond Calvert, Ph.D.
Professor
Department of Anatomy and Cell Biology
Faculty of Medicine
University of Sherbrooke
Sherbrooke, Quebec, Canada

Chris N. Conteas
Department of Internal Medicine
Wayne State University School of Medicine
Harper-Grace Hospitals
Detroit, Michigan

J. C. Dagorn, D.Sc.
Director
Unité de Recherches de Physiologie et Pathologie Digestives
INSERM U.315
Marseille, France

H. P. Elsässer, Ph.D.
Research Associate
Department of Cell Biology and Cell Pathology
University of Marburg
Marburg, FRG

D. Giorgi, Ph.D.
Unité de Recherches de Physiologie et Pathologie Digestives
INSERM U.315
Marseilles, France

F. Halter, M.D.
Professor and Head
Gastrointestinal Unit
University Hospital
Bern, Switzerland

H. F. Kern
Professor
Department of Cell Biology and Cell Pathology
University of Marburg
Marburg, FRG

Murray Korc, M.D.
Associate Professor
Section of Endocrinology
Department of Internal Medicine and Biochemistry
The University of Arizona Health Sciences Center
Tucson, Arizona

Emanuel Lebenthal, M.D.
Director
International Institute for Infant Nutrition and Gastrointestinal Disease
Children's Hospital and
Professor of Pediatrics
State University of New York at Buffalo
Buffalo, New York

Ying-kit Leung, M.D.
Assistant Professor of Pediatrics
State University of New York at Buffalo and
Attending Physician
Children's Hospital of Buffalo
Buffalo, New York

Gary M. Levine, M.D.
Head
Division of Gastroenterology and Nutrition
Albert Einstein Medical Center and
Professor of Medicine
Temple University School of Medicine
Philadelphia, Pennsylvania

Gordon D. Luk
Chief
Department of Internal Medicine
Wayne State University School of Medicine
Harper-Grace Hospitals
Detroit, Michigan

Adhip P. N. Majumdar, Ph.D., D.Sc.
Senior Research Biochemist
Research Service
Veterans Administration Medical Center
Allen Park, Michigan and
Department of Medicine and
Department of Nutrition and Food Science
Wayne State University
Detroit, Michigan

Peter P. McCann, Ph.D.
Director
Scientific Administration
Merrell Dow Research Institute and
Professor of Anatomy and Cell Biology
University of Cincinnati College of Medicine
Cincinnati, Ohio

Daniel Ménard, Ph.D.
Professor
Department of Anatomy and Cell Biology
Faculty of Medicine
University of Sherbrooke
Sherbrooke, Quebec, Canada

Jean Morisset, Ph.D.
Professor
Department of Biology
Faculty of Sciences
University of Sherbrooke
Sherbrooke, Quebec, Canada

Jeffrey A. Moshier
Department of Internal Medicine
Wayne State University School of Medicine
Harper-Grace Hospitals
Detroit, Michigan

Ann L. Silverman
Department of Internal Medicine
Wayne State University School of Medicine
Harper-Grace Hospitals
Detroit, Michigan

Pomila Singh, Ph.D.
Professor
Department of Surgery
The University of Texas Medical Branch
Galveston, Texas

Travis E. Solomon, M.D., Ph.D.
Professor
Kansas University Medical Center
Kansas City, Kansas and
Associate Chief of Staff for Research
Medical Research Service
VA Medical Center
Kansas City, Missouri

James C. Thompson, M.D.
Professor and Chairman
Department of Surgery
The University of Texas Medical Branch
Galveston, Texas

Courtney M. Townsend, Jr., M.D.
Professor
Department of Surgery
The University of Texas Medical Branch
Galveston, Texas

G. Willems, M.D., Ph.D.
Professor
Cancer Research Unit
Faculty of Medicine
Vrije Universiteit Brussel
Brussels, Belgium

TABLE OF CONTENTS

General Mechanisms of Cell Proliferation and Differentiation

Chapter 1

HORMONAL REGULATION OF CELL PROLIFERATION AND DIFFERENTIATION: CELLULAR MECHANISMS

Murray Korc

TABLE OF CONTENTS

I. INTRODUCTION

The eukaryotic cell is a very complex structure that can perform a large variety of specialized functions. There are more than 200 different types of eukaryotic cells in vertebrate animals. These range from the pluripotential hematopoietic stem cell to the highly specialized nerve cell. These different types of cells are grouped into integrated units that make up the various tissues and organs. For example, the gastrointestinal tract in humans consists of different types of absorptive and secretory epithelial cells, muscle cells, immune cells, nerve cells, endothelial cells, connective tissue cells, and endocrine cells. Many of these cell types are short lived. In order to function properly, every organism must be able to replace lost cells. This regenerative capacity is highly variable from tissue to tissue, is tightly regulated in most circumstances in order to allow the new cells to assume their proper differentiated functions, but is altogether nonexistent in certain cell types in the adult organism.

The present chapter will delineate some of the biochemical and molecular mechanisms that allow cells to proliferate in response to hormonal stimuli in a manner that leads to coordinately regulated growth and differentiation. It is not possible within the scope of a brief review chapter to cover in full detail the large number of hormones and growth factors that have been implicated in this regulation. Instead, this chapter will focus on recent advances in our understanding of the biomolecular regulation of the cell cycle and cellular proliferation, and some of the complex pathways that are activated by hormones and growth factors. Special emphasis will be placed on describing the mechanisms of action of epidermal growth factor (EGF). Other hormones and growth factors, that also act via receptors and second messengers to activate effector pathways that lead to cell proliferation and differentiation, will be mentioned briefly.

II. THE CELL CYCLE

In the healthy adult human many thousands of cells are undergoing mitosis (nuclear division) and cytokinesis (cytoplasmic division) every second. Cytoplasmic division is followed by a new cell cycle which consists of three successive stages that together are called the interphase. During the first stage of interphase (G_1), the cell exhibits a renewed high rate of protein synthesis. G_1 is followed by the S phase, during which DNA synthesis occurs. When the DNA content of the nucleus has doubled, the cell enters into the last stage of interphase (G_2). At the end of this stage the cell again undergoes mitosis and cytokinesis. Although different cell types divide at different rates, the mitotic process itself lasts only 1 to 2 h. The differences in the cell cycle time between different cell types are due mainly to the duration of the G_1 phase.[1] An elegant example of cells that are located in close proximity to each other and that exist in different stages of the cell cycle can be found in the lining of the small intestine of mammals. This lining consists of a sheet of single epithelial cells that exhibits multiple infoldings which are called crypts.[2] Cell division occurs only within a confined area at the bottom of the crypts.[3] Here, there is a small population of slowly dividing stem cells which give rise to the more rapidly dividing cells. The latter cells migrate upwards, leading to the formation of villi in which cell division does not occur.[4] The cells at the top of the villus are shed after a few days and are replenished by cells that have previously divided and which are coming up the migratory pathway.

The exact biochemical and molecular mechanisms that dictate when and why a eukaryotic cell divides are not known. It is established, however, that cells which are not actively dividing are usually growth arrested in the G_1 phase.[4] When a cell is quiescent, this phase of the cell cycle is also called the G_0 phase,[5] implying that the cell is in a sort of dormant state. Although the distinctions between G_0 and G_1 have not been clearly delineated, once a cell has gone through the late stages of the G_1 phase (the so-called restriction point), it will proceed through the other phases of the cell cycle toward increased DNA synthesis and subsequently cell division.[4,6]

It appears that both negative and positive control mechanisms modulate the ability of eukaryotic cells to traverse their restriction points. These include environmental and nutritional factors, intracellular regulatory pathways, and hormones and peptide growth factors. Examples of environmental factors include the density and ambient temperature in which the cells are growing. Examples of nutritional factors include the availability of amino acids and trace elements. Examples of intracellular pathways include the increases during the S phase of the cell cycle in the activities of many enzymes that modulate DNA synthesis.[7] These cycle-specific enzymes include such important proteins as topoisomerase I, which exhibits a 2- to 4-fold increase in activity prior to cell division,[8] and dihydrofolate reductase[9] and thymidine kinase[10] which exhibit a 10- to 20-fold increase in activity. After completion of the S phase, the activity of these enzymes decreases back to baseline levels.

The molecular basis for the periodic increase in the activity of S-phase specific enzymes is complex. In the case of thymidine kinase, an enzyme that transfers a phosphate group onto thymidine to form its monophosphate derivative, quiescent cells in culture which have been stimulated to proliferate by the addition of serum exhibit parallel increases in thymidine kinase protein and mRNA levels.[11,12] Thus, control of thymidine kinase activity appears to occur at a transcriptional level in these cells. In contrast, in actively cycling cells such as proliferating HeLa cells, the levels of thymidine kinase mRNA exhibit a 3-fold increase during each S phase of the cell cycle, whereas the corresponding protein levels increase 15-fold.[13] This marked increase in protein levels associated with increased thymidine kinase enzyme activity is due to a greater efficiency of translation of the corresponding mRNA.[13] Further, the decrease in enzyme activity that occurs following completion of DNA synthesis is due to the rapid proteolysis of the thymidine kinase polypeptide in association with a marked decrease in the rate of synthesis of the enzyme.[13] Thus, in cells that are actively cycling the increase in activity of thymidine kinase during the S phase appears to be due to post-transcriptional regulation.

Another major cell-cycled, regulated enzyme that acts to transfer a phosphate onto its target substrate is histone H1 kinase.[14] The enzyme is activated in many different kinds of dividing eukaryotic cells during their transition from the G_2 phase into mitosis.[15] The kinase activity of this enzyme preferentially phosphorylates histone H1 in a manner that may allow it to be directly involved in chromatin condensation;[14] it is dependent on magnesium, can utilize either ATP or GTP, and is independent of calcium, calmodulin, diacylglycerol, or cyclic nucleotides.[16] Histone H1 kinase appears to be directly involved in chromatin condensation[17] and initiation of mitosis[18] and may be closely related or identical to the M-phase promoting factor (MPF). The latter enzyme induces several mitosis related events, including the breakdown of the nuclear membrane, organization of the mitotic spindle, and chromosome condensation during mitosis.[14] A major component of MPF is the cdc2 complex, which is activated during the G_2/M transition and which may be responsible for initiating the process of mitosis in human cells.[19] The cdc2 complex is itself a kinase, which in human cells exists as a 34 kilodalton (kDa) protein that is complexed to a 13 kDa and 62 kDa protein.[20-22] This complex is activated in a cell cycle dependent manner as a result of the rearrangement of its subunits and the phosphorylation of the 34 kDa protein, which occurs predominantly on tyrosine residues.[20-22] Tyrosine phosphorylation is a relatively uncommon biological phenomenon that is effected physiologically by a small group of growth promoting polypeptides and that is believed to be important in the regulation of cell growth.[23] The critical role of kinases in general and tyrosine kinases in particular in many signal transduction pathways will be discussed in more detail later in this chapter.

Certain genes which exhibit an oncogenic potential may also be directly involved in the regulation of cell proliferation. The concept of protooncogenes has arisen from studies with transforming retroviruses as well as with DNA transfection experiments. These studies have pointed out the existence of a group of genes that are capable of giving rise to the malignant phenotype and in the absence of certain specific aberrations primarily have important regulatory functions in eukaryotic cells.[24,25] One gene that has been extensively studied in this regard is the c-*myc* protooncogene. In tumor cell lines, overexpression of the protein product of this cellular homologue to the avian myelocytomatosis virus (MC29) oncogene has been shown to occur as a result of gene amplification.[26-28] In contrast, following mitogenic stimulation of quiescent cells such as lymphocytes,[29] hepatocytes,[30] and fibroblasts,[31] there is a marked increase in c-*myc* mRNA levels. The increase in c-*myc* mRNA levels occurs early in the G_0/G_1 phase of the cell cycle and is observed following stimulation with such diverse agents as platelet derived growth factor (PDGF), concanavalin A, and serum.[29,32] In the basal state, c-*myc* mRNA levels are low as a result of marked degradation of this mRNA species.[33] Conversely, the marked elevation in the levels of c-*myc* mRNA following mitogenic stimulation is due mainly to an inhibition of its degradation.[34] The interactions of the c-*myc* protein product with the protein products of other oncogenes and their role in the regulation of cell proliferation and differentiation will be discussed later in this chapter.

Yet another mechanism through which genes may regulate the cell cycle relates to the ability of their protein products to inhibit cell proliferation.[35] Further, there appear to be certain genes that are active only during the G_0 phase of the cell cycle. Some of these genes may function to sustain the growth-arrested state of the cells. Others, however, may function to inhibit cell growth. The existence of growth-arrest-specific genes that inhibit cell proliferation has been recently reported.[36] The expression of these genes is inhibited following induction of cell growth, and the product of at least one of these genes appears to be cell-cycle controlled.[36]

In order for a eukaryotic cell to divide, it has to replicate its genome during the S phase of the cell cycle.[1] Several lines of evidence indicate that DNA polymerase alpha is one of the principal enzymes that is involved in the regulation of DNA replication. Thus, aphidicolin, a specific inhibitor of this enzyme, inhibits DNA replication.[37] Monoclonal antibodies against the enzyme also inhibit DNA synthesis.[38,39] Enhanced activity of the enzyme correlates with enhanced DNA synthesis.[40] A cell line that exhibits a temperature-sensitive defect in DNA replication is also a DNA polymerase alpha mutant.[41] Although there is a slight increase in the expression of this enzyme prior to the S phase, and a slight decrease through the G_2 phase, the enzyme is generally constitutively expressed throughout the cell cycle.[42] Nonetheless, the regulation of the expression of this enzyme is believed to occur at the transcriptional level.[42] Taken together, these observations indicate that enzymes which are involved in the regulation of cell proliferation are not necessarily active only during a specific phase of the cell cycle and that a variety of intracellular factors probably interact with cell cycle specific enzymes to regulate the exact timing of the proliferative response.

III. CELL DIFFERENTIATION

Differentiation is a complex process whereby a cell acquires a specific set of specialized functions. In the absence of differentiation, every cell in a particular multicellular organism would exhibit the same phenotype, have the same complement of metabolic pathways, possess the same mRNAs, and produce the same proteins as every other cell in that organism. This hypothetical organism would not have any specialized tissues and would therefore exhibit limited functional capabilities. However, the cells of multicellular organisms have developed the capacity to differentiate and to exhibit numerous specialized functions. In spite of the resultant functional and phenotypic diversity, it had been accepted that all of the cells within any given eukaryotic organism possess a complete set of genetic information which is identical with the genetic material that was originally present in the fertilized egg that gave rise to this organism. We now realize, however, that the genetic make-up of certain cells and tissues within an organism may be altered as a result of processes such as gene amplification and gene rearrangement.[43]

Our whole embryological development would not be possible without the process of differentiation. However, one of the consequences of differentiation is the loss of cellular plasticity.[44] This means that the acquisition of certain differentiated functions by a particular progenitor cell most often precludes it from acquiring other differentiated functions. For example, in the case of the red blood cell, a pluripotential stem cell divides to produce either additional stem cells or proerythroblasts. The proerythroblast, through a series of intermediate stages, is committed to become a mature red blood cell and cannot assume other pathways of differentiation.[45] In contrast to the red blood cell, many other differentiated cells retain the capacity to undergo mitosis and to produce identically differentiated daughter cells. Further, in some instances the daughter cells regain a certain degree of plasticity and exhibit new differentiated functions — a phenomenon that has been called metaplasia or transdifferentiation.[46]

In general, the phenotypic characteristics of a differentiated cell are those that allow it to perform its specialized functions. Thus, a pancreatic acinar cell exhibits the typical polarized morphology of exocrine secretory cells and synthesizes digestive enzymes that are packaged into zymogen granules which are ultimately released into the digestive tract; a nerve cell exhibits a cell body, dendrites, and an axon which serves to carry nerve impulses; an adipocyte exhibits the bag-like appearance of a cell that has accumulated numerous droplets of triacylglycerol while serving its function to store energy. In addition to their own typical morphological features, many differentiated cells exhibit certain common features. These may include the ability to form tight junctions, gap junctions, and desmosomes; to exhibit contact inhibition of cell proliferation; to express specific cell surface antigens and specific cytosolic proteins; to exhibit cell-type specific gene expression; and to be highly responsive to specific hormones and neurotransmitters as a result of the expression of genes coding for these receptors and their effector pathways. The complex processes that allow for this kind of specialized cellular differentiation to occur are believed to be dependent on the differential recruitment and activation of many diverse genes.[47]

IV. CELL PROLIFERATION

Many hormones and growth factors have been implicated in the regulation of cell proliferation and differentiation. In some instances, a hormone appears to exert its trophic effects on one or two cell types. Thus, thyroid stimulating hormone, or TSH, exerts trophic effects on the thyroid.[48] Similarly, erythropoietin acts directly on the red blood cell precursors to direct both their proliferation and differentiation toward mature erythrocytes.[49] In addition, several different hematopoietic growth factors have been described which direct progenitor cells to multiply and orchestrate their commitment toward differentiation into specific types of cells, including different subsets of lymphocytes, polynuclear cells, and macrophages.[50-52]

It is also possible for the same hormone or growth factor to regulate the proliferation of many different types of cells. For example, insulin-like growth factor I (IGF-I) has been shown to enhance the proliferation of a variety of cells.[53,54] In the case of gastrointestinal hormones, gastrin,[55] cholecystokinin (CCK),[56] and bombesin[57] are known to enhance the proliferation of specific types of cells. These same hormones and growth factors also regulate a variety of cellular processes which are not directly linked to mitogenesis, but which allow the cells to carry out their differentiated and highly specific functions.

It is likely that the major determinants of cellular proliferation and differentiation that are found in serum are polypeptide growth factors. Studies of the phenomenon whereby serum initiates DNA synthesis in quiescent cells showed that in most instances hormones such as insulin are only able to assist in this proliferative response and that this response is primarily due to the actions of growth factors that are present in serum.[58-61] In general, factors such as PDGF render a cell competent to respond to subsequent mitogenic stimuli.[58,59] Competence is achieved following a relatively brief exposure to PDGF, and the cell remains in a competent state even if the growth factor is removed from the incubation medium.[59] Many other factors that are present in serum, including epidermal growth factor (EGF) and IGF-I, are termed progression factors and allow competent cells to proceed toward initiation of DNA synthesis.[59-61] However, the distinction between competence factors and progression factors is probably less well demarcated than was initialy suggested. The remainder of this chapter will focus on EGF and the basic mechanisms through which its regulatory actions are exerted and modulated.

V. EPIDERMAL GROWTH FACTOR AS A MODEL

Epidermal growth factor, or EGF, was initially called "tooth-lid" factor because the injection of extracts from the salivary glands of newborn mice into young mice resulted in precocious incisor eruption and eyelid opening.[62] The name was subsequently changed to EGF because it was found that this heat-stable polypeptide induced epidermal proliferation.[63] EGF was also found to increase the activity of ornithine decarboxylase, the rate-limiting enzyme in the polyamine pathway.[63,64] The primary amino acid sequence of mouse EGF was reported 10 years after its discovery.[65] The protein was determined to have a molecular weight of 6045 daltons and to be trypsin resistant.[66] However, when extracted from the mouse submaxillary glands, it consists mainly of a high molecular weight form of 74 kDa that dissociates under acidic conditions into two EGF molecules and two binding proteins.[67]

EGF exerts both short and long term regulatory effects on numerous cellular processes. In some cells, EGF causes rapid elevations in the levels of cytosolic free calcium;[68] within 2 h of its addition to quiescent cells, EGF causes increased cation, sugar, and amino acid uptake;[69-71] within 12 h, EGF increases protein and RNA synthesis;[72] after 20 h, DNA synthesis is increased in preparation for mitosis.[71,73] The pleiotypic effects of EGF and its important role in regulating cell proliferation in a variety of tissues and cell types render it both a useful and an important model for analyzing the various mechanisms of action of peptide hormones and growth factors.

A. THE EGF RECEPTOR

Human EGF exhibits considerable homology with mouse EGF, mimics its biological actions, and cross-competes in radioreceptor assays.[74] Most studies of the biological actions of EGF have therefore been carried out with the more readily available mouse EGF. These studies have indicated that EGF binds to a specific cell surface receptor that is a 170 kDa glycosylated phosphoprotein.[75,76] It is autophosphorylated at tyrosine residues through a kinase that is an integral component of the receptor.[75-77] Tyrosine phosphorylation of the EGF receptor appears to be important in the mediation of the proliferative response to EGF.[78] Receptor autophosphorylation on tyrosine residues also occurs when insulin,[79] IGF-I,[80] PDGF,[81] and colony stimulating factor-I (CSF-I)[82] bind to their respective receptors.

As in the case of many other cell-surface receptors, the EGF receptor consists of an extracellular domain, a transmembrane domain, and an intracellular domain.[83] The extracellular portion of the receptor represents the ligand binding domain. It can be viewed as consisting of four segments.[84] Two of these segments are cysteine rich,[83] and the region between them appears to represent the major receptor binding region that dictates the specificity of the receptor toward EGF.[84] The transmembrane domain is highly hydrophobic and permits the receptor to anchor itself within the cell membrane.[83-85] The intracellular domain contains three tyrosine phosphorylation sites, an ATP binding domain, and the kinase region that catalyzes the autophosphorylation process.[85-87] Analysis of the primary amino acid sequence of the EGF receptor indicates that it exhibits a strong sequence homology with the protein product of the avian erythroblastosis virus v-*erb* B oncogene.[88] The latter protein, in essence, is a truncated EGF receptor that is devoid of most of the extracellular ligand-binding domain, as well as of a 34 amino acid sequence of the intracellular carboxy-terminal region.[88] This observation suggested that the extracellular binding domain may exert a negative constraint on the intracellular domain and that the potential of the v-*erb* B oncogene to induce the malignant phenotype derived from the absence of this negative constraint. However, recent studies indicate that when v-*erb* B is combined with the extracellular and transmembrane domains of the human EGF receptor, this chimer retains the ability to transform the cells to a malignant phenotype.[89] Further, the transforming ability of the chimer is increased in the presence of EGF.[89] Similarly, EGF induces transformation in NIH-3T3 cells that are transfected with the complete EGF receptor gene, but only when the number of EGF receptors is markedly increased.[90] Therefore, it would appear that the overabundance, per se, of normal EGF receptors is sufficient to induce the transformed phenotype in these transfection experiments.

Several lines of evidence suggest that overexpression of the EGF receptor is also an important oncogenic stimulus *in vivo*. An increased number of EGF receptors is associated with enhanced metastatic potential in human breast cancer[91] and enhanced tumor invasiveness in human bladder cancer.[92] Primary biopsy specimens from several different types of malignancies in humans have been reported to exhibit increased number of EGF receptors. These include glial,[93] pulmonary,[94] gastric,[95] breast,[96] and colonic[95] carcinomas. Similarly, a variety of cultured human carcinoma cell lines have been shown to have increased EGF receptor number.[97-103]

Studies with A431[77] and other cell lines[93,94] indicate that in some instances the increase in EGF receptor number is associated with an increase in the number of copies of the gene coding for this receptor. This gene has been localized to the short arm of chromosome 7 in humans.[104] Overexpression of the EGF receptor in clonal lines of A431 cells is most often, but not always, associated with a chromosomal translocation in this region.[104] These translocations have been correlated with EGF receptor gene rearrangements and with the production of aberrant EGF receptor mRNAs.[105] Structural or numerical alterations of chromosome 7 also occur in several different types of clinically important malignancies, including melanomas[97] and pancreatic[103] and pulmonary[106] carcinomas. There are many additional examples of the potential importance of clonal chromosomal abnormalities in human malignancies. These include a variety of translocations in different leukemias, lymphomas, and carcinomas.[107-109] Further, a possible role for clonal chromosomal abnormalities in solid tumors of the gastrointestinal tract is supported by the finding that colorectal cancers are monoclonal in origin,[110] that many of these tumors exhibit a loss of the short arm of chromosome 17,[110] and that an increased number of copies of chromosome 7 may occur in premalignant Barrett's esophagus mucosa.[111] These observations represent only a few of the many examples in which genetic information may be altered in some of the daughter cells of a given individual.

The above findings also serve to emphasize the importance of the EGF receptor in the regulation of normal and abnormal cellular proliferation and the potential importance of the overexpression of the EGF receptor in the etiology or growth advantage of certain malignancies. The molecular mechanisms whereby cells overexpress the EGF receptor have therefore been investigated extensively. Although these regulatory steps have not been fully elucidated, the findings to date have delineated some of the pathways that may modulate the expression of a number of important growth regulating genes. In general, the promoter regions of genes regulate the transcriptional activation of these genes and contain regions that are called TATA or CCAAT boxes.[112] These DNA sequences are important sites that function to regulate both the site of initiation of gene transcription and the rate of transcriptional synthesis.[112] However, unlike many other genes, the promoter region of the EGF receptor gene does not have the classical TATA or CCAAT boxes.[113] Instead, in the case of the EGF receptor gene, the promoter region is rich in guanine and cytosine and contains multiple transcription initiation sites.[113] There appear to be at least eight nuclear protein binding areas within these sites, and these are potentially the target of proteins that are involved in the transcriptional activation of the gene.[114] These DNA binding proteins are also called trans-acting factors because they are coded for by genes that are not necessarily in proximity to the genes whose transcription they regulate.[115] In the case of the EGF receptor, a specific 270 kDa protein has been shown to activate EGF receptor gene transcription.[113] A similar regulatory effect on EGF receptor gene transcription has also been demonstrated with several other DNA binding proteins, including the Sp-1 trans-acting factor.[113] The GC richness of the promoter region, the absence of either TATA or CCAAT boxes, and the presence of multiple transcription initiation sites is a characteristic feature of certain cellular oncogenes such as the ras genes,[116,117] housekeeping genes such as dihydrofolate reductase,[118] and DNA tumor viruses such as human hepatitis B.[119] The transcriptional regulation of the gene coding for transforming growth factor-alpha (TGF-α) also exhibits many of these features, but appears to be initiated from a single site.[120]

B. EGF BINDING

The first step in the initiation of EGF action is represented by the binding of the ligand to its specific cell-surface receptor.[121] Following binding and receptor autophosphorylation,[65-67] the EGF-receptor complex translocates through the membrane and internalizes into the cell by a process called receptor-mediated endocytosis,[122] a phenomenon which is also observed with many other peptide ligands. EGF binding is decreased by incubating cells with EGF.[123] This process of down-regulation is also observed with other hormones and growth factors. It is the consequence of the internalization and degradation of the cell-surface receptors following their occupation by the respective ligands and may represent a mechanism for dampening the actions of these ligands. The binding and subsequent endocytosis of peptides are also modulated by a variety of signals that are not related to the down-regulation phenomenon. In the case of EGF, binding is decreased by $\alpha 1$[124] and $\beta 1$[125] adrenergic agents, PDGF,[126] CCK,[127] carbachol,[127] caerulein,[127] bombesin,[127] vasopressin,[128] fibroblast-derived growth factor (FGF),[129] 1,25-dihydroxy-vitamin D_3[130] and interleukin-1.[131] The inhibitory effect of adrenergic agents is believed to be mediated through cyclic AMP (cAMP).[125] In contrast, the inhibitory action of CCK is probably mediated through a rise in cytosolic free calcium,[127] as well as through the activation of protein kinase C,[132] a calcium-activated, phospholipid-dependent enzyme.[133]

A variety of nonphysiological agents also modulate EGF binding. Foremost among these is tetradecanoyl-phorbol acetate (TPA), a compound that directly activates protein kinase C.[133-135] Presumably, protein kinase C decreases the affinity of the EGF receptor for EGF by phosphorylating the receptor on a threonine residue at position 654 close to the inner surface of the plasma membrane.[136] This hypothesis is supported by the observation that, following conversion of the threonine residue at position 654 by site-directed mutagenesis into an alanine residue, the receptor is no longer modulated by TPA.[137] However, the mutated receptor is readily down-regulated by EGF, indicating that EGF-mediated inhibition of EGF binding is not dependent on activation of protein kinase C.[137] The importance of protein kinase C in modulating hormone-mediated signal transduction is also illustrated by the ability of this enzyme to induce the phosphorylation of several other hormone receptors, including the transferrin[138] and β-adrenergic[139] receptors, and to cause the rapid internalization and degradation of muscarinic acetylcholine receptors.[140]

In contrast to the previous compounds, transforming growth factor-beta (TGF-β),[141] estrogen,[142] glucocorticoids,[143,144] thyroxine,[145] growth hormone,[146] and tumor necrosis factor-alpha (TNF-α)[147] increase EGF binding. In the case of TGF-β, the stimulatory effect on EGF binding is preceded by a transient inhibitory effect.[148] Further, in cultured human gingival fibroblasts TNF decreases EGF binding.[131] Thus, EGF receptor transmodulation can be activated by a variety of ligands in many different cell types. Further, modulation of EGF binding may result in altered responsiveness of target cells to subsequent action by EGF. However, in many instances, the exact relationship between the regulation of EGF binding and the subsequent biological response to EGF has not been completely delineated.

In addition to binding EGF with a high affinity, the EGF receptor also binds TGF-α[149] and vaccinia growth factor.[150,151] This ability of a cell-surface receptor to bind more than one ligand represents a mechanism whereby activation of cellular responses can be accomplished by more than one extracellular signal, even when these signals are acting through the same receptor. Another example of several hormones activating the same receptor is represented by the ability of insulin, IGF-I and insulin-like growth factor II (IGF-II) to bind and activate the insulin receptor.[152] Further, insulin binds and activates the IGF-I receptor.[152] In many instances, the mitogenic effects of insulin are due to the activation of the IGF-I receptor and are not mediated via the insulin receptor.[152] In contrast to the insulin/IGF-I relationship, there is no evidence for the existence of a distinct receptor for TGF-α. Instead, the biological activities of TGF-α are believed to be due exclusively to its ability to bind and activate the EGF receptor.[153,154] TGF-α is a peptide growth factor that shares only a 35% homology with EGF.[155,156] However, it possesses six cysteine residues in the same relative positions as EGF.[155,156] The three-dimensional configuration of the receptor binding domain of TGF-α is believed to be very similar to that of EGF, enabling TGF-α to readily bind to the EGF receptor and to phosphorylate it on tyrosine residues.[157]

As expected, EGF and TGF-α exert similar biological effects in a variety of cell types both *in vitro* and *in vivo*.[153-160] However, there are many exceptions to this rule. Thus, TGF-α exerts a greater stimulatory effect than EGF with respect to angiogenesis in the hamster cheek pouch model,[161] calcium mobilization from fetal rat bones,[162] arterial blood flow in the dog,[163] skin wound healing,[164] formation of keratinocyte megacolonies,[165] induction of cell ruffling,[166] and anchorage-independent growth of cultured human pancreatic carcinoma cells.[167] Further, TGF-α is a more potent inhibitor of the proliferation of RL95-2 human endometrial carcinoma cells than EGF.[168] In addition to these quantitative differences, there are some examples of qualitative differences between the actions of the two growth factors. Thus, the inhibitory effect of TGF-α on norepinephrine-induced contraction in arterial strips is not diminished following repeated exposure of the strips to TGF-α.[163] In contrast, under the same experimental conditions, the response to EGF is greatly attenuated. Similarly, in primary lung carcinoma cells TGF-α enhances, whereas EGF inhibits cell proliferation.[169] Further, the chicken EGF receptor exhibits a higher affinity toward human TGF-α than toward murine EGF, and TGF-α is 100-fold more potent than EGF in stimulating DNA synthesis in 3T3 cells that express the chicken EGF receptor than in 3T3 cells that express the human homolog.[170] Although the mechanisms underlying these various differences are not known, these observations raise the possibility that the two ligands acting via the same receptor may activate different effector pathways and second messengers.

The binding of EGF to target cells may be altered under several physiological conditions and in certain disease states. For example, the differentiation of the granulosa cell in the rat is associated with changes in EGF binding.[171] In this cell type, follicle stimulating hormone appears to maintain the increase in EGF receptor number and this effect is counterbalanced by leutinizing hormone and human chorionic gonadotropin.[171] Similarly, EGF binding is increased in placental membranes during pregnancy and this effect is maximal at term.[172,173] This increase in EGF binding is due mainly to an increase in the low affinity binding sites and is abrogated in insulin-deficient diabetes.[174] Insulin deficiency is also associated with a decrease in EGF binding in the pancreas[175] and liver.[176] However, hyperinsulinemic diabetes is also associated with decreased hepatic binding of EGF.[177] It is likely, therfore, that either the hyperglycemia per se or the associated metabolic changes cause the observed decreases in EGF binding. Presumably, in all these situations TGF-α binding is also altered, in parallel with the changes in EGF binding.

C. EGF PROCESSING

The way cells process peptide hormones and growth factors may potentially represent an important aspect of the regulation of growth control. The fate of cell-bound peptides has therefore been extensively studied in a variety of cell systems. In the case of EGF,[122] as in the case of insulin,[178] the internalized ligand is rapidly degraded within lysosomes or other acidic organelles. EGF degradation is initiated by the sequential removal of carboxy-terminal amino acids, and ultimately the ligand can no longer bind to its receptor.[179] Ostensibly, degradation proceeds very rapidly following dissociation of the truncated ligand from its receptor within the acidic milieu of the lysosomal compartment. There is an increasing number of exceptions to this general scheme. Thus, EGF is not readily degraded in normal pancreatic acini,[180] PANC-I human pancreatic carcinoma cells,[181] and Madin-Darby Canine Kidney (MDCK) cells.[182] Similarly, endothelial cells bind and internalize insulin, but do not degrade this ligand.[183] Therefore, the general dogma that peptide hormones and growth factors are always degraded following binding and internalization in target cells is no longer tenable. In the case of MDCK cells, limited EGF processing is associated with vectorial transcytosis of the ligand.[182] In the case of PANC-I cells, there is considerable recycling of the ligand between the intracellular and extracellular compartments, a phenomenon which may allow one EGF molecule to induce the internalization and degradation of numerous EGF receptors.[180] Although it is possible that the limited degradation of EGF may also allow it to exert specific intracellular functions, the evidence in this regard is very limited. In contrast, several lines of evidence indicate that insulin, once internalized into certain target cells, may exert direct effects on their nuclei.[184] Thus, there is an increasing number of exceptions to the concept that only steroid hormones act directly at the level of the nucleus and that the actions of peptide hormones are mediated solely through their cell-surface receptors.

D. MECHANISM OF ACTION OF EGF

As in the case of many other pleiotypic peptides, EGF regulates cellular growth and differentiation through a variety of mechanisms. These include rapid changes in the phosphorylation of endogenous proteins, modulation of cytosolic free calcium levels and intracellular pH, and modulation of the expression of specific genes.

1. Kinase Activity

The state of phosphorylation of many cellular proteins dictates whether these proteins are biologically active or inactive. Therefore, it is likely that alterations in the state of protein phosphorylation represent a general mechanism whereby hormones and growth factors elicit their biological actions, including their effects on cell proliferation and differentiation.[185-187] The enzymes that mediate the transfer of a phosphate group onto a protein are called kinases, whereas those that mediate the removal of a phosphate group are called phosphatases. In general, protein kinases that have been implicated in hormonally mediated regulation of cellular processes can be classified as belonging to two categories. One group of kinases phosphorylates proteins on tyrosine residues,[188] as is the case for the growth factor receptors discussed earlier. The other phosphorylates proteins on threonine and serine residues. Both groups of kinases are dependent on the availability of ATP as their phosphate source, but some kinases preferentially use GTP.

In addition to the hormone and growth factor receptors which possess tyrosine kinase activity, the protein products of certain oncogenes are known to undergo tyrosine autophosphorylation.[20] Thus, the oncogene product of the Rous sarcoma virus, $pp60^{v\text{-}src}$, is autophosphorylated at its carboxy-terminal end on tyrosine 416, and this phosphorylation is believed to be critical for the ability of the protein to exhibit protein kinase activity and to confer onto transfected cells the transformed phenotype.[189]

In contrast to the tyrosine kinases, the serine-threonine kinases are more heterogeneous and their activity is regulated by a variety of second messengers and other intracellular modulators that include cAMP, cyclic GMP (cGMP), calcium, calmodulin, arachidonic acid, polyamines, and phospholipids. There are at least five groups of calcium-calmodulin protein kinases, which have been classified as type I, II, or III, phosphorylase kinase, and myosin light-chain kinase.[190] Type III calcium-calmodulin protein kinase phosphorylates elongation factor 2 and is stimulated by a number of mitogenic hormones and growth factors.[191] Another kinase is known as the S6 kinase, which phosphorylates the S6 ribosomal protein.[192] This kinase is also activated by a variety of mitogenic hormones and growth factors, including EGF, as well as by $pp60^{v\text{-}src}$.[192]

Another important kinase is the ubiquitous calcium-phospholipid dependent protein kinase, also known as protein kinase C. This enzyme has been implicated in the regulation of gene expression and differentiated functions, cell proliferation, cellular differentiation *in vitro*, and embryonic development *in vivo*.[193-195] Sequence analysis of cloned cDNAs corresponding to the mRNAs coding for this enzyme indicate that there are several distinct but highly homologous forms of the enzyme, which are differentially expressed in various tissues.[196-198] In addition to diacylglycerol, certain unsaturated fatty acids such as arachidonic acid activate protein kinase C in either the presence or absence of phospholipid or calcium.[199] Further, lysophosphatidylcholine activates protein kinase C at low concentrations, but inhibits it at high concentrations.[200] The importance of protein kinase C in the regulation of cell growth is dramatically underscored by the observation that its overexpression may be associated with altered proliferation and enhanced tumorigenicity.[201,202]

In addition to being a substrate for protein kinase C, the EGF receptor modulates the activities of a variety of kinases. The first kinase that is activated after EGF binds to its receptor is the kinase intrinsic to the receptor.[75-77,85-87,188] This results in tyrosine autophosphorylation of the EGF receptor at position 1173, which is believed by most investigators to activate the tyrosine kinase activity of the receptor toward other substrates. This activation may be the consequence of the removal of an inhibitory constraint that in the unstimulated state prevents the receptor from binding to its substrates.[203,204] EGF receptor tyrosine autophosphorylation is calcium- and cyclic nucleotide-independent, but requires the presence of either magnesium or manganese.[205] These divalent cations enhance tyrosine autophosphorylation by lowering the K_m for the phosphate donor.[205] Manganese is somewhat more potent than magnesium in contributing to the activation of the intrinsic kinase activity of the EGF receptor.[206] However, when the EGF receptor is isolated in the presence of detergent in its monomeric form, the EGF dependence of the tyrosine kinase activity requires the presence of 0.25 *M* ammonium sulfate in addition to manganese.[207] In contrast, the intrinsic kinase activity of the monomeric EGF receptor is activated by magnesium in the absence of ammonium sulfate.[207] Although the role of ammonium sulfate has not been elucidated, these observations suggest that a single EGF receptor molecule is capable of being activated by EGF and that divalent metal cations may be directly involved in this activation process.[207] Nonetheless, EGF receptor dimerization appears to be important in the physiological activation of its kinase activity.[208,209]

In contrast to the EGF receptor, the insulin receptor exhibits an absolute requirement for manganese.[210,211] The importance of manganese in the regulation of certain phosphorylation reactions is also exemplified by the existence of specific manganese-dependent kinases in the stomach,[212] brain,[213] and HL-60 cells.[214] Further, manganese activates cytosolic casein kinase II, an enzyme that is inhibited by heparin and stimulated by spermidine and that phosphorylates a 110 kDa protein as well as other endogenous proteins.[215] Inasmuch as both EGF and insulin enhance the activity of casein kinase II,[216] these observations raise the possibility that manganese may modulate the signal transduction pathways activated by certain growth factors and hormones.[217]

Phosphorylation of the EGF receptor also occurs at other tyrosine sites, as well as on serine and threonine residues. It is believed that the resultant covalent modifications represent an important mechanism for modulating the biological actions of the receptor, as well as for dictating its intracellular distribution and ultimate fate.[218] Consequently, activation of the EGF receptor by EGF results in the phosphorylation of a number of endogenous proteins. The potential of the activated EGF receptor to act as a kinase is illustrated by the rapidity of the phosphorylation reactions,[219] as well as by the diversity of the phosphorylated proteins. In A431 cell membranes, the major protein phosphorylated by EGF is a 35 kDa protein variably known as calpactin II or lipocortin I.[220,221] In different cell types EGF also phosphorylates proteins with molecular weights of 120 kDa, 95 kDa, 42 kDa, 36 kDa, 34 kDa, and 22 kDa.[222-229]

Activation of the EGF receptor also results in the activation of protein kinase C, as well as other kinases.[230,231] In addition, the EGF receptor can phosphorylate angiotensin II,[232] the glucocorticoid receptor,[233] gastrin,[234] growth hormone,[235] and the protein product of the *erb* B-2/*neu* oncogene.[236,237] The latter protein is a 185 kDa receptor (p185^{neu}) which shares significant homology with the EGF receptor.[238,239] Its discovery represents the first example of the characterization of a receptor whose ligand is unknown. p185^{neu} exhibits tyrosine activity and appears to be very important in modulating the growth of certain cancer cells.[240,241] The ability of the EGF receptor to modulate the phosphorylation of the glucocorticoid receptor and of p185^{neu} is illustrative of the many examples in which distinct hormones and growth factors modulate the actions of each other and interact through a variety of mechanisms.

The most direct evidence that activation of the EGF receptor tyrosine kinase is a crucial step for transmitting the mitogenic signal that is initiated following the binding of EGF to its receptor is based on studies in which the receptor has been altered by site-directed mutagenesis. These studies have evaluated mutant EGF receptors possessing a point mutation at the ATP binding site of the tyrosine kinase domain (Lys-721),[242-244] an insertional mutation within the consensus sequence of this domain,[245] or point mutations at both Lys-721 and Thr-654.[246] The resultant mutated receptors do not have intrinsic tyrosine kinase activity. Further, when the EGF receptor is altered at the primary tyrosine site for *in vivo* autophosphorylation (Tyr-1173) its tyrosine kinase activity in response to low levels of EGF is markedly attenuated.[247] Cells transfected with tyrosine kinase deficient EGF receptors lose their responsiveness to EGF with respect to proliferation and with respect to the activation of a variety of cellular functions that are modulated by EGF.[218,242-246] Further, the half-life of the EGF receptor in these cells is markedly prolonged as a result of either enhanced recycling or attenuated endocytosis and degradation of the receptor.[242,248] In contrast, a mutated EGF receptor containing tyrosine kinase activity but deficient for the carboxy-terminal phosphorylation sites undergoes normal endocytosis in response to EGF.[248] Similarly, a subclone of Chinese hamster ovary (CHO) cells transfected with a variant EGF receptor that is also deficient for all three carboxy-terminal tyrosine residues still exhibits a proliferative response to EGF.[249]

Other types of experiments indicate that specific inhibitors of the EGF receptor tyrosine kinase block the ability of EGF to induce cell proliferation.[250] Further, dephosphorylation by prostatic acid phosphatase of EGF receptors obtained from DU145 prostatic cancer cells results in attenuated tyrosine kinase activity.[251] However, partial proteolysis of the EGF receptor yields a trypsin resistant 40 kDa moiety that contains the functional tyrosine kinase domain.[252,253] Taken together, these observations suggest that ligand induced endocytosis and mitogenesis occur even when the tyrosine kinase activity of the receptor is directed at tyrosine sites other than those located at the carboxy-terminal end of the receptor, that this tyrosine kinase activity may be liberated in an active form within the cell during the process of intracellular degradation of the receptor, that complete inactivation of the tyrosine kinase activity abolishes the ability of the EGF receptor to transmit its mitogenic signals, and that endogenous phosphatases may modulate EGF receptor bioactivity *in vivo*.

The tyrosine kinase deficient insulin receptor also fails to transmit insulin-activated signals.[254-256] Similarly, other mutations within different regions of a number of hormone and growth factor receptors have helped elucidate their structure-function relationship.[87] As the number of peptide hormones and growth factors that are being implicated in the regulation of cell proliferation continues to grow (Table 1), it is likely that additional tyrosine kinase dependent and independent mechanisms will be delineated. The growth-promoting roles of many of these peptides have been recently reviewed.[257-263]

2. Calcium

Calcium is an important second messenger that participates in the regulation of important cellular processes, including cell proliferation and differentiation.[264,265] The level of intracellular free calcium within a cell is dependent on basal calcium influx and efflux, on effector systems that act to modulate calcium influx and calcium mobilization from intracellular stores, and on

TABLE 1
Growth and Differentiation Regulating Peptide Factors

1.	Insulin	18.	Gastrin
2.	Insulin-like growth factor I	19.	Nerve growth factor
3.	Insulin-like growth factor II	20.	Kaposi's sarcoma heparin-bound growth factor
4.	Relaxin	21.	Erythropoietin
5.	Epidermal growth factor	22.	Thrombopoietin
6.	Vaccinia growth factor	23.	Granulocyte-macrophage colony stimulating factor
7.	Platelet-derived growth factor A	24.	Granulocyte colony stimulating factor
8.	Platelet-derived growth factor B	25.	Interleukin-1
9.	Platelet-derived growth factor AB	26.	Interleukin-2
10.	Transforming growth factor B_1	27.	Interleukin-6
11.	Transforming growth factor B_2	28.	Tumor necrosis factor-α
12.	Transforming growth factor B_3	29.	Tumor necrosis factor-β
13.	Inhibin A	30.	Interferon-α
14.	Inhibin B	31.	Interferon-β
15.	Mullerian inhibitory substance	32.	Interferon-gamma
16.	Bombesin	33.	Acidic fibroblast growth factor
17.	Cholecystokinin	34.	Basic fibroblast growth factor

the buffering capacity of intracellular organelles that act to sequester the cation.[266,267] Physiological mobilization of intracellular calcium by peptide hormones, growth factors, and neurotransmitters occurs following activation of phospholipase C.[268] This enzyme induces the hydrolysis of membrane associated phosphatidylinositol-4,5 bisphosphate, resulting in the generation of inositol trisphosphate (Ins-1,4,5-P_3) and diacylglycerol, the putative endogenous activator of protein kinase C.[269] Ins-1,4,5-P_3 raises cytosolic free calcium levels by releasing calcium from intracellular stores, most notably from the endoplasmic reticulum where it ostensibly binds to a specific receptor.[270-272] Ins-1,4,5-P_3 is either converted by a phosphomonoesterase to a metabolically inactive isomer or by Ins-1,4,5-P_3-kinase to inositol 1,3,4,5-tetrakisphosphate (Ins-1,3,4,5-P_4).[273-276] Ins-1,3,4,5-P_4 may act together with Ins-1,4,5-P_3 to enhance receptor-mediated calcium influx.[277-279] It is also possible that Ins-1,4,5-P_3 acts directly to increase calcium influx.[280-282] In addition, other phosphoinositides appear to be generated, and these compounds may also be important in modulating cellular calcium homeostasis.[283,284]

The activation of Ins-1,4,5-P_3-kinase may be modulated by calcium,[285-287] whereas the regulation of calcium deposition and release in the Ins-1,4,5-P_3-mobilizable compartment may be modulated by GTP and protein kinase C, respectively.[288-290] Further, Ins-1,4,5-P_4 may also act to sequester cytosolic calcium.[291] These observations suggest that the rise in cytosolic free calcium levels that occurs following activation of phospholipid hydrolysis may be controlled by complex feedback mechanisms and that these regulatory pathways may contribute to the oscillations in the levels of cytosolic free calcium that occur following receptor activation with a variety of ligands.[292] Irrespective of the complexity of these biochemical pathways, it is evident that the dual stimulation of two key regulatory steps (calcium and protein kinase C) that accompanies receptor-mediated phospholipid hydrolysis has the potential to initiate a cascade of intracellular reactions and modulate many cellular processes.

The role of calcium in mediating many of the effects of hormones such as cholecystokinin and bombesin has been recognized for a long time.[293-295] In contrast, the importance of this divalent cation in mediating the actions of EGF has been only recently documented in an unequivocal manner. Thus, in A431 cells and other cells that overexpress the EGF receptor, EGF has been shown to rapidly enhance $^{45}Ca^{2+}$ influx, stimulate the hydrolysis of phosphatidylinositol-4,5-bisphosphate, and induce rapid increases in the levels of Ins-1,4,5-P_3 and Ins-1,3,4,5-P_4.[68,296] The marked and prolonged activation of the phosphatidylinositol cycle in A431 cells by EGF may be dependent on the tyrosine phosphorylation of a protein which then activates phospholipase C.[297] However, there does not appear to be concomitant activation of protein kinase C in these cells.[298] To further complicate matters, a subclone of the MDA-468 human breast cancer cell line is absolutely dependent on EGF for growth, and this requirement is abrogated by pertussis toxin.[299] It is therefore likely that the effects of EGF on the growth of certain cell lines that overexpress the EGF receptor is mediated, in part, via G-proteins.[300] Inasmuch as these proteins are also believed to participate in the regulation of receptor-mediated calcium mobilization,[301] it is possible that EGF may regulate cellular calcium homeostasis by acting through G-proteins.

In addition to acting via phosphatidylinositol hydrolysis, ligands can modulate intracellular free calcium levels by inducing calcium influx as a result of their ability to activate voltage-dependent and nonvoltage-dependent calcium channels.[302,303] Both types of channels can be activated by phosphoinositides, protein kinase C, or G proteins.[304,305] Often, there are concomitant rapid stimulatory effects on the sodium-hydrogen antiport system that is located in the cell membrane, leading to increased sodium influx and increased intracellular pH.[306] EGF is known to activate the sodium-hydrogen antiport system[307-309] and this effect is mimicked by activation of protein kinase C.[310] It is believed, therefore, that EGF and growth factors induce cellular alkalinization by activating protein kinase C.[310] However, the sodium-hydrogen antiport system is also activated by a rise in cytosolic free calcium levels,[311] raising the possibility that EGF may also be acting via calcium to induce cellular alkalinization. Irrespective of the mechanisms, the actions of EGF and other growth factors on intracellular pH appear to be a component of a number of early signals generated following ligand binding that ultimately lead to initiation of cell proliferation.[312,313]

In addition to inhibiting EGF binding and endocytosis in certain cells, calcium activates an intracellular protease that can degrade the EGF receptor at neutral pH.[314] The protease acts to cleave off a portion of the cytosolic domain of the receptor, raising the possibility that it may modulate EGF receptor number and function in intact cells.[315] Further, EGF regulates prolactin gene expression via calcium[316] and retards keratinocyte differentiation through unknown mechanisms that may be modulated by calcium.[317,318] Thus, when keratinocytes are grown in the presence of low levels of extracellular calcium their rate of differentiation is slowed.[319] The cells fail to stratify and lose the capacity to form desmosomes.[320] Concomitantly, the cells exhibit a marked increase in EGF binding.[318] Taken together, the above observations suggest that some of the actions of EGF are modulated by calcium, whereas others are mediated by calcium, and that EGF-induced alterations in cellular calcium homeostasis may be dependent on the ability of EGF to increase the tyrosine kinase activity of the EGF receptor.

3. Gene Activation

It is widely recognized that steroid hormones act at the level of the nucleus to modulate gene expression.[321-323] In recent years it has become evident that many peptide hormones and growth factors also regulate the expression of specific genes by directly altering their rate of transcription. In addition, hormones and growth factors can modulate the levels of specific mRNAs indirectly, by controlling the rate of efflux from the nucleus of a particular mRNA, by increasing or decreasing the half-life of the mRNA, and by directing the splicing of a particular precursor mRNA. For example, insulin increases amylase mRNA levels in the rat exocrine pancreas,[324] enhances albumin gene transcription in primary cultures of rat hepatocytes,[325] decreases mRNA levels for phosphoenolpyruvate carboxykinase in rat hepatoma cells,[326] and increases c-*fos* mRNA levels in 3T3-L1 fibroblasts.[327] IGF-I, which regulates both cell differentiation and proliferation, acts at the transcriptional level to modulate the expression of certain genes and regulates the stability of the transcripts of other genes that are constitutively expressed.[328] PDGF, a growth factor that makes cells competent for subsequent cell division, appears to act by rapidly inducing the activity of a number of important growth regulating genes which include c-*fos*, c-*myc*, KC, JE, and JB.[329-332]

EGF and TGF-α also stimulate c-*fos* and c-*myc*.[333,334] These nuclear oncogenes appear to have a crucial role in the regulation of cell proliferation and differentiation. The viral homolog of c-*fos*, v-*fos*, was originally identified as a transforming gene in the FBJ-murine osteosarcoma virus.[335] Both v- and c-*fos* form a complex with a 39 kDa protein,[336] which is now known to be the protein product of the c-*jun* oncogene.[337-339] The latter oncogene is the normal cellular homolog of a transforming gene that was originally isolated from avian sarcoma virus 17.[340] Its name derives from the fact that 17 in Japanese is "*ju-nanna*."[340] The c-*jun* protein product has been determined to be identical to the AP-1 transcription factor.[341,342] AP-1 was originally characterized as a DNA-binding protein that recognizes a specific sequence on the human metallothionein IIA gene and SV40 virus.[342,343] This regulatory binding site, TGACTCA, is present in many other genes and is also recognized by the DNA binding domain of the yeast transcription factor GCN4.[341,342,344-346] Further, the amino acid sequence of this domain is similar to those of c-*jun* and c-*fos*.[344] All three proteins have amino terminal ends that are rich in basic amino acids and carboxy-terminal ends that have five to six leucine residues which are separated by sequences of six amino acids, indicating that these structural features are functionally important.[346] Indeed, activation of genes that are regulated by the AP-1 regulatory elements occurs following the binding of the protein products of c-*jun* and c-*fos*, which in turn is dependent on the ability of these proteins to form a dimeric complex.[346] Formation of this complex occurs as a result of an interaction between opposing leucine residues on the two proteins, a phenomenon which has been called the leucine zipper.[346-349] EGF also enhances the transcription of c-*jun*.[350] Thus, the rapid activation of c-*fos* and c-*jun* by EGF and other mitogenic growth factors and differentiation agents, which occurs by both transcriptional and post-transcriptional mechanisms, results in intranuclear regulatory interactions between their protein products.[346-358] This mechanism ostensibly allows for the precise activation of the transcription of specific growth regulating genes.[346-358]

The exact mechanisms whereby EGF induces the transcription of c-*myc* and c-*fos* are not known. It is established, however, that in Swiss 3T3 cells this effect is independent of protein kinase C activation.[359] In contrast, in the same cells, the effects of bombesin c-*myc* and c-*fos* induction appear to be mediated via protein kinase C.[359] Similarly, the induction by EGF of the proliferation-associated VL-30 gene in mouse AKR-2B cells occurs via a C-kinase independent mechanism.[360] EGF also increases the transcription of the EGF receptor gene in several cell lines.[361-364] In the liver, this induction phenomenon appears to be mediated by mechanisms that are both C-kinase dependent and independent.[363]

The actions of EGF on gene expression are most likely modulated by a variety of factors. Thus, the ability of EGF to enhance transin mRNA expression is antagonized by TGF-β.[365] Similarly, the actions of EGF on EGF receptor gene induction are modulated by gamma-interferon.[364] It is also probable that many hormones and growth factors act through a variety of mechanisms to modulate the expression of genes that are also modulated by EGF. For example, C-kinase activation can induce c-*jun* transcription,[359] thereby potentially modulating the action of EGF. Similarly, TGF-β modulates the actions of EGF on phosphatidylinositol metabolism and cellular calcium homeostasis,[366] thereby potentially altering the effects of EGF on gene transcription.

4. Differentiation/Growth Inhibition

In addition to its mitogenic effects, EGF modulates a variety of differentiated functions that range from regulation of smooth muscle contraction to the maintenance of cell survival.[163,367,368] EGF also has an important role in fetal development, as well as in the postnatal development of certain tissues. In the fetus, EGF participates in the regulation of palate growth and differentiation,[369] pulmonary development and maturation,[370,371] and gastrointestinal mucosa maturation.[372] In the postnatal period EGF may also have an important role in the further maturation of a number of tissues.[373,374] It is also possible that milk-borne EGF may be especially important during this early neonatal period.[375] The EGF ingested by the suckling may act locally on enterocyte development[376] and systemically following absorption through the immature intestine.[377] It is possible that the actions of EGF on intestinal maturation are mediated, in part, through its effects on polyamine synthesis.[378]

EGF has also been implicated in the regulation of cellular differentiation and maturation in a number of tissues and cell types.[121,317,318,379-383] In some cell types, the differentiation-inducing actions of EGF are not associated with any changes in cell proliferation, indicating that the two processess are not necessarily linked. Thus, in cultured term trophoblasts, EGF induces the fusion of cytotrophoblasts into differentiated syncytiotrophoblasts without inducing cellular proliferation.[383] Conversely, in some cell types EGF prevents the process of differentiation. Thus, in BC3H1 mouse smooth muscle cells, EGF is both a mitogen and an inhibitor of cellular differentiation.[384] In this cell line EGF also blocks the expression of muscle specific proteins.[384] The importance of EGF in the inhibition of cellular differentiation is also underscored by the observation that in some cell types EGF receptors are no longer expressed as the cells acquire a terminally differentiated state.[385] Similarly, the regenerating liver[386] and pancreas[387] exhibit a decrease in EGF binding that coincides with maximal mitotic activity, raising the possibility that in these tissues EGF is a differentiation inducing agent and that the differentiation signal is attenuated during mitogenesis as a result of the decrease in EGF binding. In support of this hypothesis, the *in vivo* administration of EGF to Sprague Dawley rats is associated with decreased pancreatic DNA synthesis and modulation of differentiated acinar cell functions.[388,389]

EGF is also known to inhibit the proliferation of many different types of cultured cells. Thus, EGF inhibits the growth of A431 human vulvar carcinoma cells,[390,391] several human mammary and squamous carcinoma cells,[392,393] RL95-2 human endometrial cancer cells,[394] a human hepatoma cell line,[395] two types of rat pituitary tumor cells,[396,397] and R1E-1 rat intestinal cells.[398] The growth inhibitory effects of EGF have been most extensively studied in A431 cells. These studies suggested that the inhibitory actions of EGF were due to the presence of a large number of EGF receptors in this cell line, which are reported to range from one to six million receptors per cell.[399] EGF appears to act at two levels in the cell cycle in A431 cells; it blocks their entry into mitosis and it impedes their progression from G_1 into the S phase of the cycle.[400] It was initially proposed that the inhibitory effect of EGF was due to excessive kinase stimulation which ostensibly depletes cellular energy stores.[399] In support of this hypothesis, concentrations of EGF that are suboptimal for full kinase activation (3 to 100 p*M*) are mitogenic in this cell line.[401] It was also suggested that the stimulatory effect of EGF in these cells, as in the case of other cell types, may be mediated by a small group of high affinity receptors.[401,402] It has been suggested that EGF-induced changes in the cytoskeleton interefere with the mitotic process, thereby causing EGF to inhibit cell growth.[403] However, several of the cell lines cited above do not overexpress the EGF receptor, and a clonal line of A431 cells undergoes marked morphological changes in the presence of EGF but is not growth-inhibited.[404] These observations suggest that the EGF receptor number per se does not necessarily dictate the biological response to EGF and that the morphological actions of EGF are not directly linked to its growth-inhibitory effects.

The growth of RL95-2 and R1E-1 is inhibited by EGF when the cells are plated at low seeding densities, but is enhanced by EGF when the cells are plated at high seeding densities.[168,398] It is therefore likely that postreceptor events, including cell density and other factors, contribute to the ultimate responsiveness of the cell to EGF. It has been suggested that EGF-mediated growth inhibition may occur following activation of a critical ratio of high- and low-affinity receptors in susceptible cell types.[168] Although both high- and low-affinity EGF receptors are expressed in transfection experiments in which a single EGF receptor cDNA was used,[218] the two classes of receptors exhibit certain functional differences. Thus, the number of high-affinity binding sites is markedly decreased by protein kinase C activators and is increased by sphingosine through protein kinase C independent mechanisms, whereas the low-affinity binding sites are not similarly regulated.[405,406] Further, the lateral mobility of the low-affinity receptors within the membrane is greater than that of the high-affinity receptors, even though this lateral mobility appears to be independent of most of the cytoplasmic domain of the receptor.[407] Therefore, it is conceivable that different effector pathways are activated by the high- and low-affinity binding sites. Irrespective of the underlying mechanisms, these findings lend further support to the hypothesis that growth factors may act as bifunctional regulators of cell growth.[257]

VI. SUMMARY

The regulation of the proliferative response of eukaryotic cells is very complex and is dependent on the activation of multiple intracellular processes by both intracellular and extracellular signals. The extracellular signals are provided numerous hormones, growth factors, neurotransmitters, and a variety of nutrients. The actions of these extracellular signals are modulated in complex ways and are modified by various components in the extracellular matrix.[408] The resultant interactions lead to the activation of many cellular processes that include the induction of phosphatidylinositol hydrolysis, regulation of ion fluxes, phosphorylation of numerous proteins, activation of a variety of genes including protooncogenes, and specific intranuclear interactions between many nuclear trans-binding proteins. The ultimate response of a eukaryotic cell to this tremendous input of complex signals is often dictated by its differentiated state and may be different from cell type to cell type. As illustrated in this review with respect to EGF, instead of relying on the activation of just one regulatory pathway, hormones and growth factors often activate a cascade of intracellular reactions the complexity of which is yet to be fully understood. In many instances this regulation is modulated further by a variety of autocrine and paracrine mechanisms, whose importance in the regulation of cell proliferation and differentiation in both normal and disease states is still being unraveled.

REFERENCES

1. **Pardee, A. B., Dubrow, R., Hamlin, J. L., and Kletzien, R. F.,** Animal cell cycle, *Annu. Rev. Biochem.*, 47, 715, 1978.
2. **Hermos, J. A., Mathan, M., and Trier, J. S.,** DNA synthesis and proliferation by villous epithelial cells in fetal rats, *J. Cell Biol.*, 50, 255, 1971.
3. **Potten, C. S., Schofield, R., and Lajtha, L. G.,** A comparison of cell replacement in bone marrow testis and three regions of surface epithelium, *Biochim. Biophys. Acta*, 560, 281, 1979.
4. **Pardee, A. B.,** A restriction point for control of normal animal cell proliferation, *Proc. Natl. Acad. Sci. U.S.A.*, 7, 1286, 1974.
5. **Brooks, R. F., Bennett, D. C., and Smith, J. A.,** Mammalian cell cycles need two random transitions, *Cell*, 19, 493, 1980.
6. **Pledger, W. J., Stiles, C. D., Antoniades, H. N., and Scher, C. D.,** An ordered sequence of events is required before BALB/c-3T3 cells become committed to DNA synthesis, *Proc. Natl. Acad. Sci. U.S.A.*, 75, 2839, 1978.
7. **Hochauser, S. J., Stein, J. L., and Stein, G. S.,** Gene expression and cell cycle regulation, *Int. Rev. Cytol.*, 71, 95, 1981.
8. **Tricoli, J. V., Sahai, B. M., McCormick, P. J., Jarlinski, S. J., Bertram, J. S., and Kowalski, D.,** DNA topoisomerases I and II activities during cell proliferation and the cell cycle in mouse embryo fibroblast, *Exp. Cell. Res.*, 158, 1, 1985.
9. **Farnham, P. J. and Schimke, R. T.,** Transcriptional regulation of mouse dihydrofolate reductase in the cell cycle, *J. Biol. Chem.* , 260, 7675, 1985.
10. **Schlosser, C. A., Steglich, C., DeWet, J. R., and Scheffler, I. E.,** Cell cycle-dependent regulation of thymidine kinase activity introduced into mouse LMTK$^-$ cells by DNA and chromatin-mediated gene transfer, *Proc. Natl. Acad. Sci. U.S.A.*, 78, 1119, 1981.
11. **Stuart, P., Ito, M., Stewart, C., and Conrad, S. E.,** Induction of cellular thymidine kinase occurs at the mRNA level, *Mol. Cell. Biol.*, 5, 1490, 1985.

12. **Coppock, D. L. and Pardee, A. B.,** Control of thymidine kinase mRNA during the cell cycle, *Mol. Cell Biol.*, 7, 2925, 1987.
13. **Sherley, J. L. and Kelly, T. J.,** Regulation of human thymidine kinase during the cell cycle, *J. Biol. Chem.*, 263, 8350, 1988.
14. **Wu, K. S., Panusz, H. T., Hatch, C. L., and Bonner, W. H.,** Histones and their modification, *CRC Crit. Rev. Biochem.*, 20, 201, 1986.
15. **Woodford, T. A. and Pardee, A. B.,** Histone H1 kinase in exponential and synchronous populations of Chinese hamster fibroblast, *J. Biol. Chem.*, 261, 4669, 1986.
16. **Cicirelli, M. F., Pelech, S. L., and Krebs, E. G.,** Activation of multiple protein kinases during the burst in protein phosphorylation that precedes the first meiotic cell division in *Xenopus* oocytes, *J. Biol. Chem.*, 263, 2009, 1988.
17. **Mueller, R. D., Yasuda, H., and Bradbury, E. M.,** Phosphorylation of histone H1 through the cell cycle of Physarum polycephalum. 24 sites of phosphorylation at metaphase, *J. Biol. Chem.*, 260, 5081, 1985.
18. **Inglis, R. J., Langan, T. A., Matthews, H. R., Hardie, D. G., and Bradbury, E. M.,** Advance of mitosis by histone phosphokinase, *Exp. Cell. Res.*, 97, 418, 1976.
19. **Arion, D., Meijer, L., Brizuela, L., and Beach, D.,** cdc2 is a component of the M phase-specific histone H1 kinase: evidence for identity with MPF, *Cell*, 55, 371, 1988.
20. **Dunphy, W. G. and Newport, J. W.,** Unraveling of mitotic control mechanisms, *Cell*, 55, 925, 1988.
21. **Draetta, G., Brizuela, L., Potashkin, J., and Beach, D.,** Identification of p34 and p13, human homologs of the cell cycle regulators of fission yeast encoded by *cdc*2$^+$ and *suc*1$^+$, *Cell*, 50, 319, 1987.
22. **Draetta, G., Piwnica-Worms, H., Morrison, D., Drukerm B., Roberts, T., and Beach, D.,** Human cdc2 protein kinase is a major cell-cycle regulated tyrosine kinase substrate, *Nature (London)*, 336, 738, 1988.
23. **Bishop, J. M.,** Cellular oncogenes and retroviruses, *Ann. Rev. Biochem.*, 52, 301, 1983.
24. **Land, H., Parada, L. F., and Weinberg, R. A.,** Cellular oncogenes and multistep carcinogenesis, *Science*, 222, 771, 1983.
25. **Duesberg, P. H.,** Activated proto-onc genes: sufficient or necessary for cancer?, *Science*, 228, 669, 1985.
26. **Collins, S. J. and Groudine, M.,** Amplification of endogenous *myc*-related DNA sequences in a human myeloid leukaemia cell line, *Nature (London)*, 298, 679, 1982.
27. **Dalla-Favera, R.,** *onc* gene amplification in promyelocytic leukaemia cell line HL-60 and primary leukaemic cells of the same patient, *Nature (London)*, 299, 61, 1982.
28. **Alitalo, K., Schwab, M., Lin, C. C., Varmus, H. E., and Bishop, J. M.,** Homogeneously staining chromosomal regions contain amplified copies of an abundantly expressed cellular oncogene (c-*myc*) in malignant neuroendocrine cells from a human colon carcinoma, *Proc. Natl. Acad. Sci. U.S.A.*, 80, 1707, 1983.
29. **Kelly, K., Cochran, B. H., Stiles, C. D., and Leder, P.,** Cell-specific regulation of the c-*myc* gene by lymphocyte mitogens and platelet-derived growth factor, *Cell*, 35, 603, 1983.
30. **Greenberg, M. E. and Ziff, E. B.,** Stimulation of 3T3 cells induces transcription of the c-*fos* proto-oncogene, *Nature (London)*, 311, 433, 1984.
31. **Makino, R., Hayashi, K., and Sugimura, T.,** c-*myc* transcript is induced in rat liver at a very early stage of regeneration or by cycloheximide treatment, *Nature (London)*, 310, 697, 1984.
32. **Campisi, J., Gray, H. E., Pardee, A. B., Dean, M., and Sonensheim, G. E.,** Cell-cycle control of c-*myc* but not c-*ras* expression is lost following chemical transformation, *Cell*, 36, 241, 1984.
33. **Dani, Ch., Blanchard, J. M., Piechaczyk, M., El Sabouty, S., Marty, L., and Jeanteur, Ph.,** Extreme instability of myc mRNA in normal and transformed human cells, *Proc. Natl. Acad. Sci. U.S.A.*, 81, 7046, 1984.
34. **Blanchard, J. M., Piechaczyk, M., Dani, C., Chambard, J. -C., Franchi, A., Pouyssegur, J., and Jeanteur, P.,** c-*myc* gene is transcribed at high rate in G_0-arrested fibroblasts and is post-transcriptionally regulated in response to growth factors, *Nature (London)*, 317, 443, 1985.
35. **Murphree, A. L. and Benedict, W. F.,** Retinoblastoma: clues to human oncogenesis, *Science*, 223, 1028, 1984.
36. **Schneider, C., King, R. M., and Philipson, L.,** Genes specifically expressed at growth arrest of mammalian cells, *Cell*, 54, 787, 1988.
37. **Ikegami, S., Taguchi, T., Ohashi, M., Oguro, K., Nagano, H., and Mano, Y.,** Aphidicolin prevents mitotic cell division by interfering with the activity of DNA polymerase-α, *Nature (London)*, 275, 458, 1978.
38. **Miller, M. R., Ulrich, R. G., Wang, T. S.-F., and Korn, D.,** Monoclonal antibodies against human DNA polymerase-α inhibit DNA replication in permeabilized cells, *J. Biol. Chem.*, 260, 134, 1985.
39. **Kaczmarek, L., Miller, M. R., Hammond, R. A., and Mercer, W. E.,** A microinjected monoclonal antibody against human DNA polymerase-α inhibits DNA replication in human, hamster, and mouse cell lines, *J. Biol. Chem.*, 261, 10802, 1986.
40. **Fry, M. and Loeb, L. A.,** *Animal Cell DNA Polymerases*, CRC Press, Boca Raton, FL, 1986, 13.
41. **Murakami, Y., Yasuda, H., Miyazawa, H., Hanaoka, F., and Yamada, M.,** Characterization of a temperature-sensitive mutant of mouse FM3A cells defective in DNA replication, *Proc. Natl. Acad. Sci. U.S.A.*, 82, 1761, 1985.

42. **Wahl, A. F., Geis, A. M., Spain, B. H., Wong, S. W., Korn, D., and Wang, T. S.-F.,** Gene expression of human DNA polymerase-α during cell proliferation and the cell cycle, *Mol. Cell. Biol.*, 8, 5016, 1988.
43. **Schimke, R. T.,** Gene amplification in cultured cells, *J. Biol. Chem.*, 263, 5989, 1988.
44. **Blau, H. M., Pavlath, G. K., Hardeman, E. C., Choy, C.-P., Silberstein, L., Webster, S. G., Miller, S. C., and Webster, C.,** Plasticity of the differentiated state, *Science*, 230, 758, 1985.
45. **Lajtha, L. G.,** Recent studies in erythroid differentiation and proliferation, *Medicine*, 43, 625, 1964.
46. **Rao, M. S., Scarpelli, D. G., and Reddy, J. K.,** Transdifferentiated hepatocytes in rat pancreas, *Curr. Top. Dev. Biol.*, 20, 63, 1986.
47. **Schmid, V. and Alder, H.,** Isolated, mononucleated, striated muscle can undergo pluripotent transdifferentiation and form a complex regenerate, *Cell*, 38, 801, 1984.
48. **Colletta, G., Cirafici, A. M., and Vecchio, G.,** Induction of the c-*fos* oncogene by thyrotropic hormone in rat thyroid cells in culture, *Science*, 233, 458, 1986.
49. **Burgess, A. W.,** Hematopoietic growth factors, in *Mediators in Cell Growth and Differentiation*, Vol. 37, Ford, R. J. and Maizel, A. L., Eds., Raven Press, New York, 1985, 159.
50. **Metcalf, D.,** The granulocyte-macrophage colony stimulating factors, *Science*, 229, 16, 1985.
51. **Metcalf, D., Begley, C. G., Johnson, G. R., Nicola, N. A., Vadas, M. A., Lopez, A. F., Williamson, D. J., Wong, G. G., Clark, S. C., and Wang, E. A.,** Biologic properties in vitro of a recombinant granulocyte-macrophage colony stimulating factor, *Blood*, 67, 37, 1986.
52. **Fung, M. C., Hapel, A. J., Yoner, S., Cohen, D. R., Johnson, R. M., Campbell, H. D., and Young, I. G.,** Molecular cloning of cDNA for murine interleukin-3, *Nature (London)*, 307, 233, 1984.
53. **Conover, C. A., Hintz, R. L., and Rosenfeld, R. G.,** Comparative effects of somatomedin C and insulin on the metabolism and growth of cultured human fibroblasts, *J. Cell. Phys.*, 122, 133, 1985.
54. **Froesch, E. R., Schmid, C., Schwander, J., and Zapf, J.,** Actions of insulin-like growth factors, *Annu. Rev. Physiol.*, 47, 443, 1985.
55. **Kusyk, C. J., McNiel, N. O., and Johnson, L. R.,** Stimulation of growth of a colon cancer cell line by gastrin, *Am. J. Physiol.*, 251, G597, 1986.
56. **Logsdon, C. D.,** Stimulation of pancreatic acinar cell growth by CCK, epidermal growth factor, and insulin *in vitro*, *Am. J. Physiol.*, 251, G487, 1986.
57. **Heldin, C.-H. and Westermark, B.,** Growth factors: mechanism of action and relation to oncogenes, *Cell*, 37, 9, 1984.
58. **Pledger, W. J., Stiles, C. D., Antoniades, H. N., and Scher, C. D.,** Induction of DNA synthesis in Balb/c 3T3 cells by serum components: re-evaluation of the commitment process, *Proc. Natl. Acad. Sci. U.S.A.*, 74, 2839, 1977.
59. **Stiles, C. D., Capone, G. T., Scher, C. D., Antoniades, H. N., Van Wyk, J. J., and Pledger, W. J.,** Dual control of cell growth by somatomedins and platelet-derived growth factor, *Proc. Natl. Acad. Sci. U.S.A.*, 76, 1279, 1979.
60. **Vogel, A., Raines, E., Kariya, B., Rivest, M. J., and Ross, R.,** Coordinate control of 3T3 cell proliferation by platelet-derived growth factor and plasma components, *Proc. Natl. Acad. Sci. U.S.A.*, 75, 2810, 1978.
61. **Leof, E. B., Wharton, W., Van Wyk, J. J., and Pledger, W. J.,** Epidermal growth factor (EGF) and somatomedin C regulate G_1 progression in competent BALB/c-3T3 cells, *Exp. Cell. Res.*, 147, 202, 1982.
62. **Cohen, S.,** Isolation of a mouse submaxillary gland protein accelerating incisor eruption and eyelid opening in the new-born animal, *J. Biol. Chem.*, 237, 1555, 1962.
63. **Stastny, M. and Cohen, S.,** Epidermal growth factor IV the induction of ornithine decarboxylase, *Biochim. Biophys. Acta*, 204, 578, 1970.
64. **Moriarity, D. M., DiSorbo, D. M., Litwack, G., and Savage, C. R., Jr.,** Epidermal growth factor stimulation of ornithine decarboxylase activity in a human hepatoma cell line, *Proc. Natl. Acad. Sci. U.S.A.*, 78, 2752, 1981.
65. **Pegg, A. E. and McCann, P. P.,** Polyamine metabolism and function, *Am. J. Physiol.*, 243, C212, 1982.
66. **Savage, C. R., Inagami, T., and Cohen, S.,** The primary structure of epidermal growth factor, *J. Biol. Chem.*, 247, 7612, 1972.
67. **Taylor, J. M., Mitchell, W. M., and Cohen, S.,** A characterization of the high molecular weight form of epidermal growth factor, *J. Biol. Chem.*, 249, 3198, 1974.
68. **Sawyer, S. and Cohen, S.,** Enhancement of calcium uptake and phosphatidylinositol turnover by epidermal growth factor in A-431 cells, *Biochemistry*, 20, 6280, 1981.
69. **Boonstra, J., Moolenaar, W. H., Harrison, P. H., Moed, P., Van der Saag, P. T., and de Laat, S. W.,** Ionic responses and growth stimulation induced by nerve growth factor and epidermal growth factor in rat pheochromocytoma (PC12) cells, *J. Cell Biol.*, 97, 92, 1983.
70. **Barnes, D. and Colowick, S. P.,** Stimulation of sugar uptake in cultured fibroblasts by epidermal growth factor (EGF) and EGF-binding arginine esterase, *J. Cell. Physiol.*, 89, 633, 1976.
71. **Hollenberg, B. D. and Cuatrecasas, P.,** Insulin and epidermal growth factor. Human fibroblast receptors related to deoxyribonucleic acid synthesis and amino acid uptake, *J. Biol. Chem.*, 250, 3845, 1975.

72. **Covelli, I., Mozzi, R., Rossi, R., and Frati, L.,** A mechanism of action of the epidermal growth factor. III. Stimulation of the uptake of labeled precursors into RNA, DNA and proteins induced by EGF in isolated tumor cells, *Hormones*, 3, 183, 1972.
73. **Adamson, E. D. and Rees, A. R.,** EGF receptors, *Mol. Cell. Biochem.*, 34, 129, 1981.
74. **Gregory, H.,** Isolation and structure of urogastrone and its relation to EGF, *Nature (London)*, 257, 325, 1975.
75. **Ushiro, H. and Cohen, S.,** Identification of phosphotyrosine as a product of epidermal growth factor-activated protein kinase in A431 cell membranes, *J. Biol. Chem.*, 255, 8363, 1980.
76. **Hunter, T. and Cooper, J. A.,** Epidermal growth factor induces rapid tyrosine phosphorylation of proteins in A431 human tumor cells, *Cell*, 24, 741, 1981.
77. **Soderquist, A. M. and Carpenter, G.,** Biosynthesis and metabolic degradation of receptors for epidermal growth factor, *J. Membrane Biol.*, 90, 97, 1986.
78. **Hunter, T. and Cooper, J. A.,** Protein tyrosine kinases, *Annu. Rev. Biochem.*, 54, 897, 1985.
79. **Rosen, O. M., Herrera, R., Olowe, Y., Petruzzelli, L. M., and Cobb, N.,** Phosphorylation activates the insulin receptor tyrosine protein kinase, *Proc. Natl. Acad. Sci. U.S.A.*, 80, 3237, 1983.
80. **Jacobs, S., Kull, F. C., Jr., Earp, H. S., Scvobada, N. E., and Van Wyk, H. H.,** Somatomedin-C stimulates the phosphorylation of the β-subunit of its own receptor, *J. Biol. Chem.*, 258, 9581, 1983.
81. **Ek, B., Westermark, B., Wasteson, A., and Heldin, C. -H.,** Stimulation of tyrosine- specific phosphorylation by platelet derived growth factor, *Nature (London)*, 295, 419, 1982.
82. **Rettenmier, C. W., Chen, J. H., Roussel, M. F., and Sherr, C. J.,** The product of the c-*fms* proto-oncogene: a glycoprotein with associated tyrosine kinase activity, *Science*, 228, 320, 1985.
83. **Ullrich, A., Coussens, L., Hayflick, J. S., Dull, T. J., Gray, A., Tam, A. W., Lee, J., Yarden, Y., Liverman, T. A., Schlessinger, J., Downward, J., Mayes, E. L. V., Whittle, N., Waterfield, M. D., and Seeburg, P. H.,** Human epidermal growth factor receptor cDNA sequence and aberrant expression of the amplified gene in A431 epidermoid carcinoma cells, *Nature (London)*, 309, 418, 1984.
84. **Lax, I., Burgess, W. H., Bellot, F., Ullrich, A., Schlessinger, J., and Givol, D.,** Localization of a major receptor-binding domain for epidermal growth factor by affinity labeling, *Mol. Cell. Biol.*, 8, 1831, 1988.
85. **Downward, J., Waterfield, M. D., and Parker, P. J.,** Autophosphorylation and protein kinase C phosphorylation of the epidermal growth factor receptor, *J. Biol. Chem.*, 260, 14538, 1985.
86. **Ruso, M. W., Lukas, T. J., Cohen, S., and Staros, J. V.,** Identification of residues in the nucleotide binding site of the epidermal growth factor receptor/kinase, *J. Biol. Chem.*, 260, 5205, 1985.
87. **Yarden, Y. and Ullrich, A.,** Molecular analysis of signal transduction by growth factors, *Biochem.*, 27, 3113, 1988.
88. **Downward, J., Yarden, Y., Mayes, E., Scrace, G., Totty, N., Stockwell, P., Ullrich, A., Schlessinger, J., and Waterfield, M. D.,** Close similarity of epidermal growth factor receptor and v-*erb*-B oncogene protein sequences, *Nature (London)*, 307, 521, 1984.
89. **Riedel, H., Massoglia, S., Schlessinger, J., and Ullrich, A.,** Ligand activation of overexpressed epidermal growth factor receptors transforms NIH3T3 mouse fibroblasts, *Proc. Natl. Acad. Sci. U.S.A.*, 85, 1477, 1988.
90. **Velu, T. J., Bequinot, L., Vass, W. C., Willingham, W. C., Merlino, G. T., Pastan, I., and Lowy, D. R.,** Epidermal growth factor-dependent transformation by a human EGF receptor proto-oncogene, *Science*, 238, 1408, 1987.
91. **Sainsbury, J., Sherbet, G., Farndon, J., and Harris, A.,** Epidermal growth factor receptors and oestrogen receptors in human breast cancer, *Lancet*, 1, 364, 1985.
92. **Neal, D., Bennett, M., Hall, R., Marsh, C., Abel, P., Sainsbury, J., and Harris, A.,** Epidermal growth factor receptors in human bladder cancer: comparison of invasive and superficial tumours, *Lancet*, 1, 366, 1985.
93. **Libermann, T., Razon, M., Bartal, A., Yarden, Y., Schlessinger, J., and Soreq, J.,** Expression of epidermal growth factor receptors in human brain tumors, *Cancer Res.*, 44, 753, 1984.
94. **Hendler, F. and Ozanne, B.,** Human squamous cell lungs cancers express increased growth factor receptors, *J. Clin. Invest.*, 74, 647, 1984.
95. **Yasui, W., Hiromichi, S., Hata, J., Kameda, T., Ochiai, A., Hisao, I., and Tahara, E.,** Expression of epidermal growth factor receptor in human gastric and colonic carcinomas, *Cancer Res.*, 48, 137, 1987.
96. **Ro, J., North, S. M., Gallick, G. E., Hortobagyi, G. N., Gutterman, J. U., and Blick, M.,** Amplified and overexpressed epidermal growth factor receptor gene in uncultured primary human breast carcinoma, *Cancer Res.*, 48, 161, 1987.
97. **Koprowski, H., Herlyn, M., Balaban, G., Parmiter, A., Ross, A., and Nowell, P.,** Expression of the receptor for epidermal growth factor correlates with increased dosage of chromosome 7 in malignant melanoma, *Somatic Cell Molec. Genet.*, 11, 297, 1985.
98. **Filmus, J., Pollak, M., Caincross, J., and Buick, R.,** Amplified overexpressed and rearranged epidermal growth factor receptor gene in a human astrocytoma cell line, *Biochem. Biophys. Res. Commun.*, 131, 207, 1985.
99. **Filmus, J., Pollak, M., Cailleau, R., and Buick, R.,** MDA-468, a human breast cancer cell line with a high number of epidermal growth factor (EGF) receptor, has an amplified EGF receptor gene and is growth inhibited by EGF, *Biochem. Biophys. Res. Commun.*, 128, 898, 1985.

100. **Merlino, G., Xu, Y.-H., Richert, N., Clark, A. J. L., Ishii, S., Schlegel, S. B., and Pastan, I.,** Elevated epidermal growth factor receptor gene copy number and expression in a squamous carcinoma cell line, *J. Clin. Invest.*, 75, 1077, 1984.
101. **Hendler, F. J. and Ozanne, B. W.,** Human squamous cell lung cancers express increased epidermal growth factor receptors, *J. Clin. Invest.*, 74, 647, 1984.
102. **Ozanne, B., Shum, A., Richards, C. S., Cassells, D., Grossman, D., Trent, J., Gusterson, B., and Hendler, F.,** Evidence for an increase of EGF receptors in epidermoid malignancies, in *Cancer Cells 3: Growth Factors and Transformation*, Feramisco, J., Ozanne, B., and Stiles, C., Eds., Cold Spring Harbor Laboratory, Cold Spring Harbor, NY, 1985, 41.
103. **Shimizu, N., Kondo., I., Gamon, S., Behzadian, M., and Shimizu, Y.,** Genetic analysis of hyperproduction of epidermal growth factor receptors in human epidermoid carcinoma A431 cells, *Somatic Cell Molec. Genet.*, 10, 45, 1984.
104. **Lin, C. R., Chen, W. S., Kruiger, W., Stolarsky, L. S., Weber, W., Evans, R. M., Verma, I. M., Gill, G. N., and Rosenfeld, M. G.,** Expression cloning of human EGF receptor complementary DNA: gene amplification and three related messenger RNA products in A431 cells, *Science*, 224, 843, 1984.
105. **Korc, M., Meltzer, P., and Trent, J.,** Enhanced expression of epidermal growth factor receptor correlates with alterations of chromosome 7 in human pancreatic cancer, *Proc. Natl. Acad. Sci. U.S.A.*, 83, 5141, 1986.
106. **Lee, J. S., Pathak, S., Hopwood, V., Tomasovic, B., Mullins, T. D., Baker, F. L., Spitzer, G., and Neidhart, J. A.,** Involvement of chromosome 7 in primary lung tumor and nonmalignant normal lung tissue, *Cancer Res.*, 47, 6349, 1987.
107. **Yunis, J.,** The chromosomal basis of human neoplasia, *Science*, 221, 227, 1983.
108. **Yang-Feng, T., Francke, U., and Ullrich, A.,** Gene for human insulin receptor: localization to site on chromosome 19 involved in pre-B-cell leukemia, *Science*, 228, 728, 1985.
109. **Klein, G.,** The role of gene dosage and genetic transpositions in carcinogenesis, *Nature (London)*, 294, 313, 1981.
110. **Fearon, E. R., Hamilton, S. R., and Vogelstein, B.,** Clonal analysis of human colorectal tumors, *Science*, 238, 193, 1987.
111. **Garewal, H., Meltzer, P., Trent, J., Prabhala, R., Sampliner, R., and Korc, M.,** Chromosome 7 alteration and overexpression of EGF receptor in Barrett's esophagus, *Clin. Res.*, 36, 170A, 1988.
112. **Breathnach, R. and Chambon, P.,** Organization and expression of eukaryotic split genes coding for proteins, *Annu. Rev. Biochem.*, 50, 349, 1981.
113. **Kageyama, K., Merlino, G. T., and Pastan, I.,** Epidermal growth factor (EGF) receptor gene transcription, *J. Biol. Chem.*, 263, 6329, 1988.
114. **Johnson, A. C., Ishii, S., Jinno, Y., Pastan, I., and Merlino, G. T.,** Epidermal growth factor receptor gene promoter, *J. Biol. Chem.*, 263, 5693, 1988.
115. **Maniatis, T., Goodbourn, S., and Fischer, J. A.,** Regulation of inducible and tissue-specific gene expression, *Science*, 236, 1237, 1987.
116. **Hoffman, E. K., Trusko, S. P., Feeman, N. A., and George, D. L.,** Structural and functional characterization of the promoter region of the mouse c-Ki-ras gene, *Mol. Cell. Biol.*, 7, 2592, 1987.
117. **Honkawa, H., Masahashi, W., Hashimoto, S., and Hashimoto-Gotoh, T.,** Identification of the principal promoter sequence of the c-H-ras transforming oncogene: deletion analysis of the 5′-flanking region by focus formation assay, *Mol. Cell. Biol.*, 7, 2933, 1987.
118. **Azizkhan, J. C., Vaughn, J. P., Christy, R. J., and Hamlin J. L.,** Nucleotide sequence and nuclease hypersensitivity of the Chinese hamster dihydrofolate reductase gene promoter region, *Biochemistry*, 25, 6228, 1986.
119. **Cattaneo, R., Will, H., Hernandez, N., and Schaller, H.,** Signals regulating hepatitis B surface antigen transcription, *Nature (London)*, 305, 326, 1983.
120. **Jakobovits, E. B., Schlokat, U., Vannice, J. L., Derynck, R., and Levinson, A. D.,** The human transforming growth factor alpha promoter directs transcription initiation from a single site in the absence of a TATA sequence, *Mol. Cell. Biol.*, 8, 5549, 1988.
121. **Carpenter, G. and Cohen, S.,** Epidermal growth factor, *Annu. Rev. Biochem.*, 48, 193, 1979.
122. **Pastan, I. H. and Willingham, M. C.,** Journey to the center of the cell: role of the receptosome, *Science*, 214, 504, 1981.
123. **Beguinot, L., Lyall, R. M., Willingham, M. C., and Pastan, I.,** Down-regulation of the epidermal growth factor receptor in KB cells is due to receptor internalization and subsequent degradation in lysosomes, *Proc. Natl. Acad. Sci. U.S.A.*, 81, 2384, 1984.
124. **Cruise, J. L., Cotecchia, S., and Michalopoulos, G.,** Norepinephrine decreases EGF binding in primary rat hepatocyte cultures, *J. Cell. Phys.*, 127, 39, 1986.
125. **Pessin, J. E., Gitomer, W., Oka, Y., Oppenheimer, C. L., and Czech, M. P.,** β-adrenergic regulation insulin and epidermal growth factor receptors in rat adipocytes, *J. Biol. Chem.*, 258, 7386, 1983.

126. **Wharton, W., Leof, E., Pledger, W. J., and O'Keefe, E. J.,** Modulation of the epidermal growth factor receptor by platelet-derived growth factor and choleragen: effects on mitogenesis, *Proc. Natl. Acad. Sci. U.S.A.*, 79, 5567, 1982.
127. **Korc, M., Matrisian, L. M., and Magun, B. E.,** Cytosolic calcium regulates EGF binding and internalization in pancreatic acini and cultured fibroblasts, *Proc. Natl. Acad. Sci. U.S.A.*, 81, 461, 1984.
128. **Rozengurt, E., Brown, K. D., and Pettican, P.,** Vasopressin inhibition of epidermal growth factor binding to cultured mouse cells, *J. Biol. Chem.*, 256, 716, 1981.
129. **Rozengurt, E., Collins, M., Brown, K. D., and Pettican, P.,** Inhibition of epidermal growth factor binding to mouse cultured cells by fibroblast-derived growth factor, *J. Biol. Chem.*, 257, 3680, 1982.
130. **Koga, M., Eisman, J. A., and Sutherland, R. L.,** Regulation of epidermal growth factor receptor levels by 1,25-dihydroxyvitamin D_3 in human breast cancer cells, *Cancer Res.*, 48, 2734, 1988.
131. **Bird, T. A. and Saklatvala, J.,** IL-1 and TNF transmodulate epidermal growth factor receptors by a protein kinase c-independent mechanism, *J. Immunol.*, 142, 126, 1989.
132. **Lee, L.-S. and Weinstein, I. B.,** Mechanism of tumor promoter inhibition of cellular binding of epidermal growth factor, *Proc. Natl. Acad. Sci. U.S.A.*, 76, 5168, 1979.
133. **Magun, B. E., Matrisian, L. M., and Bowden, G. T.,** Epidermal growth factor: ability of tumor promoter to alter its degradation, receptor affinity, and receptor number, *J. Biol. Chem.*, 255, 6373, 1980.
134. **Korc, M. and Magun, B. E.,** Residual inhibition of epidermal growth factor binding by pancreatic secreatagogues and phorbol ester in rat pancreas, *J. Cell. Physiol.*, 124, 344, 1985.
135. **Castagna, M., Takai, Y., Kaibuchi, K., Sano, K., Kikkawa, U., and Nishizuka, Y.,** Direct activation of calcium-activated, phospholipid-dependent protein kinase by tumour-promoting phorbol esters, *J. Biol. Chem.*, 257, 7847, 1982.
136. **Hunter, T., Ling, N., and Cooper, J. A.,** Protein kinase C phosphorylation of the EGF receptor at a threonine residue close to the cytoplasmic face of the plasma membrane, *Nature (London)*, 311, 480, 1985.
137. **Lin, C. R., Chen, W. S., Lazar, C. S., Carpenter, C. D., Gill, G. N., Evans, R. M., and Rosenfeld, M. G.,** Protein kinase C phosphorylation at Thr 654 of the unoccupied EGF receptor and EGF binding regulate functional receptor loss by independent mechanisms, *Cell*, 44, 839, 1986.
138. **May, W. S., Jacobs, S., and Cuatrecasas, P.,** Association of phorbol ester induced hyperphosphorylation and reversible regulation of transferrin membrane receptors in HL60 cells, *Proc. Natl. Acad. Sci. U.S.A.*, 81, 2016, 1984.
139. **Strulovici, B., Cerione, R. A., Kilpatrick, B. F., Caron, M. G., and Lefkowitz, R. J.,** Direct demonstration of impaired functionality of a purified desensitized beta-adrenergic receptor in a reconstituted system, *Science*, 225, 837, 1984.
140. **Liles, W. C., Hunter, D. D., Meier, K. E., and Nathanson, N. M.,** Activation of protein kinase C induces rapid internalization and subsequent degradation of muscarinic acetylcholine receptors in neuroblastoma cells, *J. Biol. Chem.*, 261, 5307, 1986.
141. **Assoian, R. K., Frolik, C. A., Roberts, A. B., Miller, D. M., and Sporn, M. B.,** Transforming growth factor-β controls receptor levels for epidermal growth factor in NRK fibroblasts, *Cell*, 36, 35, 1984.
142. **Mukku, V. R. and Stancel, G. M.,** Regulation of epidermal growth factor receptor by estrogen, *J. Biol. Chem.*, 260, 9820, 1985.
143. **Ivanovic, V. and Weinstein, I. B.,** Glucocorticoids and benzo(a)pyrene have opposing effects on EGF receptor binding, *Nature (London)*, 293, 404, 1981.
144. **Fanger, B. O., Viceps-Madore, D., and Cidlowski, J. A.,** Regulation of high- and low-affinity epidermal growth factor receptors by glucocorticoids, *Arch. Biochem. Biophys.*, 235, 141, 1984.
145. **Mukku, V. R.,** Regulation of epidermal growth factor receptor levels by thyroid hormone, *J. Biol. Chem.*, 259, 6543, 1984.
146. **Jansson, J.-O., Ekberg, S., Hoath, S. B., Beamer, W. G., and Frohman, L. A.,** Growth hormone enhances hepatic epidermal growth factor receptor concentration in mice, *J. Clin. Invest.*, 82, 1871, 1988.
147. **Palombell, V. J., Yamashiro, D. J., Maxfield, F. R., Decker, S. J., and Vilcek, J.,** Tumor necrosis factor increases the number of epidermal growth factor receptors on human fibroblasts, *J. Biol. Chem.*, 262, 1950, 1987.
148. **Assoian, R. K.,** Biphase effects of type β transforming growth factor on epidermal growth factor receptors in NRK fibroblasts, *J. Biol. Chem.*, 260, 9613, 1985.
149. **Massague, J.,** Epidermal growth factor-like transforming growth factor. I. Isolation, chemical characterization and potentiation by other transforming factors from feline sarcoma virus-transformed rat cells, *J. Biol. Chem.*, 258, 13606, 1983.
150. **Reisner, A. H.,** Similarity between the vaccinia virus 19 K early protein and epidermal growth factor, *Nature (London)*, 313, 801, 1985.
151. **Twardzik, D. R., Brown, J. P., Ranchalis, J. E., Todaro, G. J., and Moss, B.,** Vaccinia virus-infected cells release a novel polypeptide functionally related to transforming and epidermal growth factors, *Proc. Natl. Acad. Sci. U.S.A.*, 82, 5300, 1985.

152. **Massague, J. and Czech, M. P.,** The subunit structures of two distinct receptors for IGF-I and II and their relationship to the insulin receptor, J. *Biol. Chem.*, 257, 5038, 1982.
153. **Todaro, G. J., Fryling, C., and DeLarco, J. E.,** Transforming growth factors produced by certain human tumor cells: polypeptides that interact with epidermal growth factor receptors, *Proc. Natl. Acad. Sci. U.S.A.*, 77, 5258, 1980.
154. **Massague, J.,** Epidermal growth factor-like transforming growth factor. II. Interaction with epidermal growth factor receptors in human placenta membranes and A431 cells, *J. Biol. Chem.*, 258, 13614, 1983.
155. **Derynck, R., Roberts, A. B., Winkler, M. E., Chen, E. Y., and Goeddel, D. V.,** Human transforming growth factor-α: precursor structure and expression in *E. coli*, *Cell*, 38, 287, 1984.
156. **Marquardt, H., Hunkapiller, M. W., Hood, L. E., and Todaro, G. J.,** Rat transforming growth factor type 1 structure and relation to epidermal growth factor, *Science*, 223, 1079, 1984.
157. **Pike, L. H., Marquardt, H., Todaro, G. J., Gallis, B., Casnellie, J. E., Bornstein, P., and Krebs, E. G.,** Transforming growth factor and epidermal growth factor stimulate the phosphorylation of a synthetic tyrosine-containing peptide in a similar manner, *J. Biol. Chem.*, 257, 14628, 1982.
158. **Smith, J. M., Sporn, M. B., Roberts, A. B., Derynck, R., Winkler, M. E., and Gregory, H.,** Human transforming growth factor-α causes precocious eyelid opening in newborn mice, *Nature (London)*, 315, 515, 1985.
159. **Rhodes, J. A., Tam, J. P., Finke, U., Saunders, M., Bernanke, J., Silen, W., and Murphy, R. A.,** Transforming growth factor α inhibits secretion of gastric acid, *Proc. Natl. Acad. Sci. U.S.A.*, 83, 3844, 1986.
160. **Tam, J. P.,** Physiological effects of transforming growth factor in the newborn mouse, *Science*, 229, 673, 1985.
161. **Schreiber, A. B., Winkler, M. E., and Derynck, R.,** Transforming growth factor-α: a more potent angiogenic mediator than epidermal growth factor, *Science*, 232, 1250, 1986.
162. **Ibbotson, K. J., Harrod, J., Gowen, M., D'Souza, S., Smith, D. D., Winkler, M. E., Derynck, R., and Mundy G. R.,** Human recombinant transforming growth factor α stimulates bone esorption and inhibits formation *in vitro*, *Proc. Natl. Acad. Sci. U.S.A.*, 83, 2228, 1986.
163. **Gan, B. S., Hollenberg, M. D., MacCannell, K. L., Lederis, K., Winkler, M. E., and Derynck, R.,** Distinct vascular actions of epidermal growth factor-urogastrone and transforming growth factor-α, *J. Pharmacol. Exp. Ther.*, 242, 331, 1987.
164. **Schultz, G. S., White, M., Mitchell, R., Brown, G., Lynch, J., Twardzik, D. R., and Todaro, G. J.,** Epithelial wound healing enhanced by transforming growth factor α and vaccinia growth factor, *Science*, 235, 350, 1987.
165. **Barrandon, Y. and Green, H.,** Cell migration is essential for sustained growth of keratinocyte colonies: the roles of transforming growth factor-α and epidermal growth factor, *Cell*, 50, 1131, 1987.
166. **Myrdal, S. E., Twardzik, D. R., and Auersperg, N.,** Cell-mediated co-action of transforming growth factors: incubation of type β with normal rat kidney cells produces a soluble activity that prolongs the ruffling response to type α, *J. Cell Biol.*, 102, 1230, 1986.
167. **Smith, J. J., Derynck, R., and Korc, M.,** Production of transforming growth factor α in human pancreatic cancer cells: evidence for a superagonist autocrine cycle, *Proc. Natl. Acad. Sci. U.S.A.*, 84, 7567, 1987.
168. **Korc, M., Haussler, C. A., and Trookman, N. S.,** Divergent effects of epidermal growth factor and transforming growth factors on a human endometrial carcinoma cell line, *Cancer Res.*, 47, 4909, 1987.
169. **Siegfried, J. M.,** Detection of human lung epithelial cell growth factors produced by a lung carcinoma cell line: use in culture of primary solid lung tumors, *Cancer Res.*, 47, 2903, 1987.
170. **Lax, I., Johnson, A., Howk, R., Sap, J., Bellot, F., Winkler, M., Ullrich, A., Vennstrom, B., Schlessinger, J., and Givol, D.,** Chicken epidermal growth factor (EGF) receptor: cDNA cloning, expression in mouse cells, and differential binding of EGF and transforming growth factor alpha, *Mol. Cell. Biol.*, 8, 1970, 1988.
171. **Feng, P., Knecht, M., and Catt, K.,** Hormonal control of epidermal growth factor receptors by gonadotropins during granulosa cell differentiation, *Endocrinology*, 120, 1121, 1987.
172. **Adamson, E. D. and Meek, J.,** The ontogeny of epidermal growth factor receptors during mouse development, *Dev. Biol.*, 103, 62, 1984.
173. **Lai, W. H. and Guyda, H. J.,** Characterization and regulation of epidermal growth factor receptors in human placental cell cultures, *Clin. Endocrinol. Metab.*, 58, 344, 1984.
174. **Sissom, J. F., Stenzel, W. K., and Warshaw, J. B.,** Decreased binding of epidermal growth factor in placentas from streptozotocin-diabetic rats, *J. Clin. Invest.*, 80, 242, 1987.
175. **Korc, M., Matrisian, L. M., Nakamura, R., and Magun, B. E.,** Epidermal growth factor binding is altered in pancreatic acini from diabetic rats, *Life Sci.*, 35, 2049, 1984.
176. **Lev-Ran, A., Hwang, D. L., and Barseghian, G.,** Decreased expression of liver epidermal growth factor receptors in rats with alloxan and streptozotocin diabetes, *Biochem. Biophys. Res. Comm.*, 137, 258, 1986.
177. **Blackshear, P. J., Stumpo, D. J., Kennington, E. A., Tuttle, J. S., Orth, D. N., Thompson, K. L., Hung, M.-C., and Rosner, M. R.,** Decreased levels of hepatic epidermal growth factor receptors in obese hyperglycemic rodents, *J. Biol. Chem.*, 262, 12356, 1987.
178. **Horuk, R. and Olefsky, J. M.,** Post binding events in insulin action, *Diabetes/Metabolism Rev.*, 1, 59, 1985.

179. **Planck, S. R., Finch, J. S., and Magun, B. E.,** Intracellular processing of epidermal growth factor. II. Intracellular cleavage of the COOH-terminal region of ^{125}I-epidermal growth factor, *J. Biol. Chem.*, 259, 3053, 1984.
180. **Korc, M., Matrisian, L. M., Planck, S. R., and Magun, B. M.,** Binding of epidermal growth factor in rat pancreatic acini, *Biochem. Biophys. Res. Commun.*, 111, 1066, 1983.
181. **Korc, M. and Magun, B.,** Recycling of epidermal growth factor in a human pancreatic carcinoma cell line, *Proc. Natl. Acad. Sci. U.S.A.*, 82, 6172, 1985.
182. **Maratos-Flier, E., Kao, C. Y., Verdin, E. M., and King, G. L.,** Receptor-mediated vectorial transcytosis of epidermal growth factor by madin-darby canine kidney cells, *J. Cell Biol.*, 105, 1595, 1987.
183. **Dernovsek, K. D., Bar, R. S., Ginsberg, B. H., and Lioubin, M. N.,** Rapid transport of biologically intact insulin through cultured endothelial cells, *J. Clin. Endocrinol. Metab.*, 58, 761, 1984.
184. **Goldfine, I. D., Purrello, F., Vigneri, R., and Clawson, G. A.,** Insulin and the regulation of isolated nuclei and nuclear subfractions: potential relationship to mRNA metabolism, *Diab. Met. Rev.*, 1, 119, 1985.
185. **Greengard, P.,** Phosphorylated proteins as physiological effectors, *Science*, 199, 146, 1978.
186. **Hunter, T.,** A thousand and one protein kinases, *Cell*, 50, 823, 1987.
187. **Hanks, S. K., Quinn, A. M., and Hunter, T.,** The protein kinase family: conserved features and deduced phylogeny of the catalytic domains, *Science*, 241, 42, 1988.
188. **Yarden, Y.,** Growth factor receptor tyrosine kinases, *Ann. Rev. Biochem.*, 57, 443, 1988.
189. **Hunter, T.,** A tail of two src's: mutatis mutandis, *Cell*, 49, 1, 1987.
190. **Blackshear, P. J., Nairn, A. C., and Kuo, J. F.,** Protein kinases 1988: a current perspective, *FASEB J.*, 2, 2957, 1988.
191. **Palfrey, H. C., Nairn, A. C., Muldoon, L. L., and Villereal, M. L.,** Rapid activation of calmodulin dependent protein kinase III in mitogen-stimulated human fibroblasts, *J. Biol. Chem.*, 262, 9785, 1987.
192. **Blenis, J., Kuo, C. J., and Erickson, R. L.,** Identification of ribosomal protein S6 kinase regulated by transformation and growth-promoting stimuli, *J. Biol. Chem.*, 262, 14373, 1987.
193. **Nishizuka, Y.,** The role of protein kinase C in cell surface signal transduction and tumour promotion, *Nature (London)*, 308, 693, 1984.
194. **Takai, Y., Kaibuchi, K., Tsuda, T., and Hoshijima, M.,** Role of protein kinase C in transmembrane signaling, *J. Cell. Biochem.*, 29, 143, 1985.
195. **Otte, A. P., Koster, C. H., Snoek, G. T., and Durston, A. J.,** Protein kinase C mediates neural induction in *Xenopus laevis*, *Nature (London)*, 334, 618, 1988.
196. **Parker, P. J., Coussens, L., Totty, N., Young, S., Chen, E., Stabel, S., Waterfield, M. D., and Ulrich, A.,** The complete primary structure of protein kinase C, the major phorbol ester receptor, *Science*, 233, 853, 1986.
197. **Coussens, L., Parker, P. J., Rhee, R., Yang-Feng, T. L., Chen, E., Waterfield, M. D., Francke, U., and Ulrich, A.,** Multiple, distinct forms of bovine and human protein kinase C suggest diversity in cellular signaling pathways, *Science*, 233, 853, 1986.
198. **Huang, F. L., Yoshida, Y., Nakabayashi, H., and Huang, K.-P.,** Differential distribution of protein kinase C isozymes in the various regions of brain, *J. Biol. Chem.*, 262, 15714, 1987.
199. **McPhail, L. C., Clayton, C. C., and Synderman, R. A.,** Potential second messenger role for unsaturated fatty acids: activation of Ca^{2+}-dependent protein kinase, *Science*, 224, 622, 1984.
200. **Oishi, K., Raynor, R. L., Charp, P. A., and Kuo, J. F.,** Regulation of protein kinase C by lysophospholipids. Potential role in signal transduction, *J. Biol. Chem.*, 263, 6865, 1988.
201. **Persons, D. A., Wilkison, W. O., Bell, R. M., and Finn, O. J.,** Altered growth regulation and enhanced tumorigenicity of NIH 3T3 fibroblasts transfected with protein kinase C-I cDNA, *Cell*, 52, 447, 1988.
202. **Housey, G. M., Johnson, M. D., Hsiao, W. L. W., O'Brian, C. A., Murphy, J. P., Kirschmeier, P., and Weinstein, I. B.,** Overproduction of protein kinase C causes disordered growth control in rat fibroblasts, *Cell*, 52, 343, 1988.
203. **Weber, W., Bertics, P. J., and Gill, G. N.,** Immunoaffinity purification of the epidermal growth factor receptor, *J. Biol. Chem.*, 259, 14631, 1984.
204. **Bertics, P. J. and Gill, G. N.,** Self phosphorylation enhances the protein tyrosine kinase activity of the epidermal growth factor receptor, *J. Biol. Chem.*, 260, 14642, 1985.
205. **Carpenter, G.,** The biochemistry and physiology of the receptor-kinase for EGF, *Mol. Cell. Endocrin.*, 31, 1, 1983.
206. **Pike, L. J., Kuenzel, E. A., Casnellie, J. E., and Krebs, E. G.,** A comparison of the insulin- and epidermal growth factor-stimulated protein kinases from human placenta, *J. Biol. Chem.*, 259, 9913, 1984.
207. **Koland, J. G. and Cerione, R. A.,** Growth factor control of epidermal growth factor receptor kinase activity via an intramolecular mechanism, *J. Biol. Chem.*, 263, 2230, 1988.
208. **Schlessinger, J.,** Allosteric regulation of the epidermal growth factor receptor kinase, *J. Cell Biol.*, 103, 2067, 1986.
209. **Fanger, B. O., Stephens, J. E., and Staros, J. V.,** High-yield trapping of EGF-induced receptor dimers by chemical cross-linking, *FASEB J.*, 3, 71, 1989.

210. **White, M., Haring, H. U., Kasuga, M., and Kahn, C. R.,** Kinetic properties and sites of autophosphorylation of the partially purified insulin receptor from hepatoma cells, *J. Biol. Chem.* 259, 255, 1984.
211. **Czech, M. P.,** The nature and regulation of the insulin receptor: structure and function, *Ann. Rev. Physiol.*, 47, 357, 1985.
212. **Goldenring, J. R., Oddsdottir, M., and Modlin, I. M.,** A novel protein kinase activity in rabbit gastric gland cytosol, *Biochem. Biophys. Res. Commun.*, 142, 559, 1987.
213. **Ehrlich, Y. H., Whittemore, S. R., Lambert, R., Ellis, J., Graber, S. G., and Lenox, R. H.,** GTP-preferring protein phosphorylation systems in brain membranes: possible role in adenylate cyclase regulation, *Biochem. Biophys. Res. Commun.*, 107, 699, 1982.
214. **Elias, L. and Davis, A.,** Manganese-phospholipid-stimulated protein kinase activity of human leukemic cells, *J. Biol. Chem.*, 260, 7023, 1985.
215. **Hathaway, G. M. and Traugh, J. A.,** Casein kinase II, *Methods Enzymol.*, 99, 317, 1983.
216. **Sommercorn, J., Mulligan, J. A., Lozeman, F. J., and Krebs, E. G.,** Activation of casein kinase II in response to insulin and to epidermal growth factor, *Proc. Natl. Acad. Sci. U.S.A.*, 84, 8834, 1987.
217. **Korc, M.,** Manganese homeostasis in humans and its role in disease states, in *Essential and Toxic Trace Elements in Human Health and Disease*, Pasada, A. S., Ed., Alan R. Liss, New York, 1988, 253.
218. **Schlessinger, J.,** The epidermal growth factor receptor as a multifunctional allosteric protein, *Biochem.*, 27, 3119, 1988.
219. **Cohen, S., Carpenter, G., and King, L.,** Epidermal growth factor-receptor-protein kinase interactions. Co-purification of receptor and epidermal growth factor-enchanced phosphorylation activity, *J. Biol. Chem.*, 255, 4834, 1980.
220. **Fava, R. A. and Cohen, S.,** Isolation of a calcium-dependent 35-kilodalton substrate for the epidermal growth factor receptor/kinase from A-431 cells, *J. Biol. Chem.*, 259, 2636, 1984.
221. **Pepinsky, R. B. and Sinclair, L. K.,** Epidermal growth factor-dependent phosphorylation of lipocortin, *Nature (London)*, 321, 81, 1986.
222. **Ehrhart, J. -C., Creuzet, C., Rollet, E., and Loeb, J.,** Epidermal growth factor-stimulated phosphorylation of tyrosine residues on a 120000 dalton protein in mouse liver plasma membrane subfractions, *Biochem. Biophys. Res. Comm.*, 102, 602, 1981.
223. **Tavare, J. M., Diggle, T. A., and Denton, R. M.,** Epidermal growth factor, but not insulin, stimulates tyrosine phosphorylation of an endogenous protein of Mr 35000 in Triton extracts of human placental syncytiotrophoblast membranes, *Biochem. J.*, 244, 769, 1987.
224. **Nakamura, K. D., Martinez, R., and Weber, M. J.,** Tyrosine phosphorylation of specific proteins after mitogen stimulation of chicken embryo fibroblasts, *Mol. Cell. Biol.*, 3, 380, 1983.
225. **Cooper, J. A., Bowen-Pope, D. F., Raines, E., Ross, R., and Hunter, T.,** Similiar effects of platelet-derived growth factor and epidermal growth factor on the phosphorylation of tyrosine in cellular proteins, *Cell*, 31, 263, 1982.
226. **Alitalo, K., Ralston, R. R., and Keski-Oja, J.,** Distribution of the 36000 dalton tyrosine protein kinase substrate in drug and epidermal growth factor-treated epithelial cells, *Exp. Cell Res.*, 150, 177, 1984.
227. **Erikson, E., Shealy, D. J., and Erikson, R. L.,** Evidence that viral transforming gene products and epidermal growth factor stimulate phosphorylation of the same cellular protein with similar specificity, *J. Biol. Chem.*, 256, 11381, 1981.
228. **Ghosh-Dastidar, P. and Fox, C. F.,** Epidermal growth factor and epidermal growth factor receptor-dependent phosphorylation of a M_r = 34000 protein substrate for $pp60^{src}$, *J. Biol. Chem.*, 258, 2041, 1983.
229. **Blackshear, P. J., Nemenoff, R. A., and Avruch, J.,** Insulin and growth factors stimulate the phosphorylation of a M_r-22000 protein in 3T3-L1 adipocytes, *Biochem. J.*, 214, 11, 1983.
230. **Iwashita, S. and Fox, C. F.,** Epidermal growth factor and potent phorbol tumor promoters induce epidermal growth factor receptor phosphorylation in a similiar but distinctively different manner in human epidermoid carcinoma A431 cells, *J. Biol. Chem.*, 259, 2559, 1984.
231. **Friedman, B., Francelton, A. R., Jr., Ross, A. H., Connors, J. M., Fujiki, H., Sugimura, T., and Rosner, M. R.,** Tumor promoters block tyrosine-specific phosphorylation of the epidermal growth factor receptor, *Proc. Natl. Acad. Sci. U.S.A.*, 81, 3034, 1984.
232. **Wong, T. W. and Goldberg, A. R.,** *In vitro* phosphorylation of angiotensin analogs by tyrosyl protein kinases, *J. Biol. Chem.*, 258, 1022, 1983.
233. **Rao, K. V. S. and Fox, F. C.,** Epidermal growth factor stimulates tyrosine phosphorylation of human glucocorticoid receptor in cultured cells, *Biochem. Biophys. Res. Commun.*, 144, 512, 1987.
234. **Baldwin, G. S., Knesel, J., and Monckton, J. M.,** Phosphorylation of gastrin-17 by epidermal growth factor-stimulated tyrosine kinase, *Nature (London)*, 301, 435, 1983.
235. **Baldwin, G. S., Grego, B., Hearn, M. T. W., Knesel, J. A., Morgan, F. J., and Simpson, R. J.,** Phosphorylation of human growth hormone by the epidermal growth factor-stimulated tyrosine kinase, *Proc. Natl. Acad. Sci. U.S.A.*, 80, 5276, 1983.

236. **Stern, D. F. and Kamps, M. P.,** EGF-stimulated tyrosine phosphorylation of p185neu: a potential model for receptor interactions, *EMBO J.*, 7, 995, 1988.
237. **Kokai, Y., Dobashi, K., Weiner, D. B., Meyers, J. N., Nowell, P. C., and Greene, M. I.,** Phosphorylation process induced by epidermal growth factor alters the oncogenic and cellular *neu* (NGL) gene products, *Proc. Natl. Acad. Sci. U.S.A.*, 85, 5389, 1988.
238. **Bargmann, C. I., Hung, M.-C., and Weinberg, R. A.,** The *neu* oncogene encodes an epidermal growth factor receptor-related protein, *Nature (London)*, 319, 226, 1986.
239. **Coussens, L., Yang-Feng, T. L., Liao, Y.-C., Chen, E., Gray, A., McGrath, J., Seeburg, P. H., Libermann, T. A., Schlessinger, J., Francke, U., Levinson, A., and Ullrich, A.,** Tyrosine kinase receptor with extensive homology to EGF receptor shares chromosomal location with *neu* oncogene, *Science*, 230, 1132, 1985.
240. **Hung, M.-C., Schechter, A. L., Chevray, P. Y., Stern, D. F., and Weinberg, R. A.,** Molecular cloning of the *neu* gene: absence of gross structural alteration in oncogenic alleles, *Proc. Natl. Acad. Sci. U.S.A.*, 83, 261, 1986.
241. **Bargmann, C. I. and Weinberg, R. A.,** Increased tyrosine kinase activity associated with the protein encoded by the activated *neu* oncogene, *Proc. Natl. Acad. Sci. U.S.A.*, 85, 5394, 1988.
242. **Honegger, A. M., Dull, T. J., Felder, S., van Obberghen, E., Beilot, F., Szapary, D., Schmidt, A., Ullrich, A., and Schlessinger, J.,** Point mutation at the ATP binding site of EGF receptor abolishes protein-tyrosine kinase activity and alters cellular routing, *Cell*, 51, 199, 1987.
243. **Honegger, A. M., Szapary, D., Schmidt, A., Lyall, R., Van Obberghen, E., Dull, T. J., Ullrich, A., and Schlessinger, J.,** A mutant epidermal growth factor receptor with defective protein tyrosine kinase is unable to stimulate proto-oncogene expression and DNA synthesis, *Mol. Cell. Biol.*, 7, 4568, 1987.
244. **Moolenaar, W. H., Biermann, A. J., Tilly, B. C., Verlaan, I., Honegger, A. M., Ullrich, A., and Schlessinger, J.,** A point mutation at the ATP-binding site of the EGF-receptor abolishes signal transduciton, *EMBO J.*, 7, 707, 1988.
245. **Livneh, E., Prywes, R., Dull, T. J., Ullrich, A., and Schlessinger, J.,** Release of a phorbol ester-induced mitogenic block by mutation at Thr-654 of the epidermal growth factor receptor, *Mol. Cell. Biol.*, 8, 2302, 1988.
246. **Chen, W. S., Lazar, C. S., Poenie, M., Tsien, R. Y., Gill, G. N., and Rosenfeld, M. G.,** Requirement for intrinsic protein tyrosine kinase in the immediate and late actions of the EGF receptor, *Nature (London)*, 328, 820, 1987.
247. **Bertics, P. J., Chen, W. S., Hubler, L., Lazar, C. S., Rosenfeld, M. G., and Gill, G. N.,** Alteration of epidermal growth factor receptor activity by mutation of its primary carboxyl-terminal site of tyrosine self-phosphorylation, *J. Biol. Chem.*, 263, 3610, 1988.
248. **Glenney, J. R., Jr., Chen, W. S., Lazar, C. S., Walton, G. M., Zokas, L. M., Rosenfeld, M. G., and Gill, G. N.,** Ligand-induced endocytosis of the EGF receptor is blocked by mutational inactivation and by microinjection of anti-phosphotyrosine antibodies, *Cell*, 52, 675, 1988.
249. **Clark, S., Cheng, D. J., Hsuan, J. J., Haley, J. D., and Waterfield, M. D.,** Loss of three major auto phosphorylation sites in the EGF receptor does not block the mitogenic action of EGF, *J. Cell. Physiol.*, 134, 421, 1988.
250. **Chinkers, M. and Brugge, J. S.,** Characterization of structural domains of the human epidermal growth factor receptor obtained by partial proteolysis, *J. Biol. Chem.*, 259, 11534, 1984.
251. **Basu, M., Biswas, R., and Das, M.,** 42,000-molecular weight EGF receptor has protein kinase activity, *Nature (London)*, 311, 477, 1984.
252. **Yaish, P., Gazit, A., Gilon, C., and Levitzki, A.,** Blocking of EGF-dependent cell proliferation by EGF receptor kinase inhibitors, *Science*, 242, 933, 1989.
253. **Lin, M.-F. and Clinton, G. M.,** The epidermal growth factor receptor from prostate cells is dephosphorylated by a prostate-specific phosphotyrosyl phosphatase, *Mol. Cell. Biol.*, 8, 5477, 1988.
254. **Chou, C. K., Dull, T. J., Russell, D. S., Gherzi, R., Lebwohl, D., Ullrich, A., and Rosen, O. M.,** Human insulin receptors mutated at the ATP-binding site lack protein tyrosine kinase activity and fail to mediate postreceptor effects of insulin, *J. Biol. Chem.*, 262, 1842, 1987.
255. **Ebina, Y., Araki, E., Taira, M., Shimada, F., Mori, M., Craik, C. S., Siddle, K., Pierce, S. B., Roth, R. A., and Rutter, W. J.,** Replacement of lysine residue 1030 in the putative ATP-binding region of the insulin receptor abolishes insulin- and antibody-stimulated glucose uptake and receptor kinase activity, *Proc. Natl. Acad. Sci. U.S.A.*, 84, 704, 1987.
256. **McClain, D. A., Maegawa, H., Lee, J., Dull, T. J., Ullrich, A., and Olefsky, J. M.,** A mutant insulin receptor with defective tyrosine kinase displays no biologic activity and does not undergo endocytosis, *J. Biol. Chem.*, 262, 14663, 1987.
257. **Sporn, M. B. and Roberts, A. B.,** Peptide growth factors are multifunctional, *Nature (London)*, 332, 217, 1988.
258. **Mercola, M. and Stiles, C. D.,** Growth factor superfamilies and mammalian embryogenesis, *Development*, 102, 451, 1988.
259. **King, G. L. and Kahn, C. R.,** The growth promoting effects of insulin, in *Growth and Growth Maturation Factors*, Guroff, G., Ed., John Wiley & Sons, New York, 1984, 224.

260. **Gospodarowicz, D.,** Isolation and characterization of acidic and basic fibroblast growth factor, *Methods Enzymol.*, 147, 106, 1987.
261. **Massague, J.,** The TGF-β family of growth and differentiation factors, *Cell*, 49, 437, 1987.
262. **Cuttitta, F., Carney, D. N., Mulshine, J., Moody, T. W., Fedorko, J., Fischler, A., and Minna, J. D.,** Bombesin-like peptides can function as autocrine growth factors in human small-cell lung cancer, *Nature (London)*, 316, 823, 1985.
263. **Nathan, C. F.,** Secretory products of macrophages, *J. Clin. Invest.*, 79, 319, 1987.
264. **Berridge, M. J.,** Oncogenes, inositol lipids and cellular proliferation, *Biotechnology*, 2, 541, 1984.
265. **Poenie, M., Alderton, J., Steinhardt, R., and Tsien, R.,** Calcium rises abruptly and briefly throughout the cell at the onset of anaphase, *Science*, 233, 886, 1986.
266. **Rasmussen, W., Zawalich, W., and Kojima, I.,** Ca^{2+} and cAMP in the regulation of cell function, in *Calcium and Cell Physiology*, Marme, D., Ed., Springer-Verlag, New York, 1985, 1.
267. **Korc, M. and Schoni, M. H.,** Modulation of cytosolic free calcium levels by extracellular phosphate and lanthanum, *Proc. Natl. Acad. Sci. U.S.A.*, 84, 1282, 1987.
268. **Majerus, P. W., Connolly, T. M., Deckmyn, H., Ross, T. S., Bross, T. E., Ishii, H., Bansal, V., and Wilson, D. B.,** The metabolism of phosphoinositide-derived messenger molecules, *Science*, 234 , 1519, 1986.
269. **Berridge, M. J.,** Inositol trisphosphate and diacylglycerol as second messengers, *Biochem. J.*, 220, 345, 1984.
270. **Volpe, P., Salviati, G., di Virgilio, F., and Pozzan, T.,** Inositol 1,4,5-trisphosphate induces calcium release from sarcoplasmic reticulum of skeletal muscle, *Nature (London)*, 316, 347, 1985.
271. **Prentki, M., Biden, T. J., Janjic, D., Irvine, R. F., Berridge, M. J., and Wollheim, C. B.,** Rapid mobilization of Ca^{2+} from rat insulinoma microsomes by inositol-1,4,5-triphosphate, *Nature (London)*, 309, 562, 1984.
272. **Spat, A., Bradford, P. G., McKinney, J. S., Rubin, R. P., and Putney, J. W., Jr.,** A saturable receptor for ^{32}P-inositol-1,4,5-trisphosphate in hepatocytes and neutrophils, *Nature (London)*, 319, 514, 1986.
273. **Connolly, T. M., Bansal, V. S., Bross, T. E., Irvine, R. F., and Majerus, P. W.,** The metabolism of tris- and tetraphosphates of inositol by 5-phosphomonoesterase and 3-kinase enzymes, *J. Biol. Chem.*, 262, 2146, 1987.
274. **Hansen, C. A., Mah, S., and Williamson, J. R.,** Formation and metabolism of inositol 1,3,4,5-tetrakisphosphate in liver, *J. Biol. Chem.*, 261, 8100, 1986.
275. **Hawkins, P. T., Stephens, L., and Downes, C. P.,** Rapid formation of inositol 1,3,4,5-tetrakisphosphate and inositol 1,3,4-trisphosphate in rat parotid glands may both result indirectly from receptor-stimulated release of inositol 1,4,5-trisphosphate from phosphatidylinositol 4,5-bisphosphate, *Biochem. J.*, 238, 507, 1986.
276. **Irvine, R. F., Letcher, A. J., Heslop, J. P., and Berridge, M. J.,** The inositol tris/tetrakisphosphate pathway-demonstration of Ins(1,4,5)P_3 3-kinase activity in animal tissues, *Nature (London)*, 320, 631, 1986.
277. **Irvine, R. F. and Moor, R. M.,** Micro-injection of inositol 1, 3, 4, 5-tetrakisphosphate activates sea urchin eggs by a mechanism dependent on external Ca^{2+}, *Biochem. J.*, 240, 917, 1986.
278. **Irvine, R. F. and Moor, R. M.,** Inositol(1,3,4,5)tetrakisphosphate-induced activation of sea urchin eggs requires the presence of inositol trisphosphate, *Biochem. Biophy. Res. Comm.*, 146, 284, 1987.
279. **Morris, A. P., Gallacher, D. V., Irvine, R. F., and Peterson, O. H.,** Synergism of inositol trisphosphate and tetrakisphosphate in activation of Ca^{2+}-dependent K^+ channels, *Nature (London)*, 330, 653, 1987.
280. **Kuno, M. and Gardner, P.,** Ion channels activated by inositol 1,4,5-trisphosphate in plasma membrane of human T-lymphocytes, *Nature (London)*, 326, 301, 1987.
281. **Parker, I. and Miledi, R. F. R. S.,** Inositol trisphosphate activates a voltage-dependent calcium influx in *Xenopus* oocytes, *Proc. R. Soc. Lond.*, B231, 27, 1987.
282. **Berridge, M. J.,** Regulation of ion channels by inositol trisphosphate and diacylglycerol, *J. Exp. Biol.*, 124, 323, 1986.
283. **Vallejo, M., Jackson, T., Lightman, S., and Hanley, M. R.,** Occurrence and extracellular actions of inositol pentakis- and hexakisphosphate in mammalian brain, *Nature (London)*, 330, 656, 1987.
284. **Stephens, L., Hawkins, P. T., Carter, N., Chahwala, S. B., Morris, A. J., Whetton, A. D., and Downes, P. C.,** L-*myo*-inositol 1,4,5,6-tetrakisphosphate is present in both mammalian and avian cells, *Biochem. J.*, 249, 271, 1988.
285. **Biden, T. J. and Wollheim, C. B.,** Ca^{2+} regulates the inositol tris/tetrakisphosphate pathway in intact and broken preparations of insulin-secreting RINm5F cells, *J. Biol. Chem.*, 261, 11931, 1986.
286. **Biden, T. J., Comte, M., Cox, J. A., and Wollheim, C. B.,** Calcium-calmodulin stimulates inositol 1,4,5-trisphosphate kinase activity from insulin-secreting RINm5F cells, *J. Biol. Chem.*, 262, 9437, 1987.
287. **Zilberman, Y., Howe, L. R., Moore, J. P., Hesketh, T. R., and Metcalfe, J. C.,** Calcium regulates inositol 1,3,4,5-tetrakisphosphate production in lysed thymocytes and in intact cells stimulated with concanavalin A, *EMBO J.*, 6, 957, 1987.
288. **Mullaney, J. M., Yu, M., Ghosh, T. K., and Gill, D. L.,** Calcium entry into the inositol 1,4,5-triphosphate-releasable calcium pool is mediated by a GTP-regulatory mechanism, *Proc. Natl. Acad. Sci. U.S.A.*, 85, 2499, 1988.
289. **Cuthbertson, K. S. R. and Cobbold, P. H.,** Phorbol ester and sperm activate mouse oocytes by inducing sustained oscillations in cell Ca^{2+}, *Nature (London)*, 316, 541, 1985.

290. **Woods, N. M., Cuthbertson, K. S. R., and Cobbold, P. H.,** Phorbolester-induced alterations of free calcium ion transients in single rat hepatocytes, *Biochem. J.*, 246, 619, 1987.
291. **Hill, T. D., Dean, N. M., and Boynton, A. L.,** Inositol 1,3,4,5- tetrakisphosphate induces Ca^{2+} sequestration in rat liver cells, *Science*, 242, 1176, 1988.
292. **Berridge, M. J. and Galione, A.,** Cytosolic calcium oscillators, *FASEB J.*, 2, 3074, 1988.
293. **Putney, J. W., Jr.,** Formation and actions of calcium-mobilizing messenger, inositol 1,4,5-trisphosphate, *Am. J. Physiol.*, 252, G149, 1987.
294. **Schultz, I. and Stolze, H. H.,** The exocrine pancreas: the role of secretagogues, cyclic nucleotides, and calcium in enzyme secretion, *Ann. Rev. Physiol.*, 423, 127, 1980.
295. **Gardner, J. D. and Jensen, R. T.,** Receptors and cell activation associated with pancreatic enzyme secretion, *Ann. Rev. Physiol.*, 48, 103, 1986.
296. **Wahl, M. and Carpenter, G.,** Regulation of epidermal growth factor-stimulated formation of inositol phosphates in A-431 cells by calcium and protein kinase C, *J. Biol. Chem.*, 263, 7581, 1988.
297. **Wahl, M. I., Daniel, T. O., and Carpenter, G.,** Antiphosphotyrosine recovery of phospholipase C activity after EGF treatment of A-431 cells, *Science*, 241, 968, 1988.
298. **Thompson, D. M., Proctor, J., Grant, M., and Thomas, C.,** Epidermal growth factor stimulates phosphatidylinositol turnover for ten hours in A431 cells without activation of protein kinase C, *Biochem. Biophys. Res. Comm.*, 155, 877, 1988.
299. **Church, J. G. and Buick, R. N.,** G-protein-mediated epidermal growth factor signal transduction in a human breast cancer cell line, *J. Biol. Chem.*, 263, 4242, 1988.
300. **Neer, E. J. and Clapham, D. E.,** Roles of G protein subunits in transmembrane signalling, *Nature (London)*, 333, 129, 1988.
301. **Thomas, A. P.,** Enhancement of the inositol 1,4,5-trisphosphate-releasable Ca^{2+} pool by GTP in permeabilized hepatocytes, *J. Biol. Chem.*, 263, 2704, 1988.
302. **Matthews, E. K.,** Calcium and membrane permeability, *Br. Med. Bull.*, 42, 391, 1986.
303. **Stanfield, P. R.,** Voltage-dependent calcium channels of excitable membranes, *Br. Med. Bul.*, 42, 359, 1986.
304. **Yatani, A., Codina, J., Imoto, Y., Reeves, J. P., Birnbaumer, L., and Brown, A. M.,** A G protein directly regulates mammalian cardiac calcium channels, *Science*, 238, 1288, 1987.
305. **Brown, A. M. and Birnabaumer, L.,** Direct G protein gating of ion channels, *Am. J. Physiol.*, 254, H401, 1988.
306. **Moolenaar, W. H., Defize, L. H. K., and de Laat, S. W.,** Ionic signalling by growth factor receptors, *J. Exp. Biol.*, 124, 359, 1986.
307. **Moolenaar, W. H., Yarden, Y., de Laat, S. W., and Schlessinger, J.,** Epidermal growth factor induces electrically silent Na^{+} influx in human fibroblasts, *J. Biol. Chem.*, 257, 8502, 1982.
308. **Moolenaar, W. H., Tsien, R. Y., van der Saag, P. T., and de Laat, S. W.,** Na^{+}/H^{+} exchange and cytoplasmic pH in the action of growth factors in human fibroblasts, *Nature (London)*, 304, 645, 1983.
309. **Rothenberg, P., Glaser, L., Schlesinger, P., and Cassel, D.,** Epidermal growth factor stimulates amiloride-sensitive ^{22}Na uptake in A431 cells. Evidence for Na^{+}-H^{+} exchange, *J. Biol. Chem.*, 258, 4883, 1983.
310. **Moolenaar, W. H., Tertoolen, L. G. J., and de Laat, S. W.,** Phorbol ester and diacylglycerol mimic growth factors in raising cyoplasmic pH, *Nature (London)*, 312, 371, 1984.
311. **Rosoff, P. M. and Cantley, L. C.,** Stimulation of the T3-T cell receptor associated Ca^{2+} influx enhances the activity of the Na^{+}-H^{+} exchanger in a leukemic human T cell line, *J. Biol. Chem.*, 260, 14053, 1985.
312. **Rozengurt, E.,** Early signals in the mitogenic response, *Science*, 234, 161, 1986.
313. **Soltoff, S. P. and Cantley, L. C.,** Mitogens and ion fluxes, *Ann. Rev. Physiol.*, 50, 207, 1988.
314. **Gates, R. E. and King, L. E., Jr.,** Calcium facilitates endogenous proteolysis of the EGF receptor-kinase, *Mol. Cell. Endocrinol.*, 27, 263, 1982.
315. **Yeaton, R. W., Lipari, M. T., and Fox, C. F.,** Calcium-mediated degradation of epidermal growth factor receptor in dislodged A431 cells and membrane preparations, *J. Biol. Chem.*, 258, 9254, 1983.
316. **White, B. A. and Bancroft, F. C.,** Epidermal growth factor and thyrotropin-releasing hormone interact synergistically with calcium to regulate prolactin mRNA levels, *J. Biol. Chem.*, 258, 4618, 1983.
317. **Rheinwald, J. G. and Green, H.,** Epidermal growth factor and the multiplication of cultured human epidermal keratinocytes, *Nature (London)*, 265, 219, 1982.
318. **O'Keefe, E. J. and Payne, R. E., Jr.,** Modulation of the epidermal growth factor receptor of human keratinocytes by calcium ion, *J. Invest. Derm.*, 81, 231, 1983.
319. **Tsao, M. C., Walthall, B. J., and Ham, T. G.,** Clonal growth of normal human epidermal keratinocytes in a defined medium, *J. Cell. Physiol.*, 110, 219, 1982.
320. **Hennings, H., Michael, D., Cheng, C., Steinert, P., Holbrook, K., and Yuspa, S. H.,** Calcium regulation of growth and differentiation of mouse epidermal cells in culture, *Cell*, 19, 245, 1980.
321. **Grody, W. W., Schrader, W. R., and O'Malley, B. W.,** Activation, transformation, and subunit structure of steroid hormone receptors, *Endocr. Rev.*, 3, 141, 1982.
322. **Yamamoto, K. R.,** Steroid receptor regulated transcription of specific genes and gene networks, *Annu. Rev. Genet.*, 19, 209, 1985.

323. **Evans, R. M.,** The steroid and thyroid hormone receptor superfamily, *Science,* 240, 889, 1988.

324. **Korc, M., Owerbach, D., Quinto, C., and Rutter, W. J.,** Pancreatic islet-acinar cell interaction: amylase messenger RNA levels are determined by insulin, *Science*, 213, 351, 1981.

325. **Lloyd, C. E., Kalinyak, J. E., Hutson, S. M., and Jefferson, L. S.,** Stimulation of albumin gene transcription by insulin in primary cultures of rat hepatocytes, *Am. J. Physiol.*, 252, C205, 1987.

326. **Sasaki, K., Cripe, T. P., Koch, S. R., Andreone, T. L., Petersen, D. D., Beale, E. G., and Granner, D. K.,** Multihormonal regulation of phosphoenolpyruvate carboxykinase gene transcription, *J. Biol. Chem.*, 259, 15242, 1984.

327. **Stumpo, D. J. and Blackshear, P. J.,** Insulin and growth factor effects on c-*fos* expression in normal and protein kinase C-deficient 3T3-L1 fibroblasts and adipocytes, *Proc. Natl. Acad. Sci. U.S.A.*, 83, 9453, 1986.

328. **Zumstein, P. and Stiles, C. D.,** Molecular cloning of gene sequences that are regulated by insulin-like growth factor, *J. Biol. Chem.*, 262, 11252, 1987.

329. **Kruijer, W., Cooper, J. A., Hunter, T., and Verma, I. M.,** Platelet-derived growth factor induces rapid but transient expression of the c-*fos* gene and protein, *Nature (London)*, 312, 711, 1984.

330. **Cochran, B. H., Zullo, J., Verma, I. M., and Stiles, C. D.,** Expression of the c-*fos* gene and the *fos*-related gene as stimulated by platelet-derived growth factor, *Science*, 226, 1080, 1984.

331. **Kelly, K., Cochran, B. H., Stiles, C. D., and Leder, P.,** Cell specific regulation of the c-*myc* gene by lymphocyte mitogens and platelet-derived growth factor, *Cell,* 35, 603, 1983.

332. **Rollins, B. J., Morrison, E. D., and Stiles, C. D.,** A cell-cycle constraint on the regulation of gene expression by platelet-derived growth factor, *Science,* 238, 1269, 1987.

333. **Cutry, A. F., Kinniburgh, A. J., Twardzik, D. R., and Wenner, C. E.,** Transforming growth factor alpha (TGFα) induction of c-*fos* and c-*myc* expression in C3H 10T1/2 cells, *Biochem. Biophys. Res. Comm.*, 152, 216, 1988.

334. **Ran, W., Dean, M., Levine, R. A., Henkle, C., and Campisi, J.,** Induction of c-*fos* and c-*myc* mRNA by epidermal growth factor or calcium ionophore is cAMP dependent, *Proc. Natl. Acad. Sci. U.S.A.*, 83, 8216, 1986.

335. **Finkel, M. P., Biskis, B. O., and Jinkins, P. B.,** Virus induction of osteosarcomas in mice, *Science*, 151, 698, 1966.

336. **Curran, T., Van Beveren, C., Ling, N., and Verma, I. M.,** Viral and cellular *fos* proteins are complexed with a 39,000-dalton cellular protein, *Mol. Cell. Biol.*, 5, 167, 1985.

337. **Rauscher, F. J., III, Cohen, D. R., Curran, T., Bos, T. J., Vogt, P. K., Bohmann, D., Tjian, R., and Franza, B. R., Jr.,** Fos-associated protein p39 is the product of the *jun* proto-oncogene, *Science,* 240, 1010, 1988.

338. **Chiu, R., Boyle, W. J., Meek, J., Smeal, T., Hunter, T., and Karin, M.,** The c-Fos protein interacts with c-Jun/AP-1 to stimulate transcription of AP-1 responsive genes, *Cell,* 54, 541, 1988.

339. **Sassone-Corsi, P., Lamph, W. W., Kamps, M., and Verma, I. M.,** *fos*-associated cellular p39 is related to nuclear transcription factor AP-1, *Cell,* 54, 553, 1988.

340. **Maki, Y., Bos, T. J., Davis, C., Starbuck, M., and Vogt, P. K.,** Avian sarcoma virus 17 carries the *jun* oncogene, *Proc. Natl. Acad. Sci. U.S.A.*, 84, 2848, 1987.

341. **Bohmann, D., Bos, T. J., Admon, A., Nishimura, T., Vogt, P. K., and Tjian, R.,** Human proto-oncogene c-*jun* encodes a DNA binding protein with structural and functional properties of transcription factor AP-1, *Science*, 328, 1386, 1987.

342. **Angel, P., Allegretto, E. A., Okino, S. T., Hattori, K., Boyle, W. J., Hunter, T., and Karin, M.,** Oncogene *jun* encodes a sequence-specific *trans*-activator similar to AP-1, *Nature (London)*, 332, 166, 1988.

343. **Lee, W., Mitchell, P., and Tjian, R.,** Purified transcription factor AP-1 interacts with TPA-inducible enhancer elements, *Cell*, 49, 741, 1987.

344. **Vogt, P. K., Bos, T. J., and Doolittle, R. F.,** Homology between the DNA binding domain of the GCN4 regulatory protein of yeast and the carboxyl-terminal region of a protein coded for by the oncogene *jun*, *Proc. Natl. Acad. Sci. U.S.A.*, 84, 3316, 1987.

345. **Struhl, K.,** The DNA-binding domains of the jun oncoprotein and the yeast GCN4 transcriptional activator protein are functionally homologous, *Cell*, 50, 841, 1987.

346. **Landschulz, W. H., Johnson, P. F., and McKnight, S. L.,** The leucine zipper: a hypothetical structure common to a new class of DNA binding proteins, *Science*, 240, 1759, 1988.

347. **Halazonetis, T. D., Georgopoulos, K., Greenberg, M. E., and Leder, P.,** c-*jun* dimerizes with itself and with c-*fos* forming complexes of different DNA binding affinities, *Cell*, 55, 917, 1988.

348. **Nakabeppu, Y., Ryder, K., and Nathans, D.,** DNA binding activities of three murine *jun* proteins: stimulation by *fos*, *Cell*, 55, 907, 1988.

349. **Sassone-Corsi, P., Ransone, L. J., Lamph, W. W., and Verma, I. M.,** Direct interaction between *fos* and *jun* nuclear oncoproteins: role of the "leucine zipper" domain, *Nature (London)*, 336, 692, 1988.

350. **Quantin, B. and Breathnach, R.,** Epidermal growth factor stimulates transcription of the c-*jun* proto-oncogene in rat fibroblasts, *Nature (London)*, 334, 538, 1988.

351. **Gonda, T. J. and Metcalf, D.,** Expression of *myb*, *myc*, and *fos* proto-oncogenes during the differentiation of a murine myeloid leukaemia, *Nature (London)*, 310, 249, 1984.

352. **Lachman, H. M., Hatton, K. S., Skoultchi, A. I., and Schildkraut, C. L.,** c-*myc* mRNA levels in the cell cycle change in mouse erythroleukemia cells following inducer treatment, *Proc. Natl. Acad. Sci. U.S.A.*, 82, 5323, 1985.
353. **Vaux, D. L., Cory, S., and Adams, J. M.,** *Bcl*-2 gene promotes haemopoietic cell survival and cooperates with c-*myc* to immortalize pre-B cells, *Nature (London)*, 335, 440, 1988.
354. **Schweinfest, C. W., Fujiwara, S., Lau, L. F., and Papas, T. S.,** c-*myc* can induce expression of G_0/G_1 transition genes, *Mol. Cell. Biol.*, 8, 3080, 1988.
355. **Kaneko-Ishino, T., Kume, T. U., Sasaki, H., Obinata, M., and Oishi, M.,** Effect of c-*myc* gene expression on early inducible reaction required for erythroid differentiation *in vitro*, *Mol. Cell. Biol.*, 8, 5545, 1988.
356. **Sasson-Corsi, P., Sisson, J. C., and Verma, I. M.,** Transcriptional autoregulation of the proto-oncogene *fos*, *Nature (London)*, 334, 314, 1988.
357. **Sariban, E., Luebbers, R., and Kufe, D.,** Transcriptional and posttranscriptional control of c-*fos* gene expression in human monocytes, *Mol. Cell. Biol.*, 8, 340, 1988.
358. **Curran, T. and Franza, R. B., Jr.,** *Fos* and *jun*: the AP-1 connection, *Cell*, 55, 395, 1988.
359. **McCaffrey, P., Ran, W., Campisi, J., and Rosner, M. R.,** Two independent growth factor-generated signals regulate c-*fos* and c-*myc* mRNA levels in Swiss 3T3 cells, *J. Biol. Chem.*, 262, 1442, 1987.
360. **Rodland, K. D., Muldoon, L. L., Dinh, T.-H., and Magun, B. E.,** Independent transcriptional regulation of a single VL30 element by epidermal growth factor and activators of protein kinase C, *Mol. Cell. Biol.*, 8, 2247, 1988.
361. **Clark, A. J. L., Ishii, S., Richert, N., Merlino, G. T., and Pastan, I.,** Epidermal growth factor regulates the expression of its own receptor, *Proc. Natl. Acad. Sci. U.S.A.*, 82, 8374, 1985.
362. **Earp, H. S., Austin, K. S., Blaisdell, J., Rubin, R. A., Nelson, K. G., Lee, L. W., and Grisham, J. W.,** Epidermal growth factor (EGF) stimulates EGF receptor synthesis, *J. Biol. Chem.*, 261, 4777, 1986.
363. **Earp, H. S., Hepler, J. R., Petch, L. A., Miller, A., Berry, A. R., Harris, J., Raymond, V. W., McCune, B. K., Lee, L. W., Grisham, J. W., and Harden, T. K.,** Epidermal growth factor (EGF) and hormones stimulate phosphoinositide hydrolysis and increase EGF receptor protein synthesis and mRNA levels in rat liver epithelial cells, *J. Biol. Chem.*, 263, 13868, 1988.
364. **Black, R. J., Yu, Z.-P., Brown, D., and Chang, E.,** Modulation of oncogene expression by epidermal growth factor and gamma-interferon in A431 squamous carcinoma cells, *J. Biol. Reg. Homeo. Agents*, 2, 35, 1988.
365. **Machida, C. M., Muldoon, L. L., Rodland, K. D., and Magun, B. E.,** Transcriptional modulation of transin gene expression by epidermal growth factor and transforming growth factor beta, *Mol. Cell. Biol.*, 8, 2497, 1988.
366. **Muldoon, L. L., Rodland, K. D., and Magun, B. E.,** Transforming growth factor β modulates epidermal growth factor-induced phospoinositide metabolism and intracellular calcium levels, *J. Biol. Chem.*, 263, 5030, 1988.
367. **Berk, B. C., Brock, T. A., Webb, R. C., Taubman, M. B., Atkinson, W. J., Gimbrone, M. A., Jr., and Alexander, R. W.,** Epidermal growth factor, a vascular smooth muscle mitogen, induces rat aortic contraction, *J. Clin. Invest.*, 75, 1083, 1985.
368. **Brannon, P. M., Hirschi, K., and Korc, M.,** Effects of epidermal growth factor, insulin, and insulin-like growth factor on rat pancreatic acinar cells cultured in serum-free medium, *Pancreas*, 3, 41, 1988.
369. **Pratt, R. M.,** Involvement of hormones and growth factors in the development of the secondary palate, in *Development in Mammals*, Vol. 4, Johnson, M. H., Ed., Elsevier/North-Holland Biomedical Press, New York, 1980, 203.
370. **Catterton, W. Z., Escobedo, M. B., Sexson, W. R., Gray, M. E., Sundell, H. W., and Stahlman, M. T.,** Effect of epidermal growth factor on lung maturation in fetal rabbits, *Pediatr. Res.*, 13, 104, 1979.
371. **Messmer, T. O., Armour, R., and Holley, R. W.,** Factors influencing the growth of alveolar type II epithelial cells isolated from rat lungs, *Exp. Cell Res.*, 142, 417, 1982.
372. **Clavert, R., Beaulieu, J.-F., and Menard, D.,** Epidermal growth factor (EGF) accelerates the maturation of fetal mouse intestinal mucosa in utero, *Experientia*, 38, 1096, 1982.
373. **Barr, P. J. and Valenzuela, P.,** Epidermal growth factor genes: from molecular biology to clinical therapeutics, in *Oncogenes, Genes, and Growth Factors*, Guroff, G., Ed., John Wiley & Sons, New York, 1987, 77.
374. **Malo, C. and Menard, D.,** Infuence of epidermal growth factor on the development of suckling mouse intestinal mucosa, *Gastroenterol.*, 83, 28, 1982.
375. **Carpenter, G.,** Epidermal growth factor is a major growth-promoting agent in human milk, *Science*, 210, 198, 1980.
376. **Oka, Y., Ghisham, F. K., Green, H. L., and Orth, D. N.,** Effect of mouse epidermal growth factor/urogastrone on the functional maturation of rat intestine, *Endocrinology*, 112, 940, 1983.
377. **Pollack, P. F., Goda, T., Colony, P. C., Edmond, J., Thornburg, W., Korc, M., and Koldovsky, O.,** Effects of enterally fed epidermal growth factor on the small and large intestine of the suckling rat, *Reg. Peptides*, 17, 121, 1987.
378. **Dufour, C., Dandrifosse, G., Forget, P., Vermesse, F., Romain, N., and Lepoint, P.,** Spermine and spermidine induced intestinal maturation in the rat, *Gastroenterology.*, 95, 112, 1988.

379. **Muller-Glauser, W. and Preisig, E.,** The effect of cholera toxin and epidermal growth factor on the *in-vitro* growth of human oral epithelial cells, *Arch. Oral Biol.*, 8, 765, 1983.
380. **Stanley, M. A. and Dahlenburg,** The effect of choleragen and epidermal growth factor on proliferation and maturation *in vitro* of human ectocervical cells, *In Vitro*, 20, 144, 1984.
381. **Wu, K., Yankaskas, J., Cheng, E., Knowles, M. R., and Boucher, R.,** Growth and differentiation of human nasal epithelial cells in culture, *Am. Rev. Respir. Dis.*, 132, 311, 1985.
382. **Askanas, V. and Gallez-Hawkins, G.,** Synergistic influence of polypeptide growth factors on cultured human muscle, *Arch. Neurol.*, 42, 749, 1985.
383. **Morrish, D. W., Bhardwaj, D., Dabbagh, L. K., Marusyk, H., and Siy, O.,** Epidermal growth factor induces differentiation and secretion of human chorionic gonadotropin and placental lactogen in normal human placenta, *J. Clin. Endocrinol. Metab.*, 65, 1282, 1987.
384. **Wang, Y.-C. and Rubenstein, P. A.,** Epidermal growth factor controls smooth muscle α-isoactin expression in BC3H1 cells, *J. Cell Biol.*, 106, 797, 1988.
385. **Olwin, B. B. and Hauschka, S. D.,** Cell surface fibroblast growth factor and epidermal growth factor receptors are permanently lost during skeletal muscle terminal differentiation in culture, *J. Cell Biol.*, 107, 761, 1988.
386. **Brockenbrough, J. S., Weir, G. C., and Korc, M.,** Alterations in EGF binding to acini during pancreatic regeneration in the rat, *Int. J. Pan.*, 3, 415, 1988.
387. **Earp, H. S. and O'Keefe, E. J.,** EGF receptor number decreases during rat liver regeneration, *J. Clin. Invest.*, 67, 1580, 1981.
388. **Morisset, J., Korc, M., and Larose, L.,** Antagonistic and additive trophic effects of caerulein and EGF in rat pancreas, *Gastroenterology*, 90, 1557, 1986.
389. **Larose, L., Morisset, J., and Korc, M.,** Modulation of caerulein-mediated amylase secretion by epidermal growth factor, *Gastroenterology*, 90, 1512, 1986.
390. **Gill, G. N., Buss, J. E., Lazar, C. S., Lifshitz, A., and Cooper, J. A.,** Role of epidermal growth factor-stimulated protein kinase in control of proliferation of A431 cells, *J. Cell. Biochem.*, 19, 249, 1982.
391. **Barnes, D. W.,** Epidermal growth factor inhibits growth of A431 human epidermoid carcinoma in serum-free cell culture, *J. Cell. Biol.*, 93, 1, 1982.
392. **Imai, Y., Leung, C. K. H., Friesen, G. H., and Shiu, R. P. C.,** Epidermal growth factor receptors and effect of epidermal growth factor on growth of human breast cancer cells in long-term tissure culture, *Cancer Res.*, 42, 4394, 1982.
393. **Kamata, N., Chida, K., Rikimaru, K., Horikoshi, M., Enomoto, S., and Kuroki, T.,** Growth-inhibitory epidermal growth factor and overexpression of its receptors on human squamous cell carcinomas in culture, *Cancer Res.*, 46, 1648, 1986.
394. **Korc, M., Padilla, J., and Grosso, D.,** Epidermal growth factor inhibits the proliferation of a human endometrial carcinoma cell line, *J. Clin. Endocrinol. Metab.*, 62, 874, 1986.
395. **Knowles, A., Salas-Prato, M., and Villeala, J.,** Epidermal growth factor inhibits growth while increasing the expression of an ecto Ca^{2+} ATPase of a human hepatoma cell, *Biochem. Biophys. Res. Commun.*, 126, 8, 1985.
396. **Johnson, L. K., Baxter, J. D., Vlodavsky, I., and Gospodarowicz, D.,** Epidermal growth factor and expression of specific genes: effects on cultured rat pituitary cells are dissociable from the mitogenic response, *Proc. Natl. Acad. Sci. U.S.A.*, 77, 394, 1980.
397. **Schonbrunn, A., Krasnoff, M., Westenclorf, J. M., and Tashjian, A. H., Jr.,** Epidermal growth factor and thyrotropin-releasing hormone act similarly on a cloncal pituitary cell strain, *J. Cell. Biol.*, 85, 786, 1980.
398. **Blay, J. and Brown, K.,** Contradistinctive growth responses of cultured rat intestinal epithelial cells to epidermal growth factor depending on cell population density, *J. Cell. Physiol.*, 129, 343, 1986.
399. **Kawamoto, T., Sato, J. D., Le, A., Sato, G. H., Lazar, C. S., and Gill, G. N.,** Relation of epidermal growth factor receptor concentration to growth of human epidermoid carcinoma A431 cells, *J. Biol. Chem.*, 259, 7761, 1984.
400. **MacLeod, C. L., Luk, A., Castagnola, J., Cronin, M., and Mendelsohn, J.,** EGF induces cell cycle arrest of A431 human epidermoid carcinoma cells, *J. Cell. Phys.*, 127, 175, 1986.
401. **Kawamoto, T., Sato, J. D., Le, A., Polikoff, J., Sato, G. H., and Mendelsohn, J.,** Growth stimulation of A431 cells by epidermal growth factor: identification of high-affinity receptors for epidermal growth factor by an anti-receptor monoclonal anitbody, *Proc. Natl. Acad. Sci. U.S.A.*, 80, 1337, 1983.
402. **Schechter, Y., Hernaez, L., and Cuatrecasas, P.,** Epidermal growth factor: biological activity requires persistent occupation of high-affinity cell surface receptors, *Proc. Natl. Acad. Sci. U.S.A.*, 75, 5788, 1978.
403. **Chinkers, M., McKanna, J. A., and Cohen, S.,** Rapid rounding of human epidermoid carcinoma cells A431 induced by epidermal growth factor, *J. Cell Biol.*, 88, 422, 1981.
404. **Lifshitz, A., Lazar, C. S., Buss, J. E., and Gill, G. N.,** Analysis of morphology and receptor metabolism in clonal variant A431 cells with differing growth responses to epidermal growth factor, *J. Cell. Physiol.*, 115, 235, 1983.
405. **Shoyab, M., DeLarco, J. E., and Todaro, G. J.,** Biologically active phorbol esters specifically alter affinity of epidermal growth factor membrane receptors, *Nature (London)*, 279, 387, 1979.

406. **Livneh, E., Benveniste, M., Prywes, R., Felder, S., Kam, Z., and Schlessinger, J.,** Large deletions in the cytoplasmic kinase domain of the epidermal growth factor receptor do not affect its lateral mobility, *J. Cell Biol.*, 103, 327, 1986.

407. **Faucher, M., Girones, N., Hannun, Y. A., Bell, R. M., and Davis, R. J.,** Regulation of the epidermal growth factor receptor phosphorylation state by sphingosine in A431 human epidermoid carcinoma cells, *J. Biol. Chem.*, 263, 5319, 1988.

408. **Yamaguchi, Y. and Ruoslahti, E.,** Expression of human proteoglycan in Chinese hamster ovary cells inhibits cell proliferation, *Nature (London)*, 336, 244, 1988.

Chapter 2

POLYAMINES AND CELL AND TISSUE GROWTH

Chris N. Conteas, Ann L. Silverman, Jeffrey A. Moshier, Peter P. McCann, and Gordon D. Luk

TABLE OF CONTENTS

I. POLYAMINE METABOLISM

The polyamines — putrescine, spermidine, and spermine — are ubiquitious, low molecular weight polycations which play important roles in a variety of cellular processes, especially those associated with growth, differentiation, and neoplasia. The current understanding of polyamine biochemistry, physiology, and function has been extensively reviewed;[1-4] however, the exact *in vivo* functions of the polyamines and their cellular physiology are not yet well understood. This chapter will briefly outline our current knowledge of the role of polyamines in cell and tissue growth, particularly in the gastrointestinal tract.

The synthesis of polyamines in mammalian systems starts with the decarboxylation of the amino acid ornithine to form putrescine (Figure 1). This enzymatic conversion is carried out by the enzyme ornithine decarboxylase (ODC) and is the first and often rate-limiting step in the polyamine biosynthetic pathway. After the formation of putrescine, aminopropyl groups are added to form, respectively, spermidine and spermine. These enzymatic reactions are carried out by the aminopropyl transferases, spermidine synthase and spermine synthase, respectively, which use decarboxylated *S*-adenosylmethionine (dc-AdoMet) as the aminopropyl donor. Dc-AdoMet is formed by the action of *S*-adenosylmethionine decarboxylase (AdoMetDC).

Therefore, AdoMetDC has also been considered as a key regulatory enzyme in polyamine biosynthesis. Although the steps described are generally irreversible, some interconversion of polyamines occurs through the action of spermidine and spermine N^1-acetyltransferases and polyamine oxidase (Figure 1). Putrescine may also be degraded by the enzyme diamine oxidase to aminobutyraldehyde and subsequently to gamma aminobutyric acid (GABA). Dc-AdoMet, after donating the aminopropyl group for the formation of spermidine and spermine, is itself converted to methylthioadenosine, which is then recycled through the purine pools and methionine pools and hence may also effect methylation reactions.

In the resting cell, the basal activity of ODC is low in most tissues. However, ODC has one of the shortest half-lives known for mammalian enzymes, ranging from 7 to 20 min in different reports. The onset of cell and tissue growth is characteristically associated with marked increases in ODC activity and rapid accumulation of tissue polyamines. This increase in ODC activity is one of the earliest cellular biochemical events occurring during the transition of cells from resting states to active proliferation. Thus ODC appears to be ready to respond to proliferative stimuli and may itself trigger critical cellular processes in tissue growth.

Several pharmacologic inhibitors of polyamine biosynthesis have been developed.[4] The enzymes ODC and AdoMetDC have been the primary targets of inhibitors because of the perceived key regulatory role they play in polyamine biosynthesis. The one inhibitor that has been studied most extensively is difluoromethylornithine (DFMO, eflornithine by Merrel-Dow Pharmaceuticals, Inc.). DFMO was developed as an enzyme-activated, irreversible inhibitor of ODC. It produces both *in vitro* and *in vivo* marked inhibition of ODC activity and depletion of cellular putrescine, and to a much lesser extent spermidine and spermine. DFMO has been used extensively to evaluate the dependence of cell and tissue growth upon the increased ODC activity and accumulation of intracellular polyamines.

II. POLYAMINES AND CELL AND TISSUE GROWTH

As described above, ODC activity and polyamine content increase rapidly and markedly during cell proliferation, both in normal and neoplastic cells. In studies of tissue growth, the regenerating rat liver after partial hepatectomy has been one of the earliest and most extensively studied models. In this model and others, tissue regeneration, as well as rapid tissue growth in response to a variety of proliferative stimuli, results in marked induction of ODC activity and increases in cellular polyamine content. As will be described subsequently, the administration of the inhibitor of ODC, DFMO, will inhibit the activity of ODC, decrease the cellular content of polyamines, and often result in an inhibition or cessation of cell and tissue growth.

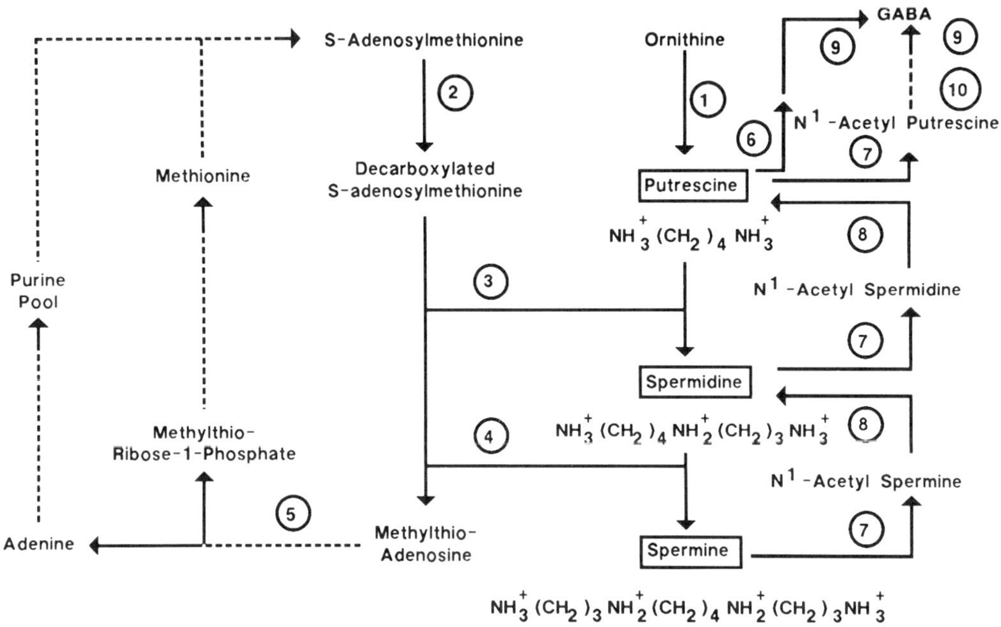

FIGURE 1. Polyamine metabolism. The enzymes indicated are: (1) ornithine decarboxylase (EC 4.1.1.17); (2) S-adenosylmethionine decarboxylase (EC 4.1.1.50); (3) spermidine synthase (EC 2.5.1.16); (4) spermine synthase (EC 2.5.1.22); (5) methylthioadenosine phosphorylase (EC 2.4.2.28); (6) diamine oxidase (EC 1.4.3.6); (7) spermidine/spermine-N[1]-acetyltransferase; (8) polyamine oxidase; (9) acetyltransferase; (10) aldehyde dehydrogenase. (From Luk, G. D. and Casero, R. A., in *Advances in Enzyme Regulation*, Vol. 26, Webber, G., Ed., Pergamon Press, England, 1987, 91. With permission.)

A. REGENERATING LIVER

In the rat partial hepatectomy model, the hepatic remnant undergoes rapid growth and regeneration, with virtual restoration of the preresection liver weight within 7 to 14 d. Hepatic regeneration is associated with marked transient increases in ODC, polyamines, and DNA synthesis.[5,6] Using competitive inhibitors of ODC, as well as DFMO, several investigators have shown that the increases in ODC may be essential for liver regeneration. These studies have shown that the inhibition of ODC prevents the accumulation of polyamines. The suppression of DNA synthesis, however, was not a universal finding. Many of those studies were affected by co-existing toxic effects of enzyme inhibitors, which resulted in weight loss and hence could have potentially inhibited the process of liver regeneration. Futhermore, in studies where there were no significant decreases in DNA and RNA synthesis, the drugs were used at low doses and might not have been effective in achieving complete inhibition of ODC and adequate depletion of intracellular polyamines.

When adequate doses of DFMO are used, and animals are given DFMO postoperatively while oral intake was low, we showed that the administration of DFMO resulted in a marked suppression of ODC activity and decreases in DNA and protein synthesis.[6] Most significantly, the process of hepatic regeneration, as measured by liver weight gain, was suppressed by greater than 80%. This study confirms that the inhibition of ODC activity and the decrease in intracellular polyamine levels by DFMO results in inhibition of DNA and protein synthesis and tissue regeneration. These results seem to indicate that in the rat liver, the process of hepatic regeneration is critically dependent upon the induction of ODC activity and the intracellular accumulation of polyamines.[6]

Furthermore, in this model, the simultaneous administration of exogenous putrescine overcame the inhibitory effect of DFMO on hepatic regeneration. This confirms that the inhibitory effect of DFMO is specifically due to the depletion of intracellular polyamines, as the administration of exogenous putrescine restores hepatic regeneration. This use of exogenous polyamines to overcome the inhibitory effect of DFMO on cell and tissue growth is important to document the specific role of polyamines and ODC in growth processes.

B. GASTRIC MUCOSAL GROWTH

The role of polyamines and ODC in gastric mucosal growth is quite controversial. Several investigators have found a dissociation between ODC activity and growth of the gastric mucosa. In the gastric mucosa, although there were changes in gastric mucosal weight with postfasting refeeding, there were no associated changes in ODC activity. Furthermore, DFMO had no effect on gastric mucosal growth in normal fed rats and only partially prevented the trophic response to refeeding.

Epidermal growth factor, a potent inducer of ODC activity in the intestinal and colonic mucosa, did not induce ODC activity in the gastric mucosa.[9] Similarly, gastrin, the trophic hormone for the gastric mucosa, also did not induce ODC activity in the gastric mucosa. However, DFMO did inhibit the trophic action of gastrin on the gastric mucosa.[7]

It thus appears that in the gastric mucosa, growth responses may depend on the polyamines in at least a few models, but increases in ODC are somehow masked. It is possible that the process of postfasting refeeding is not an adequately strong proliferative stimulus to significantly induce ODC. Evidence for the latter hypothesis comes from studies that gastric mucosal injury produced by hypertonic perfusion induced ODC activity by more than 100-fold, with a simultaneous 10-fold increase in DNA synthesis.[10] It appears that gastric mucosal growth, at least in refeeding, gastrin stimulation, and postinjury recovery models, is somewhat dependent upon ODC activity and polyamine biosynthesis and can be inhibited at least partially by inhibitors of polyamine synthesis.

Recent evidence may provide further answers for the role of ODC and polyamines in the gastric mucosa. It appears that in the gastric mucosa, the basal "ODC activity" measured biochemically is contributed primarily by nonspecific decarboxylases. The biochemical ODC activity was not inhibited by DFMO, cycloheximide, or cold ornithine.[11] However, ODC was demonstrable immunocytochemically with specific ODC antibodies in the small population of proliferating cells in the gastric mucosa. Furthermore, the immunoreactive ODC increased with pentagastrin stimulation.[11] These results suggest that in the gastric mucosa, growth processes are dependent on increases in ODC which are not demonstrable by biochemical measurements because of masking by high levels of nonspecific decarboxylases.

C. INTESTINAL MUCOSAL GROWTH

The intestinal mucosa is rapidly proliferative and can virtually completely renew its entire mass in 3 to 4 d, yet its growth is highly regulated. The polyamines appear to play an important role in regulating intestinal mucosal growth.

During mucosal growth, ODC activity and polyamine content show increases parallel with proliferation. In our own laboratory we have shown that mucosal ODC and polyamines are increased during the rapid growth phases of intestinal maturation,[12] intestinal mucosal recovery after ara-C injury,[12] and during intestinal adaptive hyperplasia during lactation.[13-15] In all of these studies, the specific ODC inhibitor DFMO was used to suppress mucosal growth. This suggested that the intestinal mucosal growth processes are specifically dependent upon the increases in ODC and polyamines.

Although the normal adult animal does not show significant toxicity when given DFMO for periods of 1 to 2 weeks, prolonged administration of DFMO will reduce the food intake and body weight of rodents.[6,13] Since weight loss alone can produce growth inhibition in animals, studies using DFMO should employ a pair-feeding regimen, especially if the drug is to be used for greater than 3 to 7 d.

As described in the section on polyamine metabolism, the enzyme diamine oxidase is the putative degradative enzyme for putrescine. When aminoguanidine, an inhibitor of diamine oxidase, is administered to rodents there is a concomitant increase in food intake and weight gain.[25] This suggests that manipulation of the polyamine pathway is possible through both the synthetic pathway (DFMO inhibition of ODC) and the degradative pathway (aminoguanidine inhibition of diamine oxidase) and can significantly affect the weight gain of the animals.

1. Intestinal Maturation

In the newborn rat, the intestinal mucosa undergoes postnatal maturation. In the first 3 weeks after birth intestinal crypt cells proliferate rapidly, the crypts elongate and the villi lengthen, cell number increases, with a concomitant increase also in the number of mature mucosal cells. The mature mucosal cells begin to express mucosal enzymes such as disaccharidases and diamine oxidase. The most rapid histologic and enzymatic maturation occurs during the third week after birth. Similarly, the most rapid increase in the activities of the disaccharidases occurs during that week. We found that mucosal ODC activity and polyamine content increased markedly and transiently just prior to the marked increase in mucosal thickness and increase in mature mucosal enzyme activities. ODC activity increased 10- to 20-fold, along with 3- to 6-fold increases in the activity of AdoMetDC and the content of putrescine, spermidine, and spermine. The enzyme activities and polyamine content returned to their low basal levels by day 24, when the intestinal mucosa had achieved its mature status.[12] These data suggest that ODC and polyamines are associated with intestinal maturation.

DFMO was then given to the nursing mothers or directly to the newborn rat pups between days 14 and 24 after birth, during which the marked transient increase in ODC activity was found. With DFMO there was a 65% inhibition of the increase in ODC and virtually complete suppression of the increases in putrescine and spermidine. These changes in ODC activity and polyamine content were associated with a delay in the maturation of the intestinal mucosa, as documented by a delay in histologic and enzymatic maturation. Mucosal disaccharidases and diamine oxidase activities did not reach normal adult levels in the DFMO treated animals until day 32, a full 8 d after untreated controls. This biochemical maturation delay was accompanied by a delay in morphometric maturation of 3 d.[12] These results suggest an important role for the increase in ODC and polyamine content in the process of proper intestinal maturation in the newborn rat.

2. Intestinal Postinjury Recovery

After the rat has achieved adult size its intestinal mucosa is in a dynamic steady state of constant cell renewal with crypt cell production of new cells balancing the loss of senescent cells into the lumen. This steady state may be altered when the crypt cells respond to a variety of growth stimuli with increased proliferative activity.

One such proliferative stimulus is the recovery from injury. Mucosal injury may be produced by the administration of antiproliferative agents, such as chemotherapy agents. Arabinosylcytosine (Ara-C) is an S-phase cell cycle specific chemotherapy agent which is markedly cytotoxic to the proliferating mucosal crypt cells. The mucosa becomes denuded of epithelial cells while the villus cells continue to mature and are lost into the lumen. When Ara-C is then stopped, the surviving crypt cells begin to proliferate at a markedly increased rate as compared to pretreatment basal cell renewal rates. New epithelial cells are thus produced to repopulate the intestinal mucosa. The most rapid crypt cell proliferation occurs between days four and seven after an appropriate dose of Ara-C.

The sloughing of the mucosa, assessed morphologically and enzymatically by measurement of mucosal thickness and by the loss of disaccharidase and diamine oxidase activities, began on day one after Ara-C and was maximal on about day four. The enzyme activities then increased rapidly and returned to normal pretreatment levels by day eight. We found that mucosal ODC activity increased 20-fold just prior to the enzymatic recovery and the maximal ODC increase was on day six. Simultaneously, there was a two- to five-fold increase in the activity of AdoMetDC activity and the content of putrescine, spermidine, and spermine. ODC activity and polyamine content rapidly returned to their low basal levels by day eight.[12]

When rats were given DFMO simultaneously with Ara-C, there was a complete suppression of the increase in ODC activity and polyamine content. These biochemical changes were accompanied by a marked delay in mucosal recovery. The activities of mucosal disaccharidases and diamine oxidase returned to normal pretreatment levels only by day 14, a full 6 d after those Ara-C treated animals not given DFMO. This delay in enzymatic recovery was also accompanied by a delay in morphologic recovery, as measured by a delay in the repopulation of the mucosa by epithelial cells and a return of mucosal thickness to pretreatment levels.[12]

D. INTESTINAL ADAPTATION

As discussed more completely in Chapter 11, intestinal adaptation can occur in response to a variety of growth stimuli. During this adaptation process, mucosal hyperplasia occurs, with increases in crypt depth and villus height, as well as increased glucose and water absorption per unit length of intestine. This morphologic and functional adaptation is most prominent in the distal small intestine. Although intraluminal factors and hyperphagia may play important roles, intestinal adaptation has been shown, at least in the lactation model, to be most dependent upon hormonal factors.[16,17]

1. Lactation

One model of intestinal adaptation is that associated with lactation. During lactation, intestinal weight, thickness, and specific enzyme activities increase markedly. By day two of lactation, intestinal segment wet weight, villus height, and crypt depth increase by 10 to 15% over nonlactating controls. This mucosal hyperplasia continues throughout the 2 to 3 weeks of lactation. Intestinal mucosal growth though is most prominent and marked during the first week with the maximal increase occurring on or about day five. We found that during lactation, mucosal ODC activity also increased, with a concomitant increase of polyamine content, with a maximal increase on day five corresponding to the time of maximal morphologic intestinal adaptation.[13] In animals given DFMO, intestinal mucosa ODC activity was inhibited, along with polyamine content, and suppression of intestinal adaptation, with marked diminishing of the adaptive increases in mucosal height and thickness, and especially of crypt depth. Again, these studies have been performed in pair-fed animals, so that the observed DFMO inhibition of mucosal growth cannot be attributed to its effect on decrease in food intake or suppression of weight gain.[13]

2. Resection

The intestinal mucosa can also respond to situations where total absorptive surface is lost. After resection of half or more of the length of the small intestine, morphologic and functional adaptation occur in the remaining intestinal remnant. This adaptation is again greater in the distal intestine after proximal resection. This jejunectomy model has been used extensively to study mucosal growth and the effect of various growth factors. This intestinal adaptation is achieved primarily by mucosal hyperplasia, with increased epithelial cell number and increased villus height and crypt depth, as well as dilation and lengthening of the intestinal segment. This adaptive hyperplasia has been shown to be dependent upon a combination of factors, including intralumenal nutrition, pancreatico-biliary secretions, and hormonal factors.[6,17]

With jejunectomy, the resection of the proximal 50% of the small intestine, marked increases of ODC and AdoMetDC by 100- to 200-fold occur in the rat between days one and four after resection. Simultaneously, there is a two- to four-fold increase in mucosal levels of putrescine and spermidine. This time period coincides and slightly precedes the period of initiation of mucosal cell hyperplasia and is characterized by increased mucosal cell proliferation, as measured morphologically and biochemically. Intestinal weight and mucosal thickness increase by 50 to 75%. Similarly mucosal DNA content and DNA synthesis increase by 70 to 100%. Increased intestinal crypt cell proliferation also occurs, with greater than 100% increase in labeling per crypt and in crypt cell production rate as measured by stathmokinetic methods.[14] In fact, the rates of mucosal cell proliferation as measured by crypt cell production rate, and the rate of lengthening of mucosal villus and crypt, are all significantly correlated with ODC activity.

To document the critical dependence of mucosal adaptive hyperplasia to induction of ODC and polyamine biosynthesis, we again used the ODC inhibitor DFMO. When DFMO was given to the animals beginning 3 days prior to jejunectomy, the increases in ODC activity and polyamine content are completely suppressed. This suppression of ODC activity is accompanied by virtually complete suppression of DNA synthesis and results in a complete abolition of intestinal weight gain, increase in mucosal thickness, and increase in crypt cell production.[15] These results indicate that the increases in ODC activity and polyamine content are critical for adaptive postresectional crypt cell proliferation *in vivo* and that the critical step mediated by polyamines in this adaptive mucosal growth process may be the onset of DNA synthesis.

3. Pancreatico-Biliary Diversion

Dowling and co-workers[16] have extended the above work to the pancreatico-biliary diversion (PBD) model. PBD is the surgical interposition of either the jejunum or the ileum between the stomach and the duodenum. As a result, the pancreatico-biliary secretions are diverted to a different segment of the intestine. This procedure results in increased circulating levels of at least two gastrointestinal peptide hormones, enteroglucagon and cholecystokinin. Subsequent studies have demonstrated that the intestinal mucosal hyperplasia is probably induced by enteroglucagon and the adaptive growth in the pancreas, also seen in this model, is induced by cholecystokinin.[16]

With PBD, increases in polyamines were seen in the intestinal mucosa. The increases in polyamines in the ileum are greater than those in the jejunum, which in turn are greater than those in the duodenum, paralleling the degree of adaptive mucosal growth in the various intestinal segments. They subsequently found that DFMO resulted in significant reductions in mucosal levels of spermidine and spermine and were associated with a marked decrease or complete prevention of the adaptive mucosal hyperplasia in both the ileum and jejunum.[16]

4. Fasting Refeeding

Although the critical dependence for ODC induction and polyamine biosynthesis on intestinal mucosal growth appears to be well proven, the exact cellular processes and the cellular physiology of the polyamines in the intestinal mucosa are still not well understood. In the basal state, ODC activity is much higher in the mature villus tip cells than the crypt cells;[18-20] however, these studies are generally done in the basal state.

Using the fasting-refeeding model, several investigators, including ourselves, have now shown that induction of ODC and polyamine biosynthesis also occur in this model.[21,22] The association of ODC and the polyamines in the refeeding model of adaptive hyperplasia is similar to the other models of intestinal mucosal adaptation. In addition, the gradient of ODC activity, being higher in the villus cells than the crypt cells, was maintained throughout starvation and refeeding. The activity of AdoMetDC, however, was highest in the crypt cells in the basal state and increased throughout the entire villus-crypt axis during refeeding and also during lactation, preserving a villus-crypt gradient opposite that of ODC.[21,22]

During refeeding hyperplasia the contents of all three polyamines increased. Putrescine was highest in the villus-tip cells, paralleling ODC activity. Spermidine and spermine content were highest in the crypt and paralleled the distribution of AdoMetDC activity. These results raised the possibility that AdoMetDC activity and spermidine and spermine content may play a more important role than ODC and putrescine in the regulation of intestinal mucosal proliferation.

Since the polyamine biosynthetic enzymes and the polyamines are intricately interrelated in the polyamine biosynthetic pathway, the exact roles of the different polyamines and their biosynthetic enzymes remain to be elucidated. It is also possible that the small increases in ODC in the crypt cells are adequate to trigger cell proliferation, whereas the higher ODC levels in the villus cells may represent an association with a differentiation of the initial epithelial cells. No matter what the exact physiological role of the polyamine it appears clear that polyamines are required for intestinal mucosal growth. In fact, continuous infusion of putrescine into the ileal-lumen for 5 d stimulated mucosal growth.[7]

E. PANCREATIC GROWTH

Using the PBD model, it was found that pancreatic polyamine levels increased in association with pancreatic hyperplastic growth.[16] However, DFMO resulted in no changes in the levels of putrescine, spermidine, or spermine in the whole pancreas and no differences in pancreatic wet weight, protein, or DNA content.[16]

Similarly, increases in pancreatic polyamine levels were also found in pancreatic hyperplastic growth induced by caerulein.[23,24] However, in the ethionine model, DFMO neither affected DNA synthesis nor prevented pancreatic hypertrophy. In the caerulein model, although DFMO was able to reduce the increase in DNA synthesis and DNA content, it did not alter pancreatic hyperplasia as measured by increases in RNA, proteins, and pancreatic enzyme activity.[23]

It thus appears that in the pancreas, similar to the gastric mucosa, the role of polyamines in hyperplastic growth of these organs is not yet firmly established. Further studies on a cellular level or organ culture level or the use of more potent inhibitors may need to be performed.

III. SUMMARY

Increases in ornithine decarboxylase activity and the biosynthesis of polyamines are characteristically associated with rapid cell and tissue growth and have been shown in numerous *in vitro* and *in vivo* systems to be critical for cell and tissue growth processes. The rapidly proliferating and maturing gastrointestinal mucosa appears to be an excellent model for studying the potential cellular roles of the polyamines in the regulation of cell and tissue growth.

At least for small intestinal mucosa, the evidence appears extensive to support the critical role of ODC and the polyamines in adaptive hyperplastic growth. This area of polyamine research has grown rapidly. New and more potent inhibitors of polyamine biosynthesis, including more potent inhibitors of ODC and new polyamine analogs, have been developed and are now being used.[4] Studies are now available to document that these newer polyamine inhibitors and analogs may allow a more complete inhibition of ODC and a more complete depletion of intracellular polyamines, particularly of spermidine and spermine.[4] These newer inhibitors may allow us to more clearly define the role of ODC and polyamines in tissue growth, particularly in the gastric mucosa and the pancreas, where the evidence supporting the critical role of polyamines is less definitive. New studies into the biochemical and molecular functions of the polyamines are also ongoing.[3,4] This increased knowledge about the biology of polyamines will no doubt allow us to understand further the potential role of polyamines in cell growth, particularly in the gastrointestinal tract.

REFERENCES

1. **Tabor, C. W. and Tabor, H.,** Polyamines, *Annu. Rev. Biochem.*, 53,749, 1984.
2. **Pegg, A. E.,** Recent advances in the biochemistry of polyamines in eukaryotes, *Biochem. J.*, 234(2), 249, 1986.
3. **Pegg, A. E.,** Polyamine metabolism and its importance in neoplastic growth and as a target for chemotherapy, *Cancer Res.*, 48, 759, 1988.
4. **McCann, P., Pegg, A. E., and Sjoerdsma, A.,** Inhibition of polyamine metabolism, in *Biological Significance and Basis for New Therapies*, Academic Press, Orlando, FL, 1987, 1.
5. **Higgins, G. M. and Anderson, R. M.,** Experimental pathology of the liver. I. Restoration of the liver of the white rat following partial surgical removal, *Arch. Pathol.*, 12, 186, 1931.
6. **Luk, G. D.,** Essential role of polyamine metabolism in hepatic regeneration, *Gastroenterology*, 90, 1261, 1986.
7. **Seidel, E. R., Tabata, K., Dembinski, A. B,. and Johnson, L. R.,** Attenuation of trophic response to gastrin after inhibition of ornithine decarboxylase, *Am. J. Physiol.*, 249, G16, 1985.
8. **Tabata, K. and Johnson, L. R.,** Ornithine decarboxylase and mucosal growth in response to feeding, *Am. J. Physiol.*, 251, 6270, 1986.
9. **Feldman, E. J., Aures, D., and Grossman, M. I.,** Epidermal growth factor stimulates ornithine decarboxylase activity in the digestive tract of the mouse, *Proc. Soc. Exp. Biol. Med.*, 159, 400, 1978.
10. **Furihata, C., Sato, Y., Hosaka, M., Matsushima, T., Furukawa, F., and Takahashi, M.,** NaCl induced ornithine decarboxylase and DNA synthesis in rat stomach mucosa, *Biochem. Biophys. Res. Comm.*, 121, 1027, 1984.
11. **Tseng, C. C., Haddox, M. K., and Johnson, L. R.,** Immunocytochemical localization of ornithine decarboxylase (ODC) in rat gastric mucosa, *Gastroenterology*, 94, A466, 1988.
12. **Luk, G. D., Marton, L. J., and Baylin, S. B.,** Ornithine decarboxylase is important in intestinal mucosal maturation and recovery from injury in rats, *Science*, 210, 195, 1980.
13. **Yang, P., Baylin, S. B., and Luk, G. D.,** Polyamines and intestinal growth: absolute requirement for ODC activity in adaptation during lactation, *Am. J. Physiol.*, 247, G553, 1984.
14. **Luk, G. D. and Baylin S. B.,** Polyamines and intestinal growth — increased polyamine biosynthesis after jejunectomy, *Am. J. Physiol.*, 245, G656, 1983.
15. **Luk, G. D. and Baylin, S. B.,** Inhibition of intestinal epithelial DNA synthesis and adaptive hyperplasia after jejunectomy in the rat by supression of polyamine biosynthesis, *J. Clin. Invest.*, 74, 698, 1984.
16. **Dowling, R. H., Hosomi, M., Stace, N. H., Lirussi, F., Miazza, B., Levan, H., and Murphy, G. M.,** Hormones and polyamines in intestinal and pancreatic adaption, *Scand. J. Gastroenterol.*, 20, 84,1985.
17. **Elias, E. and Dowling, R. H.,** The mechanism for small-bowel adaptation in lactating rats, *Clin. Sci. Mol. Med.*, 51, 427, 1976.
18. **Baylin, S. B., Stevens, S. A., and Shakir, K. M. M.,** Association of diamine oxidase and ornithine decarboxylase with maturing cells in rapidly proliferating epithelium, *Biochim. Biophys. Acta*, 541, 415, 1978.
19. **Porter, C. W. C., Dworaczyk, C., Ganis, B., and Weiser, M. M.,** Polyamines and biosynthetic enzymes in the rat intestinal mucosa and the influence of methylglyoxal-*bis* (guanylhydrazone), *Cancer Res.*, 40, 2330, 1980.
20. **Sepulveda, F. V., Burton, K. A., Clarkson, G. M., and Syme, G.,** Cell differentiation and L-ornithine decarboxylase activity in the intestine of rats fed low and high protein diets, *Biochim. Biophys. Acta*, 716, 439, 1982.
21. **Fitzpatrick, L. R., Wang, P., Eikenburg, B. E., Haddox, M. K., and Johnson, L. R.,** Effect of refeeding on polyamine biosynthesis in isolated enterocytes, *Am. J. Physiol.*, 250, G709, 1986.
22. **Luk, G. D. and Yang, P.,** Distribution of polyamines and their biosynthetic enzymes in intestinal adaptation, *Am. J. Physiol.*, 254, G194, 1988.
23. **Benrezzak, O. and Morisset, J.,** Effects of alpha-difluoromethylornithine on pancreatic growth induced by caerulein, *Regulatory Peptides*, 9, 143, 1984.
24. **Houlihan, M. J., Saluja, A., Powers, R. E., Saluja, M., Eigner, S., and Steer, M. L.,** Effects of L-difluoromethyl ornithine (DFMO) upon pancreatic regeneration, *Dig. Dis. Sci.*, 30, 975, 1985.
25. **Luk, G. D.,** unpublished results.

Chapter 3

RECEPTORS OF THE GASTROINTESTINAL HORMONES AND THEIR DEVELOPMENT

Jean Morisset

TABLE OF CONTENTS

I. INTRODUCTION

Cells of the digestive tract synthesize and secrete large amounts of various proteins and digestive enzymes, as well as fluids of different composition. These cells have the capacity to adapt to the changing needs of the organism by modulating their growth rate and adjusting their metabolism related to their specific functions. These adaptive changes are mediated by a variety of hormones and factors interacting with specific cellular receptors; the mechanisms of action involved in mediating and maintaining growth are not fully understood.

This chapter is not intended to present a comprehensive review on receptor mechanisms or the different transmembrane signaling systems involved in hormonal responses. The objectives are rather to introduce and describe the receptors of the most important gastrointestinal hormones and growth factors involved in the growth and physiological responses of many organs of the digestive tract: cholecystokinin, gastrin, secretin, insulin, and epidermal growth factor. Emphasis will be put on cellular localization, biochemical and structural characteristics, effector system, their occupancy and the resulting biological responses, modulation and finally, development with age.

II. CHOLECYSTOKININ RECEPTOR

Cholecystokinin exists in multiple molecular forms: it contains either 8 (CCK-8), 33 (CCK-33), 39 (CCK-39), or 58 (CCK-58) amino acids. It has metabolic, endocrine, vascular, and trophic actions. As a secretagogue hormone, it induces secretion of digestive enzymes and bicarbonate from the pancreas and increases the output of hepatic bile and secretion of Brunner's glands. As a motor agent, it causes contraction of the gall bladder, gastric and intestinal smooth muscles, and inhibits contraction of the lower esophageal sphincter and sphincter of Oddi. Metabolically, it stimulates pancreatic and intestinal growth, whereas its endocrine roles include stimulation of calcitonin secretion from the thyroid and pancreatic polypeptide and insulin from the endocrine pancreas.

A. LOCALIZATION

Studies on the location of hormone receptors in target cells are now possible with quantitative electron microscope autoradiographs. Such studies have been carried out with cholecystokinin on mouse [1,2] and rat[3] pancreatic acini. In acini incubated 2 min with 0.5 or 0.1 n*M* of ^{125}I-labeled BH-CCK33, 64% of the observed grains were localized over the plasma membrane. The label did not appear over the apical area of the acinar cell whereas part of labeling was observed over the endoplasmic reticulum. After 30 min of incubation, silver grains were still present in high concentration over the basolateral plasma membrane but an increased number was observed over the endoplasmic reticulum. These studies indicate that CCK binds to specific receptors on the basolateral surface of pancreatic acinar cells and suggest that after binding, the hormone or hormone-receptor complex is internalized for further biological actions or degradation.

B. BIOCHEMICAL CHARACTERIZATION

The cholecystokinin receptor has been the most extensively characterized of all the gastrointestinal hormone receptors, and the tissue of choice has been rodent pancreatic acinar plasma membrane or the acinar cell from rat, mouse, or guinea pig.

As indicated in Table 1, the CCK receptor has been biochemically characterized in rat,[4,5] mouse,[6] guinea pig,[7] bovine,[8] dog,[7] and human.[9] Most studies were performed using ^{125}I-CCK-33[10] as a ligand, but ^{125}I-CCK-8,[9] ^{125}II-E-CCK-8,[11] and ^{125}I-[Tyr^{34},Ahx^{37}] CCK(31-39)[7] were also used. The first antagonist to be developed, ^{3}H-L-364,718,[12] served to characterize the solubilized receptor from rat pancreas. These studies were performed either on dispersed acini,[5] purified plasma membranes,[13] or after receptor solubilization.[6] Apart from observations on pancreatic

TABLE 1
Biochemical Characteristics of the Cholecystokinin Receptor

Species	Ligand	Tissue	K_D (n*M*)	B_{max}	Mol wt (kDa)	Ref.
Guinea pig	^{125}I[Tyr^{34},Ahx^{37}]CCK(31-39)	Pancreatic acini	0.1 (K_H) 35.0 (K_L)	20 fmol/mg prt 500 fmol/mg prt	84	7
Dog	^{125}I[Tyr^{34},Ahy^{37}]CCK(31-39)	Pancreatic acini	0.5 (K_H) 25.0 (K_L)	25 fmol/mg prt 600 fmol/mg prt	78, 45, 28	7
Rat	^{125}I-CCK-33	Pancreatic acini	0.06 (K_H) 21.0 (K_L)	4 fmol/mg prt 640 fmol/mg prt	5	
	^{125}I-CCK-33	Pancreatic plasma membrane	1.3	204 fmol/mg prt		13
	^{3}H-L364,718	Solubilized receptor	0.5	3.1 pmol/mg prt		12
Human	^{125}I-CCK-8	Gastric smooth muscle membrane	0.07	8.4 p*M*	Major 75 Minor 200; 120, 100, 53	9
Bovine	^{125}I-CCK-33	Gallbladder muscle membrane	0.6	100 fmol/mg prt		8

Note: (K_D) Dissociation constant; (K_H) K_D of the high-affinity site; (K_L) K_D of the low-affinity site; (B_{max}) receptor concentration; (mol wt) estimated molecular weight of the receptor.

tissue, the CCK receptor was also investigated in isolated fundus gland cells[11] and membranes from gastric smooth muscle[9] and gall bladder.[8] In all these studies a single protein complex receptor was observed.

Studies performed on intact pancreatic acinar cells (Table 1) indicate the presence of two binding sites on the CCK receptor, one of high affinity (0.06 to 0.5 n*M*) with low capacity (4 to 25 fmol/mg protein) and the second of low affinity (21 to 35 n*M*) with high capacity (500 to 640 fmol/mg protein).[5,7] However, from experiments performed on pancreatic plasma membranes, only one binding site was observed and this had an affinity (0.24 to 1.3 n*M*) situated between the high and low affinities of the receptors on the dispersed acini.[6,10,13] Characterization of the solubilized receptor from pancreatic plasma membrane also revealed two binding sites, one of relatively high affinity (3.9 n*M)* and low capacity (0.12 pmol) when compared to acini and the other of lower affinity (69 nM) and higher capacity (2.9 pmol)[6] comparable to that of acini. The antagonist ^{3}H-L-364,718 was used by one group[12] to study the solubilized receptor and showed great potential as a ligand having an affinity of 0.53 n*M* which is similar to that of the purified plasma membranes.[12]

The CCK receptor characterized from different organs, from plasma membranes, acinar cells, gastric and gall bladder muscle cells exhibits affinities for the radiolabeled ligands ranging from 0.07 to 1.3 n*M* (Table 1). Such a narrow range of values suggests that a single type of receptor molecule probably exists in different forms when studied in its natural environment, the cell itself.

Two models have been presented to explain the biological effects of CCK. Collins et al.[14] present the CCK receptor as having two distinct binding sites; the first site has a high affinity for CCK and its occupation is mandatory for the occupation of the second site. Dissociation from the high affinity site 1 is rapid and that from the low affinity site 2 is also rapid when site 1 is occupied, but slow when 1 is vacant. This model was proposed to explain the residual effect of CCK on amylase release from pancreatic acinar cells. To understand the observation that amylase release and glucose uptake from pancreatic acinar cells occur at different CCK concentrations, Sankaran et al.[15] also presented a CCK receptor model. The receptor would have high-affinity sites which bind two molecules of CCK per receptor molecule. This divalent high-affinity receptor complex would be responsible for amylase release. The lower affinity receptor binds only one molecule of CCK and its occupancy would mediate glucose transport and account for inactivation of the high-affinity complex involved in amylase release.

Is it now possible to reconcile the models which implicate two CCK binding sites by using some of the structural characteristics recently obtained from covalent cross-linking studies? Analyses of photoaffinity labeled membranes by SDS/polyacrylamide gel electrophoresis reveal that the molecular weight of the major band of this receptor varies between 75 and 84 kDa with, in some cases, distinct molecular species of higher and lower molecular weights (Table 1). The larger labeled species may be oligomers of the major component[3] and the finding that the cross-linked receptor absorbed to wheat germ agglutinin and eluted with *N*-acetylglucosamine indicates that it is a glycoprotein.[16] Except for one study performed on human gastric smooth muscle membrane,[9] all the other experiments used the pancreatic acinar cell membranes prepared from guinea pig,[7] dog,[7] rat,[3] or mouse[17] to evaluate the structure of the CCK receptor. Under nonreducing conditions, the major labeled band of the cross-linked receptor was at 120 kDa; however, when membranes were reduced, the predominant labeling was at 76 kDa.[18] This suggests that the receptor is composed of two subunits, a 76 kDa subunit responsible for CCK binding and a 40 kDa subunit linked to the bigger subunit by disulfide bonds. This 40 kDa subunit resembles the 42 kDa subunit of Ns or that of 40 kDa in Ni described later as being part of the solubilized CCK receptor from rat pancreatic plasma membranes.[19] However, the Ni and Ns subunits are guanine nucleotide regulatory proteins and are not to our knowledge bound to their receptor by a disulfide bridge. Cloning of the CCK receptor gene or genes will probably determine whether the high- and low-affinity binding sites are part of one or two different molecules.

C. INTERACTION WITH GUANINE NUCLEOTIDE BINDING PROTEINS

Mechanisms by which CCK affects intracellular signal transduction in its target organs are still not fully understood. Innis and Snyder[10] were among the first to demonstrate that CCK binding to its receptor in the pancreas could be regulated in a selective fashion by guanyl nucleotides. A year later, Lambert et al.[20] presented data indicating that the guanine nucleotide regulatory units involved in adenylate cyclase activation, GTPase activity, and CCK binding had similar properties in rat pancreatic plasma membranes. This same group[19] later confirmed that, following solubilization of the CCK receptor from rat pancreatic plasma membranes, two guanine nucleotide regulatory proteins coexisted in the soluble complex. These regulatory proteins are assumed to be Ns and Ni from the molecular weights (42,000 and 40,000) established following ADP-ribosylation by cholera toxin and *Bordetella pertussis* toxin.

More recently, Merritt et al.[21] reported increased accumulation of IP_3 in response to caerulein or carbachol in rat pancreatic acinar cells. This caerulein- or carbachol-induced effect was potentiated by stable analogs of GTP and was not affected by pretreatment of the cells with pertussis or cholera toxins. These new data therefore suggest that neither Ns nor Ni are coupled to phospholipase C in pancreatic acinar cells but are bound to a different G protein. Data for the solubilized receptor[19] and the pancreatic membrane receptor[22] suggest that CCK binding induces an association of the CCK receptor with guanine nucleotide binding protein that would initiate intracellular signal transduction. This newly found G protein remains to be characterized and we also need to know more about this putative Ni regulatory protein which was copurified with the CCK receptor.[19] Yet, we do not know of any hormone inhibitors of the *in vitro* secretory effects of cholecystokinin or acetylcholine on pancreatic acinar cells.

D. ACTIVATION OF SECOND MESSENGERS

Intracellular signal transduction events stimulated by hormones are channeled through specific guanine nucleotide proteins (G proteins) which associate with their receptor as an initial step following hormone binding. Such a complex then dissociates and the alpha subunit of the G protein combines with either the catalytic subunit of adenylate cyclase or possibly also with phospholipase C according to the second messenger pathway that is activated. These two activated enzymes then synthesize the second messengers cyclic AMP, inositol trisphosphate (IP_3), and diacylglycerol (DG).

Cholecystokinin is relatively unique as it induces its physiological effects through the formation of all three second messengers. The pancreatic acinar cell has exclusively been used for these studies. Indeed, in the presence of IBMX, a phosphodiesterase inhibitor, the supramaximal concentrations of CCK (300 n*M*) and CCK-8 (above 10 n*M*) for amylase release caused a significant increase in cAMP in rat pancreatic acinar cells[23,24] and pancreatic fragments.[25] These high concentrations of CCK saturate both the high- and low-affinity binding sites of the CCK receptor[15] and such an occupation of the low-affinity sites corresponds to increased glucose absorption in these cells.[26] However, the observation that secretin and dibutyryl cAMP had no effect on 2-deoxy-D-glucose uptake[26] may indicate that there is no relation between cAMP production and glucose uptake by the acinar cells.

Recently, Merritt et al.[21] reported an increased accumulation of IP_3 in response to caerulein, a CCK analog, in permeabilized acinar cells indicating phospholipase C-mediated hydrolysis of PIP_2.[21] Previous structure and activity studies on CCK receptors in the pancreas have demonstrated a good correlation between the affinity of CCK C-terminal fragments and their potency to elicit amylase release.[27] CCK analogs ranging from the tetra- to octapeptide elicited dose-dependent increases in the breakdown of phosphoinositides in pancreatic lobules. The rank order of potency for these increases is in complete agreement with their relative potency in inducing amylase release and their affinity for the CCK high-affinity receptor sites.[28] In such a system, 70% of the maximal amount of IP accumulation can generate maximal amount of amylase release suggesting an amplification process.

These data indicate that cholecystokinin can increase IP_3 as well as cAMP in the same cell preparation. However, the strong correlation existing between the dose response curve for amylase release and IP_3 formation suggests that these two events are interrelated. Since cAMP is produced at such high concentrations of cholecystokinin, a role for this CCK-activated pathway remains to be found.

In guinea pig gastric mucosa glands, submaximal concentrations of cholecystokinin mediated an accumulation of inositol phosphate and this effect was correlated with pepsinogen secretion.[29]

E. RECEPTOR OCCUPANCY AND BIOLOGICAL FUNCTIONS

The relationship of cholecystokinin receptor occupancy to regulation of biological functions has been established primarily in pancreatic tissue. Among the known biological functions of cholecystokinin on pancreatic acini are stimulation of amylase release and regulation of the transport of sugars and amino acids. Maximal stimulation of amylase release occurred with concentrations of caerulein lower than those needed to maximally increase calcium efflux[26] and this seems to be due to the saturation of the high-affinity CCK receptors.[5] However, caerulein-stimulated maximal glucose uptake required two- to threefold higher concentrations of caerulein compared with maximal calcium efflux[26] and this appears to result from occupancy of the low-affinity CCK binding sites.[15] The occupation of these low-affinity binding sites was also associated with inhibition of amylase release from dispersed pancreatic acini.[5] In guinea pig gastric chief cells, the dose-response curve of CCK for pepsinogen secretion was superimposed on that for the binding to its high-affinity receptor, the only class identified in these cells.[30]

F. MODULATION

Activity of the pancreatic acinar cell can be regulated by many hormones and neurotransmitters, but this cell is most sensitive to cholecystokinin and its analogs. However, our knowledge of the factors that regulated CCK receptors and this hormone action is quite limited.

The CCK receptor seems quite resistant to short- and long-time exposures to its own agonist. Indeed, after 2 and 14 h of CCK-8 treatment *in vivo*, the secretory response to the agonist was slightly shifted toward higher concentrations of CCK-8 which is a sign of desensitization, whereas specific CCK binding was significantly increased only for high-affinity binding sites without change in affinity of both classes.[31] A similar CCK treatment for 7 d did not affect the CCK receptor concentration on the acinar cells nor their affinity for the ligand.[32] These data suggest that the CCK receptor can hardly be down-regulated by its own agonist.

One of the hormones known to influence exocrine pancreatic function is insulin. Its major role consists in regulating pancreatic amylase concentration[33] through an effect on the number of copies of its messenger RNA.[34] In streptozotocin-induced diabetes, altered maximal rates of amylase secretion have been demonstrated.[35] Such observations stress the importance of insulin not only in regulating specific enzyme synthesis, but also in regulating secretory processes. In a recent study,[36] Otsuki and Williams investigated the direct modulation of pancreatic CCK receptors by insulin in diabetic rats. Insulin pretreatment of acini for 2 h decreased both the affinity of the high-affinity CCK receptors and the capacity of the low-affinity sites. Since insulin potentiates the CCK response on amylase release, the insulin specific effect on the decreased number of low-affinity sites, the occupation of which corresponds to decreased amylase release, could explain how high concentrations of CCK induced greater amylase secretion. These data support a role for insulin as a positive modulator for the activation of the CCK receptor.

In suckling rats, glucocorticoid treatment resulted in pancreatic hypertrophy and hyperplasia, while after weaning, the same treatment increased the hypertrophic effect of CCK.[37,38] Since glucocorticoids increase pancreatic enzyme content upon interaction with cholecystokinin, it is possible to predict some interaction of these steroids with the CCK receptor.

In a recent study, Leung et al.[39] have indeed demonstrated that, in 2-d-old rat pups, the maximal binding to ^{125}I-BH-CCK was significantly higher in the dispersed acini from hydrocortisone treated animals when compared to the dispersed acini from the control group. Such a treatment also led to a precocious maturation of the secretory apparatus resulting in a greater responsiveness of the acini to cholecystokinin. Furthermore, in the last few years, the AR42J pancreatic acinar cells, a cell line derived from a chemically induced transplantable tumor of the exocrine pancreas, have been extensively used as replacement for a normal pancreatic acinar cell culture because the latter cannot be grown in continuous culture. This cell line has also been selected to study modulation of the CCK receptor.[40] Treatment of these cells with dexamethasone increased the specific binding of ^{125}I-CCK. This effect was time-dependent and dose-dependent, with maximal effects occurring after 48 h, at an EC_{50} of 1 n*M*. Other steroids, such as corticosterone, aldosterone, and progesterone, were also effective but less than dexamethasone, estradiol being ineffective. This cell line, compared to the normal acinar cell, presents a single class of CCK binding sites and dexamethasone treatment almost tripled the number of sites per cell without any changes in receptor affinity (K_D: 6 n*M*). In similar conditions, dexamethasone increased basal and CCK stimulated amylase secretion. These data indicate that steroids, principally glucocorticoids, can also positively modulate the CCK receptor leading to an enhanced secretory response to CCK.

G. ONTOGENY

It has been observed previously that protein and enzyme secretion in response to caerulein[41] and the muscarinic agonist carbachol[42] is absent in fetal pancreas and increases gradually after birth.[42,43] This lack of responsiveness of the pancreatic acinar cells to CCK analogs before birth may result from the absence of receptors on the cell surface or an uncoupling of the receptor to the second messenger effector system; this would impair its ability to generate both IP_3 formation and the subsequent rise in cytosolic calcium necessary for exocytosis.

To understand this failure of pancreatic acini isolated from fetal and newborn rats to respond to cholecystokinin, maturation of the CCK receptor was investigated. Although Werlin et al.[44] were unable to detect CCK receptors from membranes prepared from newborn pancreases using CCK-33 as a ligand, Chang and Jamieson[45] detected their presence with CCK-8 in the embryo. In a more detailed study, Leung et al.[46] were able to characterize two CCK binding sites in the pancreas of term fetus. One of the sites had high affinity (K_D: 75 p*M*) and low capacity (0.13 fmol/mg protein) and the other had low affinity (K_D: 8 n*M*) and high capacity (448 fmol/mg protein). The capacity of the high-affinity component gradually increased with age, whereas that of the low-affinity component remained relatively constant. According to these investigators, the lack of responsiveness of pancreatic acini to secretagogues may result from a low binding capacity of the high-affinity component of the receptor; this assumes that enzyme release correlates with occupancy of the high-affinity CCK receptor as suggested by Sankaran et al.[5]

The presence of high-affinity CCK binding sites on the acinar cell plasma membrane cannot by itself trigger amylase release in the fetal pancreas. However, the observation that Ca^{2+} efflux from both fetal and neonatal pancreas was increased in a similar fashion by CCK suggests that the cell surface receptors for CCK are functional and coupled to their second messenger effectors. However, in fetal pancreas, the Ca^{2+} ionophore A23187 stimulated Ca^{2+} extrusion without amylase release and the ionophore had no effect on amylase release in the presence of dibutyryl cAMP or phorbol dibutyrate, the protein kinase C activator. This suggests that one or more signal transduction events distal to calcium mobilization have not yet matured.

Development of the CCK receptor population after birth followed a stepwise pattern.[44] A linear increase was observed between 12 h after birth until 48 h. Total binding tripled between days 2 and 3 but remained stable until day 14 at which point the amount doubled until adult levels were reached at day 21. During these same periods, the high-affinity CCK binding sites increased almost 40-fold, from 0.13 fmol in term fetus to 4.0 fmol/mg protein in the adult. On the contrary, the low-affinity sites remained constant between 448 to 640 fmol/mg protein.[46] These studies indicate that exocytosis of the pancreatic enzymes is a complex system involving maturation of receptors as well as signal transduction events that have to be fully developed for a normal secretory response.

Finally, the early presence of CCK receptors has also been demonstrated in rat upper gastrointestinal tract.[47] In a 17-d fetus, CCK binding was present in the distal stomach, esophagus, and pylorus. The specific CCK binding in the gastric mucosa disappeared 15 d after birth. Gastric antral CCK binding peaked at day 10 and was gone by day 50. However, binding in pyloric muscle also peaked at day 10 but remained in the adult. The development pattern observed in the pylorus muscle supports a role for CCK in the control of gastric emptying.

III. GASTRIN RECEPTOR

Gastrin has been shown to exist in three different forms in the blood. For identification purposes they were called big gastrin (G34; 34 amino acids), little gastrin (G17; 17 amino acids), and minigastrin (G14; 14 amino acids). The major physiological role of gastrin is to stimulate gastric acid secretion; it is also involved in gastric motility and appears to control growth of the fundic, duodenal, and colonic mucosae.

A. LOCALIZATION

Localization of the gastrin receptor on cells of the canine and pig stomachs has been realized with a monoclonal antibody to the gastrin receptor produced in inbred BALB/C mice immunized by interperitoneal injection with 10^7 canine gastric parietal cells.[48] The IgM antibody produced inhibited labeled gastrin binding to its receptor and also reduced aminopyrine uptake in parietal cells challenged by gastrin. The antibody reacted specifically with surface membranes of canine and porcine parietal cells as evidenced by immunofluorescence. This specific reaction on the parietal cell membrane was confirmed by immunogold electron microscopy. This antibody reacted specifically to gastrin receptor on parietal cells since no reaction was observed on chief cells or kidney or liver cell suspensions.

B. CHARACTERIZATION

The first characterization of the gastrin receptor was reported from plasma membranes of antral smooth muscle[49] and binding was very low. Gastrin receptor assays performed with tritiated gastrin led to the demonstration of binding sites on gastric mucosal plasma membranes and rat fundic cells, but their affinity was not in the physiological range.[50,51] The first standardized assay for measuring gastrin binding to its receptor using an iodinated analog of gastrin was described in 1979[52] and it was performed on a rat fundus homogenate. As indicated in Table 2, the gastrin receptor has since been characterized in different species including dog,[53] pig,[54] rabbit,[55] and guinea pig.[56] Initial binding studies performed with iodinated native G17 showed nonspecific binding.[53] For this reason, iodinated gastrin analogs were used and produced ligands with reproducible high affinity. Among those, ^{125}I-leu^{15}-G17,[53] ^{125}I-Nle^{15}-G$17_{2,17}$,[54] and ^{125}I-Nle^{11}-G13[55] were used. The gastrin receptors have been characterized from gastric fundus homogenate,[52] membranes prepared from the gastric fundus,[56] duodenum, colon,[57] purified gastric parietal cells,[53,58] gastric somatostatin cells,[59] and pancreatic acini[60] (Table 2).

Whether binding experiments were performed with different ligands,[53,55] biological material from isolated cells,[53,59] total cell homogenate,[52] or membranes from different organs,[56,57] the gastrin receptor exhibits high affinities ranging from 0.07 to 0.46 n*M* (Table 2). The highest affinity was found in gastric mucosal cells,[55] while affinities of 2.6 and 1.0 n*M* were found in preparations of gastric fundus cells[55] and pancreatic acini[60] with native gastrin as radioligand.

As shown in Table 2, isolated parietal cells possess eight times more receptors than their neighbor somatostatin cells.[53,59] However, membranes prepared from gastric fundus, duode-

TABLE 2
Biochemical Characteristics of the Gastrin Receptor

Species	Ligand	Tissue	K_D (n*M*)	B_{max}	Mol wt (kDa)	Ref.
Dog	^{125}I-Leu^{15}-G17	Gastric parietal cells	0.46	24,000 sites/cell	74	53, 58
	^{125}I-Leu^{15}-G17	Gastric somatostatin cells	0.41	3,000 sites/cell	—	59
Rat	^{125}I-G-17-1	Gastric fundus membrane	0.20	5.1 fmol/mg prt	—	57
	^{125}I-G-17-1	Duodenum membrane	0.27	3.0 fmol/mg prt	—	57
	^{125}I-G-17-1	Colon membrane	0.44	4.2 fmol/mg prt	—	57
Rabbit	^{125}I-Nle^{11}-G13	Gastric mucosal cells	0.07	200 sites/cell	—	55
Pig	^{125}I-Nle^{15}-G2-17	Gastric solubilized receptor	2000	—	78	54

Note: (K_D) Dissociation constant; (B_{max}) receptor concentration; (mol wt) estimated molecular weight of the receptor.

num, and colon each have comparable numbers of binding sites, between 3 and 5.1 fmol/mg protein.[57] In all these studies, the gastric receptor showed only one population of sites. The gastrin receptor was also found in the mouse colon cancer cell line MC26 and in the human colon cell line LoVo.[57] The affinities of both cell line receptors for the radio-ligand (0.5 and 0.25 n*M*) are comparable to those observed in the rat fundus mucosa (0.4 n*M*). This demonstration of high-affinity specific gastrin receptors on colonic cancer cells MC-26 supports the hypothesis that the observed trophic effects of gastrin on these cells *in vivo* may be mediated by these receptors.[61]

The first attempt to elucidate the structure of the gastrin receptor protein was performed in porcine gastric mucosal membranes.[54] The gastrin binding protein was solubilized in 0.1% Triton X-100; this treatment, however, produced a gastrin receptor of very low affinity since 2 μ*M* gastrin was required to inhibit 50% of the cross-linking of gastrin 2,17 to its receptor (Table 2). A milder solubilization in 0.4% digitonin, 0.08% cholate lowered this concentration to 0.3 μ*M*. Under these conditions the molecular weight of the purified receptor was estimated at 78,000 and was not changed under reducing conditions. Later, this same 78 kDa gastrin receptor protein was immunoprecipitated by a monoclonal antibody raised in mice immunized with gastric parietal cells prepared from canine stomach.[48] Most recently, a 74 kDa gastrin receptor was also characterized from canine gastric parietal cells.[58] It is of interest to point out that this gastrin receptor of M_r between 74 to 78,000 resembles the cholecystokinin receptor previously described (Table 1). However, despite their similarity in size, the cholecystokinin receptor of pancreatic acini has minimal affinity for gastrin,[5] whereas the parietal cell gastrin receptor does not discriminate between cholecystokinin and gastrin.[53]

C. ACTIVATION OF SECOND MESSENGERS

Any relationship between neurotransmitter or hormonal stimulation of HC1 secretion by the gastric parietal cells and the formation of second messengers remains unknown. There are no studies yet published on any possible involvement of guanine nucleotide regulatory proteins with the gastrin receptor. Recently, however, evidence was presented indicating that phosphatidylinositol breakdown was involved in HCl production. Indeed, in rabbit gastric fundic mucosal cells, the concentrations of carbachol leading to half-maximal inhibition of N-[^{3}H]methylscopolamine binding and half-maximal inositol phosphate accumulation were similar.[62] In rat gastric mucosal cells, carbachol induced inositol phosphate formation with an EC_{50} of 10 μ*M*, identical to that needed to stimulate aminopyrine uptake.[63] Recent data indicate that CCK-8 and gastrin can increase parietal cell intracellular calcium concentration probably from an intracellular store and that this mobilization may be mediated by agonist-activated inositol trisphosphate formation.[64] It seems reasonable to presume that gastrin stimulated HCl secretion from gastric parietal cells by increasing the turnover of phosphatidylinositol and thus operates through the phosphoinositol second messenger system.

D. MODULATION

Contrary to most receptors, the concentration of gastrin receptors present on gastric mucosal membranes seems to be directly regulated by serum gastrin levels. Fasting, feeding a liquid diet, and antrectomy significantly decreased serum gastrin and down-regulated the number of gastrin receptors; this phenomenon was prevented by pentagastrin treatment. Vagotomy, a surgical procedure known to raise gastrinemia, also increased the number of gastric mucosal gastrin receptors. Under these conditions, the affinity of the gastrin receptor remained unaltered.[65] This control of gastrin receptor concentration on gastric mucosa cells is directly dependent on protein synthesis. Indeed, cycloheximide treatment prevented the rise in gastrin binding capacity both in response to vagotomy and in response to refeeding following a prolonged fasting period.[66] Furthermore, regulation of the gastrin receptor concentration seems to be sex-related. In male and female rats, the concentration of gastric mucosa gastrin receptor remained the same until puberty. However, by day 40 of life, males exhibited more receptors than females. While castration did not affect gastrin receptor development in males, ovariectomy increased the number of receptors in females to a value comparable to that in males; this increase corresponded to increased food intake and higher serum gastrin levels in these ovariectomized rats. A further increase in gastrin receptor was observed following adrenalectomy of these ovariectomized females.[67] These data indicate that estrogen hormones present in ovaries and adrenals exert a negative feedback control on the expression of the gastrin receptor in gastric mucosa.

E. ONTOGENY

Newborn humans and dogs, with high serum gastrin levels, do not secrete acid basally and are unresponsive to exogenous gastrin despite the presence of parietal cells.[68,69] These findings suggest the absence of gastrin receptors on the target tissues. The only study on gastrin receptor development was performed on rat gastric mucosa and indicates that the gastrin receptor first appeared on day 20 after birth.[70] Such a late development coincided with the first response of stimulated acid secretion and increased DNA synthesis to pentagastrin on days 20 and 25, respectively. The development of these receptors was reduced by the prevention of weaning through day 25, but this treatment did not prevent or delay the onset of their appearance.[70] Under normal growth, the gastric mucosa gastrin receptors reached their adult level concentration at 60 days.

This late development of the gastrin receptor may result from the absence of another hormonal control that becomes functional around the weaning period. Indeed, many of the changes in gastrointestinal function occurring around weaning have been postulated to be partly mediated by the adrenal corticosteroids. Intestinal enzymes and gastric pepsinogen exhibited precocious increases in response to hydrocortisone.[71,72] Similarly, a single treatment of rats with corticosterone on day 7 after birth resulted in precocious development of the gastric mucosa gastrin receptors to the extent that significant gastrin binding could be detected on day 10.[73] The early appearance of these receptors also coincided with increases in antral gastrin content and increased responses of acid secretion to exogenous pentagastrin.[73] These data suggest that maturation of the gastrin receptor in the stomach is dependent on the action of glucocorticoids just before and immediately after the onset of weaning.

IV. SECRETIN RECEPTOR

Secretin, the hormone which opened the field of endocrinology, has a wide spectrum of effects on the different organs of the gastrointestinal tract. It has been shown to stimulate pepsin output in the stomach *in vitro*[74] and to stimulate net fluid and electrolyte secretion from small intestinal mucosa.[75] In the pancreas the predominant response to secretin is a water and bicarbonate hypersecretion from centroacinar and duct cells[76] and enzyme secretion from dispersed acini.[77]

A. LOCALIZATION

Although secretin has never been localized at the ultrastructural level in different cell types by autoradiography, its subcellular distribution has been evaluated in the pancreas by specific binding sites studies.[78] Maximum specific binding sites were found in the plasma membrane fraction, and the distribution pattern of the secretin receptor follows that of 5′-nucleotidase, a plasma membrane marker.

B. BIOCHEMICAL CHARACTERIZATION

The first biochemical characterization of the secretin receptor was realized from a cat pancreatic homogenate preparation.[79] Later, it was studied in pancreatic plasma membranes of rat[78,80] and in pancreatic acini of guinea pig[81,82] and rat.[83] It has also been examined in guinea pig gastric chief cells,[74] purified plasma membrane from antral and fundic parts of the rat gastric mucosa,[84] and intestinal epithelial cells[85] (Table 3).

All these studies were performed with iodinated synthetic secretin or its Tyr^1-analog. The biochemical characterization of the secretin receptor encountered some difficulties because of the structural similarities existing between secretin and vasoactive intestinal peptide (VIP); indeed, these two agonists interact with the other receptor. As indicated in Table 3, rat pancreatic plasma membranes[78] and acini[83] possess two distinct classes of receptor which interact with secretin and VIP. The first class prefers VIP and thus has higher affinity for VIP and a lower affinity for secretin; the other class prefers secretin and has a high affinity for secretin and lower affinity for VIP. The receptor with high affinity for secretin shows a K_D of 1 n*M* and the low-affinity binding sites have K_Ds between 28 and 80 n*M*.[78,83] More recently, binding studies with iodinated secretin on guinea pig pancreatic acini demonstrated another class of high-affinity secretin-preferring receptors with a K_D of 6.8 n*M*[82] different from the low-affinity VIP receptor. High-affinity (K_D: 1 n*M*) and low-affinity (K_D: 1 μ*M*) secretin binding sites were also observed in gastric chief cells.[74] Specific secretin binding in antral membranes gave two sets of binding sites, one of high affinity (K_D: 0.24 n*M*) and low capacity (158 fmol/mg protein) and the other of low affinity (K_D: 7.9 n*M*) and high capacity (1380 fmol/mg protein). Although secretin receptors are present in fundic membranes, the high nonspecific binding of the hormone (68%) made it difficult to characterize the receptor in this part of the stomach.[84] With iodinated VIP as a marker, the presence of low-affinity secretin binding sites has also been identified from competition studies in guinea pig enterocytes;[85] to our knowledge, similar studies with enterocytes were not performed with secretin as the radioligand.

Using the cross-linking reagent dithiobis succinimidyl proprionate (DSP), the structure of the secretin receptor has very recently been presented.[84] Two radiolabeled peptides of 62 and 33 kDa (Table 3), connected by disulfide bonds, were identified. Furthermore, affinity labeling of the secretin binding proteins was abolished by secretin but not by VIP, illustrating the high specificity of the receptor for secretin.

C. ACTIVATION OF SECOND MESSENGERS

Over the past 10 years, it has been well established that secretin stimulated enzyme secretion and increases cellular cAMP in pancreas *in vitro*. In rat and guinea pig pancreatic acini,[81-83] occupation of secretin high-affinity receptor caused maximal cAMP generation with minimal[83] or no effect[82] on amylase release. Hootman et al.[86] have presented evidence that this high-affinity class of receptor may be involved in Na^+-K^+-ATPase activity regulation with a role in electrolyte secretion by the acinar cells.

TABLE 3
Biochemical Characteristics of the Secretin Receptor

Species	Ligand	Tissue	K_D (n*M*)	B_{max}	Mol wt (kDa)	Ref.
Cat	^{125}I-Secretin	Pancreas homogenate	4.0	1 pmol/mg prt	—	79
Rat	^{125}I-Secretin	Pancreatic acini	1.0 (K_H) 28.0 (K_L)	0.7 pmol/mg prt	—	83
Guinea pig	^{125}I-Secretin	Pancreatic acini	6.8	5,900 sites/cell	—	82
Rat	^{125}I-Tyr^1-Secretin	Pancreatic plasma membrane	1.0 (K_H) 80.0 (K_L)	0.67 pmol/mg prt 16.8 pmol/mg prt	—	78
Guinea pig	^{125}I-Secretin	Gastric chief cells	1.0 (K_H) 1000 (K_L)	6,000 sites/cell 600,000 sites/cell	—	74
Rat	^{125}I-Secretin	Gastric antral plasma membrane	0.24 (K_H) 7.9 (K_L)	158 fmol/mg prt 1,380 fmol/mg prt	62, 33	84

Note: (K_D) Dissociation constant; (K_H) K_D of the high-affinity site; (K_L) K_D of the low-affinity site; (B_{max}) receptor concentration; (mol wt) estimated molecular weight of the receptor.

Occupation of the lower-affinity class of receptor was related to maximal amylase release from the acinar cells by a non-cAMP-mediated mechanism.[83] Recent data presented by Trimble et al.[87] indicate that such a mechanism may involve phosphoinositides hydrolysis. Indeed, at concentrations of secretin above 10 n*M*, there was evidence of phosphatidylinositol bisphosphate hydrolysis concomitant with rapid increases in inositol trisphosphate (IP_3), cytosolic free calcium, and DG content of rat pancreatic acini. Increases in Ins(1,4,5)P_3 and Ins(1,3,4,5)P_4 contents were very rapid (5 s), whereas Ins(1,3,4)P_3, the inactive IP_3, increased more slowly.[88] This very rapid formation of the active IP_3 isomer is compatible with its second-messenger role in the initial elevation of intracellular calcium. Activation of this pathway by secretin may be related to its secretory role in enzyme release. By the use of different secretin analogs, Trimble et al.[89] were also able to dissociate the stimulatory effects of secretin on cAMP and inositol phosphate formation. With [Tyr^{10,13}] secretin, a reduction in IP_3 formation was observed without any effect on cAMP production. Furthermore, treatment with cholera toxin caused an increase in basal cAMP which was maximal after 30 min while stimulation of the formation of IP_3 was reduced only after a much longer exposure (150 to 240 min) to the toxin. These data suggest that there would be two different mechanisms involved in the activation of adenylate cyclase and of phospholipase C by secretin. Each of them may involve two different substrates for cholera toxin such as GTP binding proteins.

In gastric chief cells[74] occupation of the high-affinity secretin receptor also increased maximal cAMP accumulation, but unlike the pancreatic acinar cells, such a cyclic nucleotide production was correlated with pepsin secretion. It is not known yet if secretin can induce phosphoinositide hydrolysis in these chief cells or any other target cells in the gut.

Although the development of secretin has been studied in rat and guinea pig intestine,[90] the ontogeny of the secretin receptor in any organ of the digestive tract remains to be investigated. However, from a study documenting the developmental pattern of the receptor-cAMP systems sensitive to secretin, it was possible, although indirectly, to establish the presence and the development of the secretin receptor in the rat stomach. Already on day 20 of gestation, cAMP was produced in response to secretin. At birth a 50% reduction in cAMP accumulation in response to secretin was observed. Thereafter, there was no significant variation in the cAMP response to secretin between day 5 and 20. The final maturation of the secretin response occurred between days 20 to 27 and remained constant until adulthood at 65 days.[91] Such a developmental pattern of the cAMP response to secretin parallels that of the mucosal gastrin receptor and biological sensitivity to gastrin.[70]

The secretin receptor still remains the least understood receptor among those of the gastrointestinal hormones. Even though we had a first report on its structure recently, its association with G binding regulatory proteins, its modulation by secretin itself or other hormones, and finally its development in various gastrointestinal organs still remain to be investigated.

V. INSULIN RECEPTOR

Insulin is an anabolic hormone with a wide range of effects on metabolism *in vivo*, including stimulation of the synthesis and inhibition of the breakdown of glycogen, proteins, and lipids. In addition to its well known effects on metabolism, insulin has long been known to stimulate the growth and proliferation of cells *in vitro*.[92]

Among specific actions of insulin on gut organs, the hormone stimulates the brush border enzymes activities such as lactase and sucrase, intervenes in the ontogenesis of intestinal sucrase, and seems to play a role in intestinal maturation[93,94] and organ growth.[95]

The interaction of insulin with the exocrine pancreas has been known for many years already. Indeed, insulin has been associated with the regulation of amylase[33] and protein synthesis,[96] growth of acinar[97] and islet[98] cells in culture, as well as being a potentiating agent of cholecystokinin[99] and acetylcholine[100] induced enzyme secretion. Such actions may be regulated through the activity of the insulo-acinar axis[101] and specific contacts between endocrine β cells and acinar cells.[102] These two modes of access of insulin to the exocrine pancreatic cells may explain the morphological and biochemical partition of the exocrine pancreatic tissue into peri- and tele-insular regions.[102]

A. LOCALIZATION

In a tissue-screening survey of specific insulin receptor location, Bergeron et al.[103] have demonstrated substantial binding of insulin to hepatocytes of liver, acinar cells of the pancreas, and epithelial cells of the stomach, duodenum, jejunum, ileum, and colon. In the pancreas, the exocrine cell plasmalemma[104,105] seems to be the richest source of insulin receptors with sites also observed on duct cells.[106] The presence of specific binding sites for insulin on nuclei and nuclear envelopes has also been documented and characterized in liver[107] and pancreatic acinar cells.[108]

B. BIOCHEMICAL CHARACTERIZATION

In the gut the insulin receptor has mostly been characterized in the mouse,[108] rat,[109-111] guinea pig,[112] dog,[113] and rabbit.[114]. As indicated in Table 4, these studies were performed either on pancreatic acini,[108,109,112] enterocytes,[110,114] baso-lateral membranes from intestinal mucosa,[113] or liver plasma and nuclear membranes.[111]

Similar insulin binding sites have been described in dispersed pancreatic acini isolated from mouse,[108] rat,[109] or guinea pig.[112] In all three species, the data are consistent with the presence of two classes of binding sites with relative affinities of 1 n*M* and 80 to 998 n*M* with corresponding capacities of 53 and 1735 fmol/mg protein in the rat[109] or 9300 and 760,000 sites per cell in the guinea pig[112] (Table 4). However, Williams and Goldfine[106] reported that when degradation of ^{125}I-insulin is prevented, the apparent low-affinity insulin binding disappears, suggesting that these sites are artifacts. The reality of these artifacts remains to be established in other tissues.

The insulin receptors present on rabbit intestinal mucosa[114] or dog baso-lateral membranes[113] seem to have higher affinities for both of its classes than those on pancreatic acini. Indeed K_Ds of 0.08 to 0.4 n*M* for the high-affinity class and 2.7 to 36 n*M* for the low-affinity class have been reported. Their capacities however were lower with 1600 and 2900 sites per cell.[114] The rat enterocytes, however, possess insulin receptors comparable to those characterized in the pancreas.[110] The highest concentration of receptor has been found in rat liver plasma and nuclear membranes (2 nmol/mg protein) with the presence of two classes of binding sites.[111] The relative

TABLE 4
Biochemical Characteristics of the Insulin Receptor

Species	Ligand	Tissue	K_D (n*M*)	B_{max}	Mol wt (kDa)	Ref.
Mouse	^{125}I-Insulin	Pancreatic acini	1.6 (K_H) 83.0 (K_L)	—	—	108
Rat	^{125}I-Insulin	Pancreatic acini	1.6 (K_H) 998.0 (K_L)	53 fmol/mg prt 1735 fmol/mg prt	—	109
Guinea pig	^{125}I-Insulin	Pancreatic acinar cells	1.3 (K_H) 790.0 (K_L)	93000 sites/cell 760,000 sites /cell	—	112
Mouse	^{125}I-Insulin	Pancreatic acini	1.0		130 90	116
Rabbit	^{125}I-Insulin	Ileal enterocytes	0.4 (K_H) 2.7 (K_L)	1600 sites/cell 2900 sites/cell	—	114
Rat	^{125}I-Insulin	Jejuno-ileal enterocytes	2.2 (K_H) 560.0 (K_L)	2000 sites/cell	—	110
Dog	^{125}I-Insulin	Enterocyte baso-lateral membranes	0.09 (K_H) 36.0 (K_L)	0.4 pmol/mg prt 4.7 pmol/mg prt	130	113
Rat	^{125}I-Insulin	Liver plasma membranes	0.5 (K_H) 10.0 (K_H)	2 nmol/mg prt	—	111
	^{125}I-Insulin	Liver nuclear membranes	5.6 (K_H) 65.0 (K_L)	2 nmol/mg prt	—	111

Note: (K_D) Dissociation constant; (K_H) K_D of the high-affinity site; (K_L) K_D of the low-affinity site; (B_{max}) receptor concentration; (mol wt) estimated molecular weight of the receptor.

affinities of the receptor plasma membrane, 0.5 and 10 n*M*, are higher than that of the nuclear membrane established at 5.6 and 65 n*M* (Table 4).

In liver and other tissues the insulin receptor is presented as an oligomeric protein with an M_r of 350,000 which upon reduction is separated into two pairs of identical subunits of 130,000 and 90,000 kDa.[115] In purified pancreatic membranes these same two subunits were also identified using the cross-linker dissuccinimidyl suberate under reducing conditions.[116] The 130,000 kDa α-subunit was also reported in dog baso-lateral intestinal mucosa.[113]

C. ACTIVATION OF SECOND MESSENGERS

In intact cells, insulin is known to modulate a long list of metabolic functions, but the biochemical pathways linking insulin receptor binding to its regulatory effects on cellular metabolism are unclear. Since the insulin receptor possesses tyrosine kinase activity, it is possible that activation of this receptor enzyme may control the transmembrane signalling system of insulin. Insulin action does not involve adenylate cyclase activation since the hormone has been reported as an inhibitor of this enzyme.[117] The recent observation that insulin does not affect phosphoinositide metabolism in rat liver plasma membrane suggests that it does not operate through the activation of phospholipase C either, at least in this tissue.[118] Recently, however, it has been reported that insulin interaction with purified plasma membranes could lead to the generation of a chemical mediator.[119] This substance was tentatively identified[120] as a 1000 to 1500 peptide-like molecule existing as a portion of a larger molecular weight membrane protein released upon activation of a protease which cleaves off this mediator following insulin-receptor interaction.[121] There is no evidence indicating that insulin operates on gut organs through the formation of such a mediator.

D. RECEPTOR OCCUPANCY AND BIOLOGICAL FUNCTIONS

Studies to determine if insulin exerts its biological effect directly on its target organs have been hindered by the lack of *in vitro* cell preparations sensitive to the hormone; therefore, establishing correlations between insulin receptor occupancy and the biological effects of the hormones has been difficult.

The exocrine pancreas is an excellent model used to study the biological effects of insulin due to the presence of a portal system that carries blood from the islets of Langerhans directly to the acinar cells.[106] Furthermore, insulin has been shown to regulate glucose transport,[122] protein synthesis,[123] amylase messenger RNA levels[34] in pancreatic acini, and to maintain their differentiated functions in a short-term culture system.[97]

Using this dispersed pancreatic acini model it has been demonstrated that greater than 90% of the high-affinity insulin receptor had to be occupied to elicit maximum increase in protein synthesis in diabetic rats.[109,123] Similarly, 50% of 2-deoxyglucose uptake by pancreatic acini from diabetic mice required full occupancy of the high-affinity insulin binding sites, the other half needing saturation of low-affinity receptors.[122] In this system the insulin action was not immediate, but required a preincubation of 90 min. Such a delay suggests that insulin could regulate glucose transport in pancreatic acini by translocating glucose transport carriers from an intracellular site to the plasma membrane. A similar transport system, dependent on insulin, may also exist for amino acid uptake because increases in protein synthesis in response to insulin also needed a lag period.[123]

Considerable evidence has accumulated over the last few years indicating that insulin may exert many of its effects by regulating specific protein phosphorylation and dephosphorylation reactions. According to recent studies, it seems now evident that insulin exerts its physiological effect not only at the plasma membrane level, but also on the nuclear envelope. Indeed, high-affinity insulin receptors have been characterized on the liver nuclear envelope of diabetic rats and their occupation by insulin leads to increased messenger RNA (mRNA) efflux and nucleotide triphosphatase (NTPase) activity.[107] Furthermore, the direct addition of insulin to nuclear envelope produced a decrease in the incorporation of ^{32}P into the protein of the nuclear envelope through the activation of a protein phosphatase.[124] In that study, mRNA efflux, NTPase activity, and protein dephosphorylation were all insulin-dose dependent with a maximal effect at 10 p*M* insulin. Because the insulin receptor on the nuclear envelope has a K_D of 5 n*M* for its high-affinity site,[107] whereas a maximal effect of insulin on nuclear functions occurs at 10 p*M*,[124] it seems that only a very small fraction of these receptors has to be occupied to produce these reactions. These studies give us new insights concerning the action of insulin on the cell nucleus, but the major question still remains on how insulin gets to the nucleus.

E. MODULATION

The number of insulin receptors on cells can be regulated by the concentration of insulin in their environment. When endogenous insulin is reduced in diabetic rats, the number of insulin receptors on isolated pancreatic acini increased; in these same acini from diabetic animals, their culture *in vitro* in the presence of exogenous insulin resulted in a decreased number of receptors. In both conditions of up and down regulation, receptor affinity was not modified.[125] In AR42J cultured rat pancreatic acinar cells, it was recently demonstrated that dexamethasone, a glucocorticoid analog, increased insulin receptor mRNA levels. Furthermore, such treatment led to an increase in the biosynthesis of insulin-receptor precursor in another cell line, the IM-9 cells.[126] It is therefore possible that glucocorticoid may counteract the down-regulating effect of insulin on its own receptor.

Cellular resistance to the actions of physiologic concentrations of insulin may reflect either decreased sensitivity or decreased responsiveness to insulin, or both. Decreased sensitivity is associated with a decrease in the number or affinity of receptors and the dose-response curve of the biological effect of the hormone is rightward shifted without a change in the maximal effect. On the other hand, decreased responsiveness results in a decrease in the maximal response to a hormone and is believed to involve post-receptor defect. Such a distinction in insulin effect has been investigated in HTC hepatoma cells in order to determine whether insulin has a preferential effect on its receptor or at postreceptor loci.[127]

Indeed, pre-exposure of hepatoma cells to insulin caused a temperature, time, and concentration-dependent decrease of 50 to 60% in insulin binding without apparent change in receptor affinity. Concomitantly, insulin pre-exposure resulted in a total loss of amino acid transport and induction of tyrosine aminotransferase activity in response to a subsequent stimulation by insulin.[127] So a discrepancy exists between partial loss of binding sites and total loss of biological responsiveness. Therefore, down-regulation of the insulin receptor is not sufficient to account for the complete loss of cellular responsiveness to the hormone. Furthermore, the time course of recovery of hormonal responsiveness and binding are dissociable. Indeed, while full responsiveness to insulin has been restored within 2 h after its removal from the culture medium, recovery of insulin binding had to wait 24 h. Such a study indicates that insulin receptor down-regulation is not the sole mechanism for regulation of insulin responsiveness and that the phenomenon of insulin desensitization may involve post-receptor mechanisms that still remain to be identified.

F. ONTOGENY

The ontogeny of the insulin receptor has not been extensively studied over the years and the information available comes from studies performed in liver. Indeed, in the rabbit, hepatic insulin binding increased rapidly during the last 3 d before birth, reaching a peak at birth. This burst in total insulin receptor at birth was followed by a significant fall within 2 h after birth; thereafter, the receptor concentration seems to have stabilized until adulthood.[128] A similar pattern of development was also observed in human and rat fetal liver.[129] Unfortunately, to our knowledge no data are available on the development of the insulin receptor in any other organ of the digestive tract.

VI. EGF RECEPTOR

Epidermal growth factor (EGF) is a 53 amino acid polypeptide, M_r 6045, first characterized for its ability to stimulate mitogenesis in epidermal and epithelial cells.[130] Such effects have also been observed on organs of the gastrointestinal tract. In the stomach, EGF causes significant increases in oxyntic gland mucosal DNA in both adult[131] and suckling rats.[132] EGF also accelerates the maturation of fetal mouse intestinal mucosa *in utero*[133] and influences its development in the suckling mouse.[134] In the pancreas, *in vivo* treatment with EGF has been reported to increase DNA synthesis and total DNA content.[135] *In vitro*, EGF potentiates the action of cholecystokinin-octapeptide (CCK_8) on protein synthesis in mouse pancreatic acini maintained in tissue culture,[136] increases cellular viability of rat pancreatic acinar cells when cultured in serum-free medium,[137] and enhances protein synthesis in these cells at a concentration of 42 p*M*.[138] These observations clearly indicate that this growth factor is actively implicated in maintaining growth and metabolic functions in many gut organs.

A. LOCALIZATION

The distribution of EGF receptors was studied in the intestine[139] and the pancreas[140] using light microscopic autoradiography performed at different times after an intravenous injection of ^{125}I-EGF into the adult rat. In the small intestine, labeling was mostly observed in the columnar epithelial cells of villi and crypts and was also detected in goblet cells. In the exocrine pancreas, labeling occurred over the acini and cells lining the intercalated ducts. Substantial binding was also observed to cells of the islets of Langerhans. Internalization of the growth factor with or without its receptor is suggested from the localization of silver grains within the cytoplasm between 7 and 60 min after the injection. Whether or not this process is related to the delayed biological effects of EGF in animals and in cell culture remains to be seen.

B. BIOCHEMICAL CHARACTERIZATION

As indicated in Table 5, the EGF receptor has been biochemically characterized in rat,[141]

TABLE 5
Biochemical Characterization of the EGF Receptor

Species	Ligand	Tissue	K_D (n*M*)	B_{max}	Ref.
Rat	^{125}I-EGF	Pancreatic acini	4.9	—	141
Mouse	^{125}I-EGF	Pancreatic acini	0.4	40,000 sites/cell	136
Human	^{125}I-EGF	Panc-1 pancreatic carcinoma cells	—	—	147
Guinea pig	^{125}I-EGF	Gastric antrum	0.42 (K_H)	68 fmol/mg prt	
			2.2 (K_L)	182 fmol/mg prt	142
		Gastric fundus	0.47 (K_H)	61 fmol/mg prt	
			3.4 (K_L)	234 fmol/mg prt	
Rat	^{125}I-EGF	Isolated intestinal epithelial cells	1.0	3000 sites/cell	143
Mouse	^{125}I-EGF	Isolated intestinal epithelial cells	0.19 (K_H)	21,605 site/cell	144
			1.9 (K_L)	42,485 sites/cell	
Rabbit	^{125}I-EGF	Isolated ileal epithelial cells	0.8 (K_H)	1000 sites/cell	114
			4.1 (K_L)	1180 sites/cell	
Mouse	^{125}I-EGF	Isolated colonic epithelial cells	KA 0.38×10^9 M^{-1}	323 pmol/mg prt	146
			KA 0.05×10^9 M^{-1}	1570 pmol/mg prt	
Human	^{125}I-EGF	Isolated fetal colonic epithelial cells	KA 1.79×10^8 M^{-1}	126 pmol/mg prt	
			KA 0.014×10^9 M^{-1}	1394 pmol/mg prt	160
		Isolated fetal intestinal epithelial cells	KA 1.90×10^9 M^{-1}	61 pmol/mg prt	
			KA 0.03×10^9 M^{-1}	487 pmol/mg prt	
Rat	^{125}I-EGF	Liver plasma membrane	0.4	1540 fmol/mg prt	161

Note. (K_D) Dissociation constant, (K_H) K_D of the high-affinity site, (K_L) K_D of the low-affinity site, (B_{max}) receptor concentration.

mouse,[136] guinea pig,[142] and rabbit[114] using ^{125}I-EGF purified from mouse submaxillary gland. These studies were performed either in dispersed acini,[136,141] isolated gastric antrum and fundus glands,[142] intestinal cells,[114,143,145] or colon epithelial cells.[146]

Studies performed on intact pancreatic acinar cells (Table 5) indicate the presence of one binding site on the EGF receptor with affinities of 4.9 n*M* in the rat[141] and 0.4 n*M* in the mouse.[136] A concentration of 40,000 sites per acinar cell has been reported in the mouse.[136] The presence of EGF receptors has also been demonstrated, but not characterized in Panc-1 human pancreatic carcinoma cells.[147] Kinetic properties of the EGF receptor were similar in guinea pig gastric antrum and fundus[142] and indicate two orders of binding sites (K_H 0.42 to 0.47 n*M*; K_L 2.2 to 3.4 n*M*). In younger animals, the binding capacity was lower in the fundus (B_{max} 32 and 129 fmol/mg protein) than in the antrum (59 and 169 fmol/mg protein), but no difference was observed in mature animals (Table 5). In enterocytes, characterization of the EGF receptor presents either one[143] or two classes of binding sites in the rat[145] and two classes in the mouse[144] and rabbit.[114] The single class receptor presents an affinity of 1 n*M* with a capacity of 3000 sites per cell.[143] Affinities of 0.19 and 1.9 n*M* with capacities of 21 and 42,000 sites per cell have been reported in the mouse intestinal cells.[144] In the rabbit ileal enterocytes, affinities of 0.8 and 4.1 n*M* have been reported with capacities of 1000 and 1180 sites per cell for both populations.[114] Further studies are obviously needed to clarify these discrepancies.

Although the EGF receptor has not yet been purified from any gut organs, the receptor-kinase was isolated as a 170 kDa protein from A-431 membrane vesicles;[148] a 150 kDa receptor was also observed and seems to be the product of a Ca^{2+}-dependent neutral protease. The major functional difference between these two forms appears to be the ability of the 170 kDa molecule to autophosphorylate at a much greater rate than the smaller unit.[149] The schematic representation of the receptor structure indicates that the receptor has one hydrophobic domain constituting a typical transmembrane sequence. The external domain is rich in carbohydrate residues, while the internal domain contains the tyrosine kinase activity. The primary sites of autophosphorylation occur on three tyrosyl residues while phosphorylation by protein kinase C activation happens on a threonine residue.[150]

C. INTERACTION WITH GUANINE NUCLEOTIDE BINDING PROTEINS

The recent demonstration that EGF stimulates the rapid accumulation of inositol trisphophate and a rise in cytosolic calcium in A431 cells[150,151] suggests the activation of phospholipase C. In several instances a guanine nucleotide regulatory protein has been shown to be involved in the transduction of the hormonal signal to the phospholipase C. However, in the above instance, if a G protein is involved in the transduction of the EGF signal it does not serve as a substrate for pertussis toxin since the toxin did not interfere with the production of inositol phosphates.[151] So far, no G protein has been directly implicated in the mechanism of action of EGF on its target cells.

D. ACTIVATION OF SECOND MESSENGERS

From data presented in the literature, it is not evident that the EGF receptor interacts with intrinsic membrane proteins (G protein) to activate second messenger systems. Recent information, however, indicates that the tyrosine kinase activity of the receptor is essential for the diverse biochemical effects of EGF such as rapid changes in intracellular calcium, intracellular pH, activation of gene transcription, and stimulatory effects on cell proliferation.[152]

E. RECEPTOR OCCUPANCY AND BIOLOGICAL FUNCTIONS

Any relationship of EGF receptor occupancy to regulation of cellular functions has been established in few cell types. Occupation of the high-affinity sites of the EGF receptor in human fibroblasts led to a three-fold increase in cell number after 5 d of culture, while higher concentrations of EGF were associated with inhibition of cell proliferation.[153] Similarly, saturation of the EGF high-affinity sites more than doubled the DNA content per culture plate of Panc-1 cells after 6 d of culture.[147] In pancreatic acini maintained in culture for 24 h, the potency of EGF on leucine incorporation into proteins is somewhat less than that for amylase secretion, but in both cases, the EGF concentration needed for half maximal response is slightly less than that required for half maximal inhibition of EGF binding to its receptor.[136] All these data, therefore, suggest that EGF may have a physiological role in regulating cellular functions.

F. MODULATION

As discussed earlier in this section, one of the characteristics of the EGF receptor is that it exhibits EGF-stimulable protein tyrosine kinase activity, which upon EGF stimulation causes the receptor to undergo self-phosphorylation as part of its activation. In addition to self-phosphorylation, this receptor can also be phosphorylated on threonine residues by the calcium- and phospholipid-dependent protein kinase C. This latter phosphorylation results in a suppression of its intrinsic tyrosine kinase and EGF binding activities.[154]

The binding properties of EGF to its receptor can also be affected and modified by hormones which do not interact with the EGF receptor. Evidence indicates that such a modulation may involve calcium mobilization and activation of protein kinase C by these hormones. In the pancreas, Korc et al.[155] were the first to demonstrate that physiological concentrations of CCK-8 inhibited EGF binding to its receptor, probably via calcium mobilization by reducing either the clustering or internalization of the EGF-receptor complex. The role of calcium was later confirmed with the demonstration that other pancreatic secretagogues or ionophore operating on the acinar cells through Ca^{2+} (bombesin, carbachol, gastrin, and the ionophore A23187) also inhibited EGF binding by reducing EGF internalization.[156] The implication of protein kinase C activation in this process was demonstrated when inhibition of EGF internalization occurred in the presence of phorbol esters, an effect potentiated by the ionophore A23187.[157]

In human diploid foreskin cells in culture, dexamethasone enhanced their mitogenic response to EGF. Paralleling this increased responsiveness to EGF growth stimulation, dexamethasone treated cells also exhibited an increased ability to bind EGF. Moreover, the increase in EGF binding exhibited a dexamethasone dose dependence similar to that for the enhancement of EGF mitogenesis, suggesting a relationship between the steroid effect on binding and growth.[158] In liver plasma membranes from hypothyroid rats, EGF binding was reduced by 60 to 70%. This reduction was due to a decreased number of receptors and specifically the 170,000 receptor molecule as evaluated by a cross-linking study and receptor kinase activity. These changes were restored by a single treatment with T_3. These data support a role for thyroid hormones in regulating the EGF receptor.[159]

G. ONTOGENY

The implication of epidermal growth factor in fetal and neonatal development of the gastrointestinal tract has led to some studies on the presence of specific EGF receptors in digestive organs during fetal development and different periods after birth.

Under identical optimal binding conditions, the binding capacity of mouse colonic cells remained relatively stable between 2 and 22 d of age, while that of isolated intestinal cells was much lower than that observed for colonic cells. Even though the concentration of intestinal EGF receptor increased from day 2 to 22, specific binding remained constantly below values obtained for the colonic cells at all ages studied. Furthermore, during the whole suckling period, EGF binding in the distal colon revealed a constant twofold higher concentration than in the proximal part; these regional differences disappeared in the adult.[146] Most recently, Pothier and Ménard[160] demonstrated for the first time an important concentration of EGF binding sites in isolated epithelial cells of human fetal small intestine and colon. In both organs, the pattern of EGF binding shows a higher concentration of receptors in younger fetuses (12 weeks) than in older (17 weeks). As previously observed in the mouse after birth,[146] fetal colonic cells also exhibit a constant much higher binding capacity (1.5 to 2.5 times) than the intestinal cells (Table 5). In another study,[145] EGF binding capacity was found higher in rat enterocytes from small intestine of term fetuses and declined rapidly after birth; thereafter, the EGF receptor concentration remained fairly stable until adulthood. This decline in receptor concentration immediately after birth may result from the presence of EGF present in the milk causing internalization of the receptor. Such a possibility was raised following the observation that pre-exposure of fetal enterocytes with rat milk led to a dose-dependent decrease in EGF binding.

In guinea pig gastric gland,[142] the EGF binding capacity of young animals (6 to 8 weeks) was about 50% lower in the fundus than in the antrum, whereas the kinetic properties of the receptor were similar. In the adult, the receptor concentrations were similar in both parts of the stomach although higher than in younger animals especially in the fundus. In the rat liver,[161] basolateral plasma membranes prepared from 21-d fetuses exhibit EGF binding capacity comparable to that found in membranes prepared from adult animals. However, the EGF receptor from fetuses shows a decreased affinity of binding with a higher K_D, 0.68 n*M* as compared to 0.39 n*M* in the adult. During the suckling period, age 7 and 14 d, a threefold decrease in the EGF binding capacity was observed when compared to fetal and adult age. The affinity of the receptor, however, was comparable to that found in the adult. The reduced EGF binding capacity observed in the suckling rats is consistent with the hypothesis of milk-derived EGF causing either down-regulation of receptors or masking of tracer binding; further support also comes from the observation that exogenous EGF given parenterally decreased markedly the receptor concentration of suckling rat basolateral membranes,[161] as did pre-exposure of fetal enterocytes to mature milk.[145]

In all these experiments performed on different organs, the presence of EGF receptors has been established as well as their biochemical characterization. However, the relationship between EGF binding and physiological functions especially related to growth remains to be elucidated.

VII. CONCLUSION

This review summarizes our present knowledge on the localization, biochemical and structural characterization, modulation, and ontogeny of hormones and growth factor receptors that we believed to be involved in mediating the trophic or adaptive responses of their specific target organs. However, since stimulated growth is usually a delayed phenomenon when compared to immediate secretion or contraction in response to a specific stimulus, we still have to understand how receptors initiate trophic responses.

In the near future, we shall be able to establish the intracellular signals involved in the growth process; are they the same messengers that activate the immediate hormonal responses such as cAMP, cGMP, calcium, inositol trisphosphate, and diacylglycerol? Furthermore, we will have to understand the receptor dynamics during a trophic response, especially the internalization phenomenon. Indeed, with the discovery of hormone receptors on the nuclear envelope and specific protein kinase C bound to intracellular organelles other than the plasma membrane, there exists a new field of research which needs to be explored. These are some of the important problems that we will have to investigate in order to understand how hormones induce and sustain organ growth.

REFERENCES

1. **Goldfine, I. D. and Williams, J. A.,** Receptors for insulin and CCK in the acinar pancreas: relationship to hormone action, *Int. Rev. Cytol.*, 85, 1, 1983.
2. **Williams, J. A., Sankaran, H., Roach, E., and Goldfine, I. D.,** Quantitative electron microscope autoradiographs of ^{125}I-cholecytokinin in pancreatic acini, *Am. J. Physiol.*, 243, G291, 1982.
3. **Rosenzweig, S. A., Miller, L. J., and Jamieson, J. D.,** Identification and localization of cholecystokinin-binding sites on rat pancreatic plasma membranes and acinar cells: a biochemical and autoradiography study, *J. Cell Biol.*, 96, 1288, 1983.
4. **Christophe, J., De Neef, P., Deschodt-Lanckman, M., and Robberecht, P.,** The interaction of caerulein with rat pancreas. II. Specific binding of ^{3}H caerulein on dispersed acinar cells, *Eur. J. Biochem.*, 70, 31, 1978.
5. **Sankaran, H., Goldfine, I. D., Deveney, C. W., Wong, K. Y., and Williams, J. A.,** Binding of cholecystokinin to high affinity receptors on isolated rat pancreatic acini, *J. Biol. Chem.*, 255, 1849, 1980.
6. **Szecowka, J., Goldfine, I. D., and Williams, J. A.,** Solubilization and characterization of CCK receptors from mouse pancreas, *Regul. Peptides*, 10, 71, 1985.
7. **Fourmy, D., Zahidi, A., Fabre, R., Guidet, M., Pradayrol, L., and Ribet, A.,** Receptors for cholecystokinin and gastrin peptides display specific binding properties and are structurally different in guinea pig and dog pancreas, *Eur. J. Biochem.*, 165, 683, 1987.
8. **Steigerwalt, R. W., Goldfine, I. D., and Williams, J. A.,** Characterization of cholecystokinin receptors on bovine gallbladder membranes, *Am. J. Physiol.*, 247, G709, 1984.
9. **Miller, L. J.,** Characterization of cholecystokinin receptors on human gastric smooth muscle tumors, *Am. J. Physiol.*, 247, G402, 1984.
10. **Innis, R. B. and Snyder, S. H.,** Distinct cholecystokinin receptors in brain and pancreas, *Proc. Natl. Acad. Sci. U. S. A.*, 77, 6917, 1980.
11. **Praissman, M., Walden, M. E., and Pellecchia, C.,** Identification and characterization of a specific receptor for cholecystokinin on isolated fundic glands from guinea pig gastric mucosa using a biologically active ^{125}I-CCK-8 probe, *J. Receptor Res.*, 3, 647, 1983.
12. **Chang, R. S. L., Lotti, V. J., and Chen, T. B.,** Characterization of [^{3}H] (±) L-364,718 binding to solubilized cholecystokinin (CCK) receptors of rat pancreas, *Biochem. Pharmacol.*, 36, 1709, 1987.
13. **Steigerwalt, R. W. and Williams, J. A.,** Characterization of cholecystokinin receptors on rat pancreatic membranes, *Endocrinology*, 109, 1746, 1981.
14. **Collins, S. M., Abdelmoumene, S., Jensen, R. T., and Gardner, J. D.,** Cholecystokinin-induced persistent stimulation of enzyme secretion from pancreatic acini, *Am. J. Physiol.*, 240, G459, 1981.
15. **Sankaran, H., Goldfine, I. D., Bailey, A., Licko, V., and Williams, J. A.,** Relationship of cholecystokinin receptor binding to regulation of biological functions in pancreatic acini, *Am. J. Physiol.*, 242, G250, 1982.

16. **Sakamoto, C., Goldfine, I. D., and Williams, J. A.,** Pancreatic CCK receptors: characterization of covalently labeled subunits, *Biochem. Biophys. Res. Commun.*, 118, 623, 1984.
17. **Sakamoto, C., Williams, J. A., Wong, K. Y., and Goldfine, I. D.,** The CCK receptor on pancreatic plasma membranes: binding characteristics and covalent cross-linking, *FEBS Lett.*, 151, 63, 1983.
18. **Sakamoto, C., Goldfine, I. D., and Williams, J. A.,** Characterization of cholecystokinin receptor subunits on pancreatic plasma membranes, *J. Biol. Chem.*, 258, 12707, 1983.
19. **Lambert, M., Svoboda, M., Furnelle, J., and Charistophe, J.,** Solubilization from rat pancreatic plasma membranes of a cholecystokinin (CCK) agonist-receptor complex interacting with guanine nucleotide regulatory proteins coexisting in the same macromolecular system, *Eur. J. Biochem.*, 147, 611, 1985.
20. **Lambert, M., Deschodt-Lanckman, M., Furnelle, J., and Christophe, J.,** Similar characteristics of guanine nucleotide regulatory sites involved in adenylate cyclase activation, specific GTPase activity, and cholecystokinin binding in rat pancreatic plasma membranes, *J. Cyclic Nucleotide Res.*, 7, 385, 1981.
21. **Merritt, J. E., Taylor, C. W., Rubin, R. P., and Putney, J. W.,** Evidence suggesting that a novel guanine nucleotide regulatory protein couples receptors to phospholipase C in exocrine pancreas, *Biochem. J.*, 236, 337, 1986.
22. **Williams, J. A. and McChesney, D. J.,** Cholecystokinin induces the interaction of its receptor with a guanine nucleotide binding protein, *Regul. Peptides*, 18, 109, 1987.
23. **Kempen, H. J. M., DePont, J. J. H. H. M., and Bonting, S. L.,** Pancreas adenylate cyclase. V. Its presence in isolated rat pancreatic acinar cells, *Biochim. Biophys. Acta*, 496, 521, 1977.
24. **Gardner, J. D., Sutliff, V. E., Walker, M. D., and Jensen, R. T.,** Effects of inhibitors of cyclic nucleotide phosphodiesterase on actions of cholecystokinin, bombesin, and carbachol on pancreatic acini, *Am. J. Physiol.*, 245, G676, 1983.
25. **Renckens, B. A. M., Van Emst-DeVries, S. E., DePont, J. J. H. H. M., and Bonting, S. L.,** Rat pancreas adenylate cyclase. VII. Effect of extracellular calcium on pancreozymin-induced cyclic AMP formation, *Biochim. Biophys. Acta*, 630, 511, 1980.
26. **Korc, M., Williams, J. A., and Goldfine, I. D.,** Stimulation of the glucose transport system in isolated mouse pancreatic acini by cholecystokinin and analogs, *J. Biol. Chem.*, 254, 7624, 1979.
27. **Jensen, R. T., Lenip, G. F., and Gardner, J. D.,** Interaction of COOH-terminal fragments of cholecystokinin with receptors on dispersed acini from guinea pig pancreas, *J. Biol. Chem.*, 257, 5554, 1982.
28. **Lin, C. W., Bianchi, B. R., Grant, D., Miller, T., Danaher, E. A., Tufano, M. D., Kopecka, H., and Nadzan, A. M.,** Cholecystokinin receptors: relationships among phosphoinositide breakdown, amylase release and receptor affinity in pancreas, *J. Pharmacol. Exp. Ther.*, 236, 729, 1986.
29. **Chang, R. S. L., Lotti, V. J., and Chen, T. B.,** Cholecystokinin receptor mediated hydrolysis of inositol phospholipids in guinea pig gastric glands, *Life Sci.*, 36, 965, 1985.
30. **Noguchi, M., Adachi, H., Sato, S., Honda, T., Ohnishi, S., Aoki, E., and Torizuka, K.,** Cholecystokinin-stimulated pepsinogen secretion and cholecystokinin receptors on gastric chief cells in guinea pigs, *Endocrinol. Jpn.*, 34, 727, 1987.
31. **Otsuki, M., Okabayashi, Y., Ohki, A., Hootman, S. R., Baba, S., and Williams, J. A.,** Amylase secretion by isolated pancreatic acini after acute cholecystokinin treatment *in vivo*, *Am. J. Physiol.*, 246, G419, 1984.
32. **Otsuki, M. and Williams, J. A.,** Amylase secretion in isolated pancreatic acini after chronic treatment with cholecystokinin, *Am. J. Physiol.*, 244, G683, 1983.
33. **Soling, H. D. and Unger, K. O.,** The role of insulin in the regulation of α-amylase synthesis in the rat pancreas, *Eur. J. Clin. Invest.*, 2, 199, 1972.
34. **Korc, M., Owerbach, D., Quinto, C., and Rutter, W. J.,** Pancreatic islet-acinar cell interaction: amylase messenger RNA levels are determined by insulin, *Science*, 213, 351, 1981.
35. **Otsuki, M. and Williams, J. A.,** Effect of diabetes mellitus on the regulation of enzyme secretion by isolated rat pancreatic acini, *J. Clin. Invest.*, 70, 148, 1982.
36. **Otsuki, M. and Williams, J. A.,** Direct modulation of pancreatic CCK receptors and enzyme secretion by insulin in isolated pancreatic acini from diabetic rats, *Diabetes*, 32, 241, 1983.
37. **Morisset, J. and Jolicoeur, L.,** Effect of hydrocortisone on pancreatic growth in rats, *Am. J. Physiol.*, 239, G95, 1980.
38. **Morisset, J., Jolicoeur, L., Genik, P., and Lord, A.,** Interaction of hydrocortisone and caerulein on pancreatic size and composition in rat, *Am. J. Physiol.*, 241, G637, 1981.
39. **Leung, Y. K., Jirapinyo, P., Lebenthal, E., and Lee, P. C.,** Effect of hydrocortisone on the maturation of cholecystokinin (CCK) binding and CCK stimulated amylase release in pancreatic acini of neonatal rats, *Pancreas*, 2, 73, 1987.
40. **Logsdon, C. D.,** Glucocorticoids increase cholecystokinin receptors and amylase secretion in pancreatic acinar AR42J cells, *J. Biol. Chem.*, 261, 2096, 1986.
41. **Doyle, C. M. and Jamieson, J. D.,** Development of secretagogue response in rat pancreatic acinar cells, *Dev. Biol.*, 65, 11, 1978.

42. **Larose, L. and Morisset, J.,** Acinar cell responsiveness to urecholine in the rat pancreas during fetal and early postnatal growth, *Gastroenterology*, 73, 530, 1977.
43. **Werlin, S. L. and Grand, R. J.,** Development of secretory mechanism in rat pancreas, *Am. J. Physiol.*, 236, E446, 1979.
44. **Werlin, S. L., Colton, D. G., Harb, J., Reynolds, E., Hoffman, R. G., and Williams, J. A.,** Ontogeny of secretory function and cholecystokinin binding capacity in immature rat pancreas, *Life Sci.*, 40, 2237, 1987.
45. **Chang, A. and Jamieson, J. D.,** Stimulus-secretion coupling in the developing exocrine pancreas: secretory responsiveness to cholecystokinin, *J. Cell Biol.*, 103, 2353, 1986.
46. **Leung, Y. K., Lee, P. C., and Lebenthal, E.,** Maturation of cholecystokinin receptors in pancreatic acini of rats, *Am. J. Physiol.*, 250, G594, 1986.
47. **Robinson, P. H., Moran, T. H., Goldrich, M., and McHugh, P. R.,** Development of cholecystokinin binding sites in rat upper gastrointestinal tract, *Am. J. Physiol.*, 252, G529, 1987.
48. **Mu, F. T., Baldwin, G., Weinstock, J., Stockman, D., and Toh, B. H.,** Monoclonal antibody to the gastrin receptor on parietal cells recognizes a 78-kDa protein, *Proc. Natl. Acad. Sci. U. S. A.*, 84, 2698, 1987.
49. **Baur, S. and Bacon, V. D.,** A specific gastrin receptor on plasma membranes of antral smooth muscle, *Biochem. Biophys. Res. Commun.*, 73, 928, 1976.
50. **Lewin, M., Soumarmon, A., Bali, J. P., Bonfils, S., Girman, J. P., Morgat, J. H., and Fromageot, P.,** Interaction of ^{3}H-labelled synthetic human gastrin I with rat gastric plasma membranes. Evidence for the existence of biologically reactive gastrin receptor sites, *FEBS Lett.*, 66, 168, 1976.
51. **Soumarmon, A., Cheret, A. M., and Lewin, M.,** Localization of gastrin receptors in intact isolated and separated rat fundic cells, *Gastroenterology*, 73, 900, 1977.
52. **Takeuchi, K., Speir, G. R., and Johnson, L. R.,** Mucosal gastrin receptor. I. Assay standardization and fulfillment of receptor criteria, *Am. J. Physiol.*, 237, E284, 1979.
53. **Soll, A. H., Amirian, D. A., Thomas, L. P., Reedy, T. J., and Elashoff, J. D.,** Gastrin receptors on isolated canine parietal cells, *J. Clin. Invest.*, 73, 1434, 1984.
54. **Baldwin, G. S., Chandler, R., Scanlon, D. B., and Weinstock, J.,** Identification of a gastrin binding protein in porcine gastric mucosal membranes by covalent cross-linking with iodinated gastrin 2-17, *J. Biol. Chem.*, 261, 12252, 1986.
55. **Magons, R. and Bali, J. P.,** High-affinity binding sites for gastrin on isolated rabbit gastric mucosal cells, *Eur. J. Pharmacol.*, 82, 47, 1982.
56. **Praissman, M. and Walden, M.,** The binding characteristics of ^{125}I-gastrin and ^{125}I-CCK-8 to guinea pig fundic gastric glands differ: is there more than one binding site for peptides of the CCK-gastrin family?, *Biochem. Biophys. Res. Commun.*, 123, 641, 1984.
57. **Singh, P., Rae-Venter, B., Townsend, C. M., Khalil, T., and Thompson, J. C.,** Gastrin receptors in normal and malignant gastrointestinal mucosa: age-associated changes, *Am. J. Physiol.*, 249, G761, 1985.
58. **Matsumoto, M., Park, J., and Yamada, T.,** Gastrin receptor characterization: affinity cross-linking of the gastrin receptor on canine gastric parietal cells, *Am. J. Physiol.*, 252, G143, 1987.
59. **Soll, A. H., Amirian, D. A., Thomas, L. P., Park, J., Elashoff, J. D., Beaven, M. A., and Yamada, T.,** Gastrin receptors on nonparietal cells isolated from canine fundic mucosa, *Am. J. Physiol.*, 247, G715, 1984.
60. **Yu, D. H., Zhou, Z. C., Noguchi, M., Gardner, J. D., and Jensen, R. T.,** Identification of gastrin receptors on pancreatic acini, *Gastroenterology*, 90, 1707, 1987.
61. **Winsett, O. E., Townsend, C. M., Glass, E. J., Raeventer, B., and Thompson, J. C.,** Gastrin stimulates growth of colon cancer, *Surg. Forum*, 33, 384, 1982.
62. **Baudière, B., Guillon, G., Bali, J. P., and Jard, S.,** Muscarinic stimulation of inositol phosphate accumulation and acid secretion in gastric fundic mucosal cells, *FEBS Lett.*, 198, 321, 1986.
63. **Pfeiffer, A., Paumgartner, G., and Herz, A.,** Muscarinic M2-receptors enhance polyphosphoinositol release in rat gastric mucosal cells, *FEBS Lett.*, 204, 352, 1986.
64. **Chew, C. S. and Brown, M. R.,** Release of intracellular Ca^{2+} and elevation of inositol trisphosphate by secretagogues in parietal and chief cells isolated from rabbit gastric mucosa, *Biochim. Biophys. Acta*, 888, 116, 1986.
65. **Takeuchi, K., Speir, G. R., and Johnson, L. R.,** Mucosal gastrin receptor. III. Regulation by gastrin, *Am. J. Physiol.*, 238, G135, 1980.
66. **Speir, G. R., Takeuchi, K., Peitsch, W., and Johnson, L. R.,** Mucosal gastrin receptor. VII. Up and downregulation, *Am. J. Physiol.*, 242, G243, 1982.
67. **Johnson, L. R., Peitsch, W., and Takeuchi, K.,** Mucosal gastrin receptor. VIII. Sex-related differences in binding, *Am. J. Physiol.*, 243, G469, 1982.
68. **Euler, A. R., Byrne, W. J., Cousins, L. M., Ament, M. F., Leaks, R. D., and Walsh, J. H.,** Increased serum gastrin concentration and gastric acid hyposecretion in the immediate newborn period, *Gastroenterology*, 72, 1271, 1977.
69. **Euler, A. R., Byrne, W. J., Meis, P. J., Leaks, R. D., and Ament, M. E.,** Basal and pentagastrin stimulated acid secretion in newborn human infants, *Pediatr. Res.*, 13, 36, 1979.

70. **Takeuchi, K., Peitsch, W., and Johnson, L. R.,** Mucosal gastrin receptor. V. Development in newborn rats, *Am. J. Physiol.*, 240, G163, 1981.
71. **Galand, G. and Forstner, G. G.,** Isolation of microvillus plasma membranes from suckling rat intestine. The influence of premature induction of digestive enzymes by injection of cortisolacetate, *Biochem. J.*, 144, 293, 1974.
72. **Kumegawa, M., Takuma, T., Hosada, S., Kunii, S., and Kanda, Y.,** Precocious induction of pepsinogen in the stomach of suckling mice by hormones, *Biochim. Biophys. Acta*, 543, 243, 1978.
73. **Peitsch, W., Takeuchi, K., and Johnson, L. R.,** Mucosal gastrin receptor. VI. Induction by corticosterone in newborn rats, *Am. J. Physiol.*, 240, G442, 1981.
74. **Sutliff, V. E., Raufman, J. P., Jensen, R. T., and Gardner, J. D.,** Actions of vasoactive intestinal peptide and secretin on chief cells prepared from guinea pig stomach, *Am. J. Physiol.*, 251, G96, 1986.
75. **Hicks, T. and Turnberg, L. A.,** The influence of secretin on ion transport in the human jejunum, *Gut*, 14, 485, 1973.
76. **Folsch, U. R. and Creutzfeldt, W.,** Pancreatic duct cells in rats: secretory studies in response to secretin, cholecystokinin-pancreozymin, and gastrin in vivo, *Gastroenterology*, 73, 1053, 1977.
77. **Deschodt-Lanckman, M., Robberecht, P., De Neef, P., Labrie, F., and Christophe, J.,** In vitro interactions of gastrointestinal hormones on cyclic adenosine 3′5′-monophosphate levels and amylase output in the rat pancreas, *Gastroenterology*, 68, 318, 1979.
78. **Okumura, K., Iwakawa, S., Inui, K. I., and Hori, R.,** Specific secretin binding sites in rat pancreas, *Biochem. Pharmacol.*, 32, 2689, 1983.
79. **Milutinovic, S., Schulz, I., and Rosselin, G.,** The interaction of secretin with pancreatic membranes, *Biochim. Biophys. Acta*, 436, 113, 1976.
80. **Robberecht, P., Waelbroeck, M., Noyer, M., Chatelain, P., De Neef, P., Konig, W., and Christophe, J.,** Characterization of secretin and vasoactive intestinal peptide receptors in rat pancreatic plasma membranes using native peptides, secretin-(7-27) and five secretin analogues, *Digestion*, 23, 201, 1982.
81. **Jensen, R. T., Charlton, C. G., Adachi, H., Jones, S. W., O'Donohue, T. L., and Gardner, J. D.,** Use of ^{125}I-secretin to identify and characterize high affinity secretin receptors on pancreatic acini, *Am. J. Physiol.*, 245, G186, 1983.
82. **Zhou, Z. C., Gardner, J. D., and Jensen, R. T.,** Receptors for vasoactive intestinal peptide and secretin on guinea pig pancreatic acini, *Peptides*, 8, 633, 1987.
83. **Bissonnette, B. M., Collen, M. J., Adachi, H., Jensen, R. T., and Gardner, J. D.,** Receptors for vasoactive intestinal peptide and secretin on rat pancreatic acini, *Am. J. Physiol.*, 246, G710, 1984.
84. **Bawab, W., Gespach, C., Marie, J. C., Chastre, E., and Rosselin, G.,** Pharmacology and molecular identification of secretin receptors in rat gastric gland, *Life Sci.*, 42, 791, 1987.
85. **Binder, H. J., Lemp, G. F., and Gardner, J. D.,** Receptors for vasoactive intestinal peptide and secretin on small intestinal epithelial cells, *Am. J. Physiol.*, 238, G190, 1980.
86. **Hootman, S. R., Ernst, S. A., and Williams, J. A.,** Secretagogue regulation of Na^+-K^+ pump activity in pancreatic acinar cells, *Am. J. Physiol.*, 245, G339, 1983.
87. **Trimble, E. R., Bruzzone, R., Biden, T. J., and Ferese, R. V.,** Secretin induces rapid increases in inositol trisphosphate, cytosolic Ca^{2+} and diacylglycerol as well as cyclic AMP in rat pancreatic acini, *Biochem. J.*, 239, 257, 1986.
88. **Trimble, E. R., Bruzzone, R., Meehan, C. J., and Biden, T. J.,** Rapid increases in inositol 1,4,5-trisphosphate, inositol 1,3,4,5-tetrakisphosphate and cytosolic free Ca^{2+} in agonist-stimulated pancreatic acini of rat, *Biochem. J.*, 242, 289, 1987.
89. **Trimble, E. R., Bruzzone, R., Biden, T. J., Meehan, C. J., Andrew, D., and Merrifield, R. B.,** Secretin stimulates cyclic AMP and inositol trisphosphate production in rat pancreatic acinar tissue by two fully independent mechanisms, *Proc. Natl. Acad. Sci. U. S. A.*, 84, 3146, 1987.
90. **Paquette, T. L., Shulman, D. F., Alpers, D. H., and Jaffe, B. M.,** Postnatal development of intestinal secretin in rat and guinea pigs, *Am. J. Physiol.*, 243, G511, 1982.
91. **Gespach, C., Cherel, Y., and Rosselin, G.,** Development of sensitivity to cAMP-inducing hormones in the rat stomach, *Am. J., Physiol.*, 247, G231, 1984.
92. **Straus, D. S.,** Effects of insulin on cellular growth and proliferation, *Life Sci.*, 29, 2131, 1981.
93. **Mahmood, A., Pathak, R. M., and Agarwal, N.,** Effect of chronic alloxan diabetes and insulin administration on intestinal brush border enzymes, *Experientia.*, 34, 741, 1978.
94. **Ménard, D. and Malo, C.,** Insulin-evoked precocious appearance of intestinal sucrase activity in suckling mice, *Dev. Biol.*, 69, 661, 1979.
95. **Cezard, J. P., Forgue-Lafitte, M. E., Chamblier, M. C., and Rosselin, G. E.,** Growth-promoting effect, biological activity, and binding of insulin in human intestinal cancer cells in culture, *Cancer Res.*, 41, 1148, 1981.
96. **Kramer, M. F. and Poort, C.,** The effect of insulin on amino acid incorporation into exocrine pancreatic cells of the rat, *Horm. Metab. Res.*, 7, 389, 1975.

97. **Logsdon, C. D. and Williams, J. A.,** Pancreatic acini in short-term culture: regulation by EGF, carbachol, insulin, and corticosterone, *Am. J. Physiol.*, 244, G675, 1983.
98. **McEvoy, R. C.,** Fetal rat pancreas in organ culture: effect of exogenous insulin on the development of the islet cells, *Horm. Metab. Res.*, 13, 5, 1981.
99. **Saito, A., Williams, J. A., and Kanno, T.,** Potentiation of cholecystokinin-induced exocrine secretion by both exogenous and endogenous insulin in isolated and perfused rat pancreata, *J. Clin. Invest.*, 65, 777, 1980.
100. **Saito, A., Williams, J. A., and Kanno, T.,** Potentiation by insulin of the acetylcholine-induced secretory response of the perfused rat pancreas, *Biomed. Res.*, 1, 101, 1980.
101. **Hellman, B., Wallgren, A., and Peterson, B.,** Cytological characteristics of the exocrine pancreatic cells with regards to their position in relation to the islets of Langerhans, *Acta Endocrinol.*, 39, 465, 1962.
102. **Bendayan, M.,** Contacts between endocrine and exocrine cells in the pancreas, *Cell Tissue Res.*, 222, 227, 1982.
103. **Bergeron, J. J. M., Rachubinski, R., Searle, N., Borts, D., Sikstrom, R., and Posner, B. I.,** Polypeptide hormone receptors *in vivo*: demonstration of insulin binding to adrenal gland and gastrointestinal epithelium by quantitative radioautography, *J. Histochem. Cytochem.*, 28, 824, 1980.
104. **Bergeron, J. J. M., Rachubinski, R., Searle, N., Sikstrom, R., Borts, D., Bastian, P., and Posner, B. I.,** Radioautographic visualization of *in vivo* insulin binding to the exocrine pancreas, *Endocrinology*, 107, 1069, 1980.
105. **Goldfine, I. D., Kriz, B. M., Wong, K. Y., Hradek, G., Jones, A. L., and Williams, J. A.,** Insulin action in pancreatic acini from streptozotocin-treated rats. III. Electron microscope autoradiography of ^{125}I-insulin, *Am. J. Physiol.*, 240, G69, 1981.
106. **Williams, J. A. and Goldfine, I. D.,** The insulin-pancreatic acinar axis, *Diabetes*, 34, 980, 1985.
107. **Goldfine, I. D., Clawson, G. A., Smuckler, E. A., Purrello, F., and Vigneri, R.,** Action of insulin at the nuclear envelope, *Molec. Cell. Biochem.*, 48, 3, 1982.
108. **Korc, M., Sankaran, H., Wong, K. Y., Williams, J. A., and Goldfine, I. D.,** Insulin receptor in isolated mouse pancreatic acini, *Biochem. Biophys. Res. Commun.*, 84, 293, 1978.
109. **Sankaran, H., Iwamoto, Y., Korc, M., Williams, J. A., and Goldfine, I. D.,** Insulin action in pancreatic acini from streptozotocin-treated rats. II. Binding of ^{125}I-insulin to receptors, *Am. J. Physiol.*, 240, G63, 1981.
110. **Forgue-Lafitte, M. E., Marescot, M. R., Chamblier, M. C., and Rosselin, G.,** Evidence for the presence of insulin binding sites in isolated rat intestinal epithelial cells, *Diabetologia*, 19, 373, 1980.
111. **Goldfine, I. D., Purrello, F., Clawson, G. A., and Vigneri, R.,** Insulin binding sites on the nuclear envelope: potential relationship to mRNA metabolism, *J. Cell. Biochem.*, 20, 93, 1982.
112. **Sjodin, L., Holmberg, K., and Lyden, A.,** Insulin receptors on pancreatic acinar cells in guinea pigs, *Endocrinology*, 115, 1102, 1984.
113. **Gingerich, R. L., Gilbert, W. R., Comens, P. G., and Gavin, J. R.,** Identification and characterization of insulin receptors in basolateral membranes of dog intestinal mucosa, *Diabetes*, 36, 1124, 1987.
114. **Zimmerman, T. W., Reinprecht, J. T., and Binder, H. J.,** Peptide binding to intestinal epithelium: distinct sites for insulin, EGF and VIP, *Peptides*, 6, 229, 1985.
115. **Czech, M. P.,** Insulin action, *Am. J. Med.*, 70, 142, 1981.
116. **Goldfine, I. D. and Williams, J. A.,** Receptors for insulin and CCK in the acinar pancreas: relationship to hormone action, *Int. Rev. Cytol.*, 85, 1, 1983.
117. **Hepp, K. D. and Renner, R.,** Insulin action on the adenylyl cyclase system, *FEBS Lett.*, 20, 191, 1972.
118. **Taylor, D., Uhing, R. J., Blackmore, P. F., Prpic, V., and Exton, J. H.,** Insulin and epidermal growth factor do not affect phosphoinositide metabolism in rat liver plasma membranes and hepatocytes, *J. Biol. Chem.*, 260, 2011, 1985.
119. **Seals, J. R. and Jarett, L.,** Activation of pyruvate dehydrogenase by direct addition of insulin to an isolated plasma membrane mitochondria mixture: evidence for generation of insulin's second messenger in a subcellular system, *Proc. Natl. Acad. Sci. U. S. A.*, 77, 77, 1980.
120. **Larner, J., Galasko, G., Cheung, K., DePaoli-Roach, A. A., Huang, L., Daggy, P., and Kellog, L.,** Generation by insulin of a chemical mediator that controls protein phosphorylation and dephosphorylation, *Science*, 206, 1408, 1979.
121. **Seals, J. R. and Czech, M. P.,** Evidence that insulin activates an intrinsic plasma membrane protease in generating a secondary chemical mediator, *J. Biol. Chem.*, 255, 6529, 1981.
122. **Williams, J. A., Bailey, A. C., Preissler, M., and Goldfine, I. D.,** Insulin regulation of sugar transport in isolated pancreatic acini from diabetic mice, *Diabetes*, 31, 674, 1982.
123. **Korc, M., Iwamoto, Y., Sankaran, H., Williams, J. A., and Goldfine, I. D.,** Insulin action in pancreatic acini from streptozotocin-treated rats. I. Stimulation of protein synthesis, *Am. J. Physiol.*, 240, G56, 1981.
124. **Purrello, F., Burnham, D. B., and Goldfine, I. D.,** Insulin regulation of protein phosphorylation in isolated rat liver nuclear envelopes: potential relationship to mRNA metabolism, *Proc. Natl. Acad. Sci.* U. S. A., 80, 1189, 1983.
125. **Mossner, J., Logsdon, C. D., Goldfine, I. D., and Williams, J. D.,** Regulation of pancreatic acinar cell insulin receptors by insulin, *Am. J. Physiol.*, 247, G155, 1984.

126. **McDonald, A. R., Maddux, B. A., Okabayashi, Y., Wong, K. Y., Hawley, D. M., Logsdon, C. D., and Goldfine, I. D.,** Regulation of insulin receptor mRNA levels by glucocorticoids, *Diabetes*, 36, 779, 1987.
127. **Heaton, J. H. and Gelehrter, T. D.,** Desensitization of hepatoma cells to insulin action. Evidence for a post-receptor mechanism, *J. Biol. Chem.*, 256, 12257, 1981.
128. **Ganguli, S., Sinha, M. K., Sterman, B., Harris, P., and Sperling, M. A.,** Ontogeny of hepatic insulin and glucagon receptors and adenylate cyclase in rabbit, *Am. J. Physiol.*, 244, E624, 1983.
129. **Neufeld, N. D., Scott, M., and Kaplan, S. A.,** Ontogeny of the mammalian insulin receptor. Studies of human and rat fetal liver plasma membranes, *Dev. Biol.*, 78, 151, 1980.
130. **Carpenter, G. and Cohen, S.,** Epidermal growth factor, *Ann. Rev. Biochem.*, 48, 193, 1979.
131. **Johnson, L. R. and Guthrie, P. D.,** Stimulation of rat oxyntic gland mucosal growth by epidermal growth factor, *Am. J. Physiol.*, 238, G45, 1980.
132. **Dembinski, A. B. and Johnson, L. R.,** Effect of epidermal growth factor on the development of rat gastric mucosa, *Endocrinology*, 116, 90, 1985.
133. **Calvert, R., Beaulieu, J. F., and Ménard, D.,** Epidermal growth factor (EGF) accelerates the maturation of fetal mouse intestinal mucosa in utero, *Experientia*, 38, 1096, 1982.
134. **Malo, C. and Ménard, D.,** Influence of epidermal growth factor on the development of suckling mouse intestinal mucosa, *Gastroenterology*, 83, 28, 1982.
135. **Dembinski, A., Gregory, H., Konturek, S. J., and Polanski, M.,** Trophic action of epidermal growth factor on the pancreas and gastroduodenal mucosa in rats, *J. Physiol.*, 325, 35, 1982.
136. **Logsdon, C. D. and Williams, J. A.,** Epidermal growth factor binding and biologic effects on mouse pancreatic acini, *Gastroenterology*, 85, 339, 1983.
137. **Brannon, P. M., Orrison, B. M., and Kretchmer, N.,** Primary cultures of rat pancreatic acinar cells in serum-free medium, *In Vitro*, 21, 6, 1985.
138. **Brannon, P. M., Demarest, A. S., Sabb, J. E., and Korc, M.,** Dietary modulation of epidermal growth factor action in cultured pancreatic acinar cells of the rat, *J. Nutr.*, 116, 1306, 1986.
139. **Chabot, J. G., Guy, J., St-Arnaud, R., Walker, P., and Pelletier, G.,** Autoradiographic localization of epidermal growth factor (EGF) receptors in different tissues of the rat, *J. Cell Biol.*, 95, 179a, 1982.
140. **Chabot, J. G., Walker, P., and Pelletier, G.,** Demonstration of epidermal growth factor binding sites in the adult rat pancreas by light microscopic autoradiography, *Pancreas*, 2, 653, 1987.
141. **Korc, M., Matrisian, L. M., Planck, S. R., and Magun, B. E.,** Binding of epidermal growth factor in rat pancreatic acini, *Biochem. Biophys. Res. Commun.*, 111, 1066, 1983.
142. **Forgue-Lafitte, M. E., Kobari, L., Gespach, C., Chamblier, M. C., and Rosselin, G.,** Characterization and repartition of epidermal growth factor-urogastrone receptors in gastric glands isolated from young and adult guinea pigs, *Biochim. Biophys. Acta*, 798, 192, 1984.
143. **Forgue-Lafitte, M. E., Laburthe, M., Chamblier, M. C., Moody, A. J., and Rosselin, G.,** Demonstration of specific receptors for EGF-urogastrone in isolated rat intestinal epithelial cells, *FEBS Lett.*, 114, 243, 1980.
144. **Gallo-Payet, N. and Hugon, J. S.,** Epidermal growth factor receptors in isolated adult mouse intestinal cells: studies *in vivo* and in organ culture, *Endocrinol*ogy, 116, 194, 1985.
145. **Toyoda, S., Lee, P. C., and Lebenthal, E.,** Interaction of epidermal growth factor with specific binding sites of enterocytes isolated from rat small intestine during development, *Biochim. Biophys. Acta*, 886, 295, 1986.
146. **Ménard, D., Pothier, P., and Gallo-Payet, N.,** Epidermal growth factor receptors during postnatal development of the mouse colon, *Endocrinol*ogy, 121, 1548, 1987.
147. **Korc, M. and Magun, B. E.,** Binding and processing of epidermal growth factor in Panc-1 human pancreatic carcinoma cells, *Life Sci.*, 36, 1849, 1985.
148. **Cohen, S.,** Epidermal growth factor, *In Vitro*, 23, 239, 1987.
149. **Carpenter, G. and Zendegui, J. G.,** Epidermal growth factor, its receptor, and related proteins, *Exp. Cell Res.*, 164, 1, 1986.
150. **Hepler, J. R., Nakahata, N., Lovenberg, T. W., DiGuiseppi, J., Herman, B., Earp, H. S., and Harden, T. K.,** Epidermal growth factor stimulates the rapid accumulation of inositol (1,4,5)-trisphosphate and a rise in cytosolic calcium mobilized from intracellular stores in A431 cells, *J. Biol. Chem.*, 262, 2951, 1987.
151. **Pike, L. J. and Eakes, A. T.,** Epidermal growth factor stimulates the production of phosphatidylinositol monophosphate and the breakdown of polyphosphoinositides in A431 cells, *J. Biol. Chem.*, 262, 1644, 1987.
152. **Chen, W. S., Lazar, C. S., Poenie, M., Tsien, R. Y., Gill, G. N., and Rosenfeld, M. G.,** Requirement for intrinsic protein tyrosine kinase in the immediate and late actions of the EGF receptor, *Nature*, 328, 820, 1987.
153. **Kawamoto, I., Mendelsohn, J., Le, A., Sato, G. H., Lazar, C. S., and Gill, G. N.,** Relation of epidermal growth factor receptor concentration to growth of human epidermoid carcinoma A431 cells, *J. Biol. Chem.*, 259, 7761, 1984.
154. **Bertics, P. J., Weber, W., Cochet, C., and Gill, G. N.,** Regulation of the epidermal growth factor receptor by phosphorylation, *J. Cell Biochem.*, 29, 49, 1985.
155. **Korc, M., Matrisian, L. M., and Magun, B. E.,** Direct modulation of epidermal growth factor binding by cholecystokinin, *Life Sci.*, 33, 561, 1983.

156. **Korc, M., Matrisian, L. M., and Magun, B. E.,** Cytosolic calcium regulates epidermal growth factor endocytosis in rat pancreas and cultured fibroblasts, *Proc. Natl. Acad. Sci. U. S. A.,* 81, 461, 1984.
157. **Logsdon, C. D. and Williams, J. A.,** Intracellular Ca^{2+} and phorbol esters synergistically inhibit internalization of epidermal growth factor in pancreatic acini, *Biochem. J.*, 223, 893, 1984.
158. **Baker, J. B., Barsh, G. S., Carney, D. H., and Cunningham, D. D.,** Dexamethasone modulates binding and action of epidermal growth factor in serum-free cell culture, *Proc. Natl. Acad. Sci. U. S. A.,* 75, 1882, 1978.
159. **Mukka, V. R.,** Regulation of epidermal growth factor receptor levels by thyroid hormone, *J. Biol. Chem.*, 259, 6543, 1984.
160. **Pothier, P. and Ménard, D.,** Presence and characteristics of epidermal growth factor receptors in human fetal small intestine and colon, *FEBS Lett.*, 228, 113-117, 1988.
161. **Hoath, S. B., Pickens, W. L., Bucuvalas, J. C., and Suchy, F. J.,** Characterization of hepatic epidermal growth factor receptors in the developing rat, *Biochim. Biophys. Acta*, 930, 107, 1987.

Exocrine Pancreas

Chapter 4

FETAL AND NEONATAL DEVELOPMENT OF THE EXOCRINE PANCREAS

Emanuel Lebenthal and Ying-kit Leung

TABLE OF CONTENTS

I. INTRODUCTION

The exocrine pancreas undergoes characteristic developmental changes at the tissue, cellular, and subcellular levels during the perinatal period. Numerous ontogenic studies have been carried out, and animal models, particularly rats, have been used predominantly. However, human data are scanty and extrapolation of animal data to humans has to be done with care and reservation. Our rudimentary understanding of the underlying mechanisms derived from animal models is often confounded owing to interspecies differences in gestational periods, qualities of milk and weaning diet, regulatory mechanisms such as hormonal regulation and preprogrammed, genetically determined species-specific developmental patterns.

It is the aim of this chapter to focus on recent developments in certain areas of interest, although a comprehensive synopsis on the functional and morphological development of the exocrine pancreas will also be given.

II. EMBRYONIC DEVELOPMENT [1,2]

In humans, the pancreatic exocrine and endocrine tissues arise from the dorsal and ventral pancreatic anlagen which are first detectable as evaginations of the primitive foregut during the fifth week of gestation. By the seventh week, as a consequence of rotation and fixation of the gut, the two anlagen fuse to form a single glandular mass. The dorsal anlage will ultimately contribute the tail and body of the mature gland, while the ventral anlage contributes the head and uncinate process. The main pancreatic duct is formed by the fusion of the ductular network of the two anlagen; the accessory pancreatic duct, which often persists, is derived from the dorsal pancreatic duct proximal to the usual site of fusion. It may either become a tributary of the main duct or retain its own duodenal outlet.[3]

Histologically, the fetal pancreas at 7 weeks of gestation consists of tubules and masses of undifferentiated epithelial cells within a loose connective tissue matrix. Zymogen granules are not detected at this age. By 12 to 14 weeks, the tubules have become a network of interconnected ductules, and the separation of endocrine and exocrine tissue has occurred. Discrete lobules and acini can be identified by 14 to 16 weeks. By 22 weeks, the acinar cells have a basophilic cytoplasm with numerous zymogen granules.[4]

On electron microscopy, more discrete changes can be detected during development. At 9 weeks the polarity of exocrine cells is poorly established and the basolateral membranes are smooth. Few intracellular organelles associated with exportable proteins, i.e., Golgi apparatus and secretory granules, are detectable. By 12 weeks, however, well-defined Golgi complexes and unusual immature secretory granules are present. In contrast to the mature granules, these are electron dense and either elongated or elliptical in shape. These granules persist up to approximately 20 weeks. By that time the acinar cells are well polarized with many mature zymogen granules in the apical cytoplasm. The cells have abundant rough endoplasmic reticulum and the basolateral membranes are interdigitated with well-developed apical junctional complexes. The apical membrane has more prominent microvilli that are weakly positive for alkaline phosphatase when studied histochemically.[5]

The mature pancreatic acinar cell is pyramidal in shape and has a central or basal nucleus. The perinuclear region has a well-developed Golgi apparatus, endoplasmic reticulum, and immature zymogen granules (condensing vacuoles). The apical region contains mature zymogen granules and a well-developed network of microvilli. The basolateral membranes are simple with few processes, interdigitation, and junctional complexes. By term, the morphological appearance of human pancreas is similar to that of the mature pancreas.

III. FUNCTIONAL DEVELOPMENT OF THE PANCREAS

The development of any organ involves both morphogenesis and cytodifferentiation.[6]

TABLE 1
Morphological Development of the Human Exocrine Pancreas

Event	Developmental age (weeks)
Dorsal and ventral anlagen detectable	5
Fusion of anlagen	7
Differentiation of endocrine and exocrine tissue	12—14
Lobules and acini formed	14—16
Polarity of acinar cells established	20
Zymogen granules present	20—22
Adult appearance of gland (histological)	40

Morphogenesis, described above, consists of mesoderm-endoderm interaction, mitotic growth, and attainment of the characteristic architecture of the pancreas. Cyto-differentiation, on the other hand, involves the acquisition of the ability to synthesize zymogen enzymes and the cytoplasmic structures required for transport and secretion. Cyto-differentiation has been studied in the fetal rat pancreas by use of a two-dimensional protein electrophoresis technique.[7] From day 14 to day 18 of the 22-d gestation period, protein profiles obtained from intestine, lung, mesenchyme, and pancreas are very similar. However, after the 18th day, the synthesis of embryonic proteins diminishes and is replaced by secretory transport proteins, i.e., the digestive enzymes. The surface properties of the differentiating pancreatic cell have also been studied by lectin binding techniques.[8,9] Acinar tissues seem to have less selective binding than endocrine and centroacinar cells. In addition, it has been shown that the lectin-binding characteristics of pancreatic tissue change during development in the rat. It has been demonstrated that the mesoderm-endoderm interaction seems to be quite important in both pancreatic morphogenesis and cytodifferentiation. In a fetal mouse model, protodifferentiated pancreatic epithelial cells (11- to 15-d gestation) were isolated from mesoderm and cultured *in vitro*.[10] Cytodifferentiation continued in this model, as exportable protein synthesis followed the expected time course. However, no functional units in the form of acini were formed in the absence of mesoderm. The process of cyto-differentiation, the acquisition of characteristic biochemical and intracellular organelles of the acinar cells, starts on day 15 of gestation in the rat and is essentially complete by day 19.[1,11,12] In man, it begins around the 12th week and is complete around term. (Table 1).

A. ROLE OF MESENCHYME IN DEVELOPMENT

The prenatal phase of pancreatic development is characterized by rapid exponential growth of both protodifferentiated and already differentiated acinar cells.[13] This growth is dependent on mesodermal tissue, which increases proportionally in mass. Mesoderm from various embryonic sources is effective.[14] A mesenchymal factor found in particulate fractions of homogenates (particles sedimenting between 4000 and 60,000 $\times g$) of midgestational rat or chicken embryos stimulates DNA synthesis and amylase accumulation in young rat or mouse pancreatic rudiments stripped of their mesenchymal component.[15-17] Older rat embryo or adult organ extracts are not effective. This mesenchymal factor appears to act via an interaction with the plasma membrane of pancreatic epithelial cells. One of the effects of this factor is to stimulate the endogenous production of cyclic AMP, because periodate-inactivated mesenchymal factor can be reactivated by the addition of a stable derivative of cyclic AMP. The pattern of mesenchymal factor production by surrounding mesenchyme may control pancreatic morphogenesis.

Mesenchymal tissue also appears to have an effect on the kinds of cells appearing during pancreatic development. The predominant cell types that developed when 12-d-old embryonic rat epithelium is cultured in the absence of mesenchyme are endocrine cells, primarily glucagon cells, since epithelial cells could not develop under these circumstances. However, if mesenchymal factor is supplied to the culture system, normal development of acinar cells is observed.

TABLE 2
Function Development of the Human Exocrine Pancreas

Event	Developmental age
Zymogens detectable in pancreatic homogenates	20 weeks
Lipase detectable in pancreatic homogenates	24 weeks
Enterokinase detectable in intestinal content	26 weeks
Response to secretagogues	1 month (postnatal)
Amylase detectable in pancreatic homogenates	2 months (postnatal)

B. PRENATAL DEVELOPMENT OF PANCREATIC DIGESTIVE ENZYMES (TABLE 2)

Accumulation of exocrine enzymes also occurs during pre- and postnatal life. In the rat, exocrine pancreatic development is considered to be a triphasic process, characterized by different rates of enzyme accumulation, in addition to an early predifferentiation period during which time no exocrine enzymes are detected. A primary transition occurs around days 10 to 11 of gestation, about which time the pancreas begins to accumulate low levels of exocrine enzymes. This is followed by a protodifferentiation period during which little or no increase in enzymes occurs.[17,18] An abrupt and dramatic increase of enzyme activities that characterizes the secondary transitional phase takes place at 16 to 19 d of gestation.[2,17,18] A tertiary transitional phase then follows. This mainly occurs after parturition and involves further modulation of the exocrine enzyme activities and exponential growth of the pancreas.[19,20]

Individual enzymes of the pancreas do not develop simultaneously and appear at different times. Compared to postnatal development, the signal that governs the orderly pattern of prenatal development is still not known, although the genetic makeup obviously plays a very important role in the orderly appearance of the individual enzymes. In humans, proteolytic activity is first detectable in the fetal intestinal lumen at 26 weeks of gestation, the time at which enterokinase is first detectable. The concentrations of the pancreatic endo- and exopeptidases increase thereafter so that, by term, the newborn has approximately 10 to 60% of the adult chymotrypsin and carboxypeptidase concentrations in pancreatic homogenates or meconium.[21,22] We have studied a group of 15 infants and found that duodenal fluid trypsin is present at an activity level 90 to 100% of that of children aged 2 years, chymotrypsin is present at a 50 to 60% activity level, and carboxypeptidase B at a 10 to 25% level.[22] Other workers have similarly found activities of the endoproteases trypsin and chymotrypsin to be 10 to 60% of the adult values.[21,23] The activation of the proenzymes by enterokinase is necessary to produce chemically active enzymes, and it has been demonstrated that enterokinase first appears in the intestine at 26 weeks of gestation and reaches 10 to 20% of the mature concentration by term.[24] It is not known whether this amount of enterokinase is sufficient to activate the pancreatic proenzymes at birth.

Pancreatic lipase activity can be detected as early as 16 weeks of gestation in pancreatic homogenates, and at term approximately 10% of the specific activity demonstrated in homogenates obtained from older children or adults is present.[22] Lipase activity in duodenal fluid, however, in preterm or full-term infants is very low.[22,23] Colipase and phospholipase, each of which promotes lipid hydrolysis, have not been fully evaluated in the human. The limiting factor of fat digestion in the neonatal period, however, is the low concentrations of bile acids in duodenal fluid, and the critical concentration of bile acid for micelle formation may not be reached in the premature infant.[25,26]

Amylase can be detected in amniotic fluid beginning at about 16 weeks of gestation.[27] It seems that both salivary and pancreatic isoenzymes are present during gestation but the salivary form is predominant. Work with fetal pancreatic homogenates has shown that the amylase content is extremely low. Negligible amylase activity was found in duodenal fluid in preterm

TABLE 3
Effects of Various Secretagogues on Amylase Release Expressed as Percentage of Control (=100%) in Dispersed Pancreatic Acini of Newborn and Adult Rats

Secretagogues	Concentration (*M*)	Amylase release (% of basal activity) Newborn	2-d-old	Adult
TPA	10^{-6}	93 ± 16	168 ± 15	208 ± 36
TPA	10^{-7}	91 ± 20	143 ± 13	—
CCK-8	3×10^{-10}	100 ± 0	237 ± 30	304 ± 43
A23187	3×10^{-6}	139 ± 15	167 ± 18	219 ± 40
Dibutyryl cAMP	10^{-3}	129 ± 20	—	—

Modified from Shimizu, K., Lebenthal, E., and Lee, P. C., *Biochim. Biophys. Acta*, 968, 186, 1988.

and full-term infants.[22] Despite this absence of pancreatic amylase, the newborn tolerates moderate amounts of starch due to alternative pathways of amylase digestion, mainly by glucoamylase present in the small intestine and partly by salivary amylase.[28]

C. POSTNATAL DEVELOPMENT OF PANCREATIC DIGESTIVE ENZYMES

After birth, specific activities of the exocrine enzymes develop independently. The changes in specific activities coincide with certain specific changes in exogenous (nutritional) and endogenous (hormonal) factors. Such changes can be manipulated experimentally and it is through these manipulations that we can gain insight into the controlling mechanisms of postnatal exocrine pancreatic development.

1. Development of the Secretagogue Response

Despite the relatively mature morphology of the pancreas, the newborn infant has partial pancreatic insufficiency.[29,30] It has been demonstrated by Doyle and Jamieson that in term fetal and neonatal rats, exocrine pancreatic response to secretagogues is immature in spite of a mature appearance of the tissue both histologically and ultrastructurally.[29] Carbachol and cholecystokinin-octapeptide (CCK-8) and analogues, which are potent secretagogues for the mature exocrine pancreas, fail to stimulate secretion of amylase by the term fetal rat pancreas. The degree of stimulated amylase release shows an age-dependent increase as the pancreas matures postnatally. In the human, the pancreas is also refractory to stimulation by secretagogues in the neonatal period.[31]

Secretagogues such as CCK and carbachol act by first binding to specific receptors on the surface of the acinar cells. After binding, a series of events collectively termed "stimulus-secretion coupling" take place which finally result in secretion of enzymes from the acinar cells by exocytosis. Pancreatic exocrine secretion is stimulated by secretagogues that act primarily through either the adenylate cyclase system (secretin and vasoactive intestinal polypeptide) or the phosphatidylinositol pathway (CCK and cholinergic agents). The binding of cholecystokinin and cholinergic agents to their respective receptors results in the release of inositol trisphosphate and 1,2-diacylglycerol. Inositol trisphosphate induces calcium mobilization and activates calmodulin-dependent protein kinases. At the same time, 1,2-diacyglycerol activates and translocates protein kinase C from a cytosolic to a membraneous site. Taken together, protein kinase C activation, calcium mobilization and subsequent activation of calmodulin dependent protein kinase are viewed as important intermediary steps leading to secretion by the exocrine pancreas.[32]

The lack of responsiveness of the neonatal pancreas to secretagogues thus could occur at various steps and has been the focus of many investigations (Table 3). Initial studies indicated

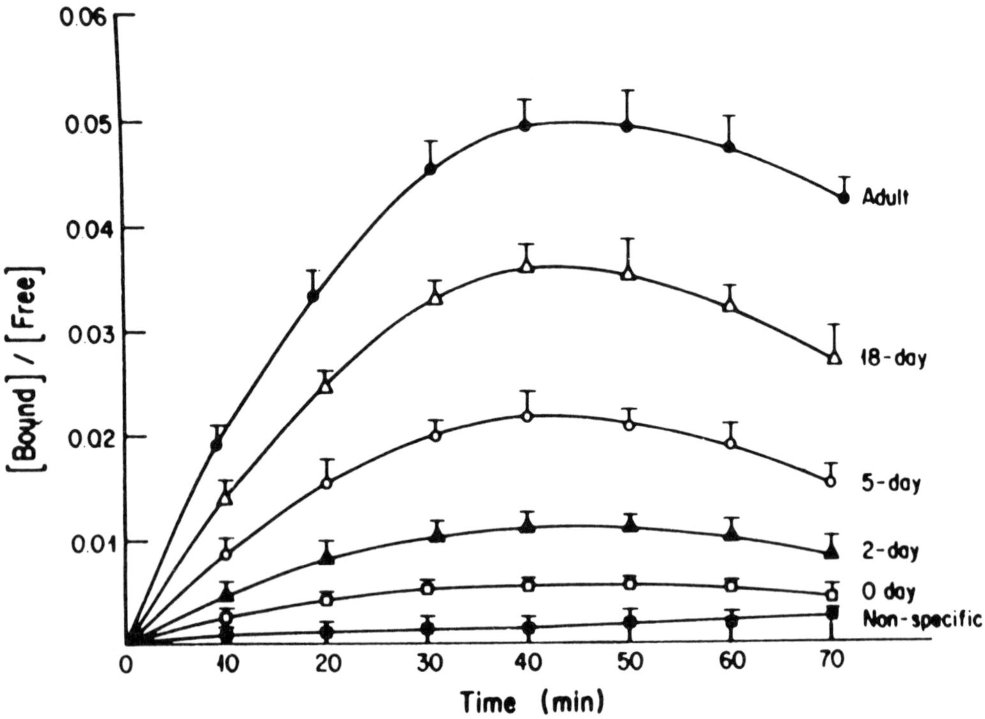

FIGURE 1. Scatchard analysis of radiolabeled CCK-8 binding to pancreatic acini of rats of different ages. (From Leung, Y. K., Lee, P. C., and Lebenthal, E., *Am. J. Physiol.*, 250, G594, 1986. With permission.)

that the immature pancreas, in spite of its unresponsiveness to cholinergic agents and CCK, responded to the calcium ionophore, A23187, and dibutyryl cAMP.[29,30] This suggests that steps distal to calcium mobilization are perhaps functioning physiologically and the dysfunction probably occurs prior to this step. We have demonstrated in a study using radiolabeled CCK as ligand that the high-affinity binding component of the CCK receptors in fetal and neonatal rats was relatively low in capacity. An age-related increase in the capacity of the high-affinity binding component with an accompanying increase in secretory response to CCK was also observed (Figure 1).[30] The maturation of the CCK receptor in addition has been shown to be inducible by hydrocortisone.[33] Dumont et al. have demonstrated that the nonresponsiveness of neonate pancreatic acini to cholinergic agents is also accompanied by a deficiency of muscarinic receptors.[34]

The relative lack of specific receptors for CCK and cholinergic agents in the neonatal pancreas can only partly explain the unresponsiveness to these agents, since measurable amounts of CCK and cholinergic receptors in the neonates are present.[34,35] Chang and Jamieson showed that the postnatal acquisition of secretory responsiveness in rat pups was not correlated with the increase in number of CCK receptors on the cell surface.[36] These results therefore point to possible additional defects in the postbinding events or factors. Recent results provide evidence in support of this hypothesis. Shimizu et al. examined the ontogenic changes of protein kinase C (PKC) in the developing exocrine pancreas in rat pups and found low levels of PKC in newborn rats.[37] A sharp increase in PKC levels was seen by day 2 after birth. TPA (tetradecanoyl-phorbol-13-acetate), a PKC activating agent shown to be a potent stimulant for pancreatic secretion in adult animals, was without effect on the newborn rat pancreas. Acinar cell responsiveness to TPA was observed at 2 days of age, however, a time when high PKC activity was demonstrable (Figure 2). These results suggest that the low levels of pancreatic PKC in newborn rats are at least in part responsible for the lack of response of pancreatic acini to CCK.

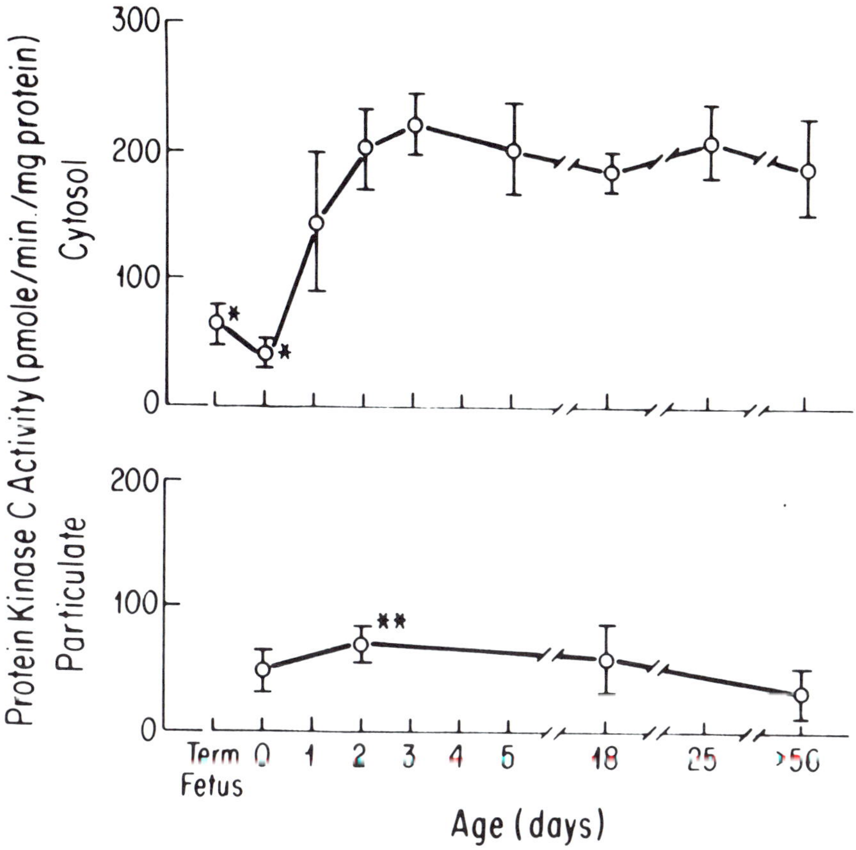

FIGURE 2. Protein kinase C activities in the cytosol and particulate fractions of rat pancreata during postnatal development. (From Shimizu, K., Lebenthal, E., and Lee, P. C., *Biochim. Biophys. Acta*, 968, 186, 1988. With permission.)

Another study on calmodulin-dependent protein kinase development yielded strikingly similar results; calmodulin-dependent protein kinase was relatively low in early life but increased at later times.[38] Its developmental pattern correlates with the responsiveness to CCK and cholinergic agents by the pancreas. These data suggest that part of the defect in the immature pancreas is a limitation of calmodulin-dependent protein kinase. It is possible that a defect in this postreceptor mechanism is one of the causes of the insensitivity of the newborn rat pancreas to secretagogue stimulation.

2. Nutrition as a Regulatory Factor for Pancreatic Development

a. Diet

In rats, as in humans and other mammals, postnatal development is accompanied by a major change in diet occurring at weaning. Rat milk is high in fat but lower in carbohydrate content. In contrast, laboratory chow has a much lower fat but higher carbohydrate content and contains twice the amount of protein as rat milk. Weaning therefore leads to drastic changes in dietary composition, especially carbohydrates, and this coincides with a rapid increase in pancreatic hydrolase accumulation, particularly amylase. The developmental changes of pancreatic enzymes with the onset of weaning have been considered to be under dietary control partly because of the possible modulation of pancreatic enzymes by diet as demonstrated in adult rats.[39-42] It has been shown experimentally that prolonged nursing of suckling rat pups resulted in a delayed increase in pancreatic amylase content,[42] and weaning pups onto a high fat, low carbohydrate chow resulted in a lower pancreatic amylase content. Early weaning led to premature increases in the specific activities of amylase, chymotrypsinogen, and trypsinogen.[43,44] The role of diet in the regulation of postnatal increases in pancreatic enzymes is not clear; in rats which had undergone prolonged nursing, a delayed but continual increase of amylase was observed. This suggests an inherent programming ultimately responsible for the increase of amylase activity, while dietary composition serves as a modulating factor. Both lipase and trypsinogen also exhibited inherent developmental changes superimposed upon changes resulting from the modulating effect of diet. Moreover, food deprivation or acute malnutrition also causes precocious changes in enzyme activities and this effect might possibly result from a stress-induced surge in corticosterone levels due to food deprivation.

Owing to limited data available in humans, whether pancreatic enzymes are altered in response to changes in diet has not been fully demonstrated. The limited human studies available agree with findings in animals. Synthetic and secretory responses to dietary variations have been reported. Premature infants fed a high-starch diet had a 10-fold increase in luminal amylase concentration, but because of an extremely low initial concentration, no meaningful improvement in amylase level in duodenal aspirate and no change in clinical parameters (i.e., starch digestion and absorption) were noted.[45] It was shown by the same investigators that a high-protein diet resulted in an increase in trypsin and chymotrypsin concentration in the duodenal aspirate. We have previously demonstrated that in premature infants, feeding with a soy-based formula resulted in higher pancreatic lipase and trypsin secretion upon stimulation by secretagogues as compared to age-matched infants fed milk-based formulas.[46]

The digestion of fat, protein, and starch in infants depends on the degree of maturation of the exocrine pancreas, which in turn can be modified by exogenous factors. Attempts to enhance the process of maturation by various interventions during critical periods of infancy may be of particular importance for further investigations. Such manipulations may result in changes in the current practices in feeding the premature and compromised infant.

b. Malnutrition

The effects of undernourishment on pancreatic development have also been the focus of considerable research. Inadequate nutrition in the form of total or protein deprivation occurring pre- or postnatally results in altered pancreatic function in adult animals or humans.[47] However, little is known about the effect of malnutrition on pancreatic development in infants. Rat pups born to dams receiving a protein-deficient diet during pregnancy had lower pancreatic enzyme contents. When these pups were nursed by dams fed a normal diet and then weaned to regular chow, the pancreatic enzyme contents returned to normal.[48] In the immediate postnatal period, malnutrition induced in rat pups starting at birth resulted in marked reduction in pancreatic weight and enzyme content.[49,50] The effect of malnutrition under these conditions is reversible since nutritional rehabilitation after weaning resulted in full recovery of pancreatic enzymes within 2 weeks.[50] It can be concluded from these data that malnutrition has detrimental effects on pancreatic development.

The effect of pre- and postnatal malnutrition of the mother on pancreatic development in her offspring has not been well studied. In one study, pre- and postnatal malnutrition in dams was produced by restriction of food consumption during pregnancy and/or lactation and the relative effects were compared by cross-fostering pups to control and malnourished dams. Results revealed that malnutrition in dams prenatally or postnatally led to a general reduction in pancreatic enzymes of the offspring. Pancreatic enzymes were lower in the group of pups born to normal mothers but nursed by malnourished dams than in those born to and nursed by malnourished dams.[51] These findings suggest possible adaptation by pups malnourished *in utero* to survive under those conditions. The nature and mechanism of these adaptive changes are unknown. It is anticipated that experiments designed along these lines would provide new approaches to evaluate the modulating role of malnutrition in pancreatic ontogeny and the design of interventional strategies for proper nutritional rehabilitation.

Malnutrition, when occurring at certain periods of development, has more serious and detrimental effects on ontogeny. This observation is in agreement with the concept of "critical periods" in development, which hypothesizes that the developmental process can be divided into well-defined stages, some of which are more important than others. Changes occurring during these periods determine subsequent outcome.[52] It has been suggested that malnutrition occurring during a "critical period" would be more harmful, long-lasting, and less likely to be reversible than at other times. Since this concept is difficult to prove experimentally it remains hypothetical and caution should be applied when it is used to explain the irreversibility of certain detrimental effects of malnutrition.

c. The Role of Hormones on Pancreatic Development

Regulation of growth and development of the pancreas is dependent on general metabolic hormones such as insulin, growth hormone, thyroxine, and the glucocorticoids and also by a host of factors brought into play by the ingestion of food, notably the gastrointestinal peptides which act as hormonal, neurocrine, or paracrine agents affecting the growth of the pancreas.

i. Nongastrointestinal Hormones

In most mammalian species, several hormones undergo developmental changes which precede the changes in enzyme levels in the pancreas. These hormones, such as glucocorticoids and thyroxine, have been considered to have important roles in the control of exocrine pancreatic development.

The glucocorticoids seem to play a pivotal role in the development of the gastrointestinal tract in general, and the pancreas in particular. Many studies have supported the importance of glucocorticoids as a modulator for pancreatic development. A temporal relationship has been shown to exist between the increase in pancreatic amylase activity and serum corticosterone (the active glucocorticoid in rats) level during postnatal development of the rat.[53,54] A close relationship was found to exist between the changing pattern of the pancreatic digestive enzymes in the developing pup and serum glucocorticoid levels. Pharmacological doses of steroids stimulate the precocious development of the pancreas.[33,55,56] Hydrocortisone, when injected into pregnant rats, induces premature maturation of fetal pancreas.[57] Glucocorticoid administration accelerates pancreatic differentiation in suckling mice and prematurely induces pancreatic amylase in suckling rats.[58] We have also demonstrated that it causes premature maturation of cholecystokinin receptors in neonatal rats.[33]

Suppression of corticosterone production in rats either chronically by adrenalectomy[59] or transiently by treatment with an inhibitor of steroidogenesis, aminoglutethimide,[60] delayed the developmental accumulation of pancreatic exocrine enzymes. These effects are reversed by hydrocortisone treatment. Receptors for the glucocorticoids have been demonstrated in the pancreas. Although specific binding for dexamethasone was demonstrable in pancreatic cytosol of newborn rat pups, total binding was very low when compared to other age groups. The glucocorticoid binding capacity gradually increased postnatally, but remained relatively low before weaning.[53] There was no significant difference in glucocorticoid binding capacity between 10-d-old and newborn rats. A significant rise was not seen until after the 15th day of life. Binding capacity was greatest at weaning (from the 3rd to 5th weeks after birth) with a peak at day 25. After this period, receptor binding activity declined rapidly for several weeks and then more gradually toward the adult level, which remained higher than that of newborns (Figure 3). It is apparent from this study that the level of glucocorticoid receptors correlates well with the accumulation of pancreatic enzymes.

The dependence of pancreatic development on glucocorticoids is not absolute. Complete abolition of corticosterone by adrenalectomy delayed but did not abolish the developmental accumulation of pancreatic enzymes.[59] Therefore, pancreatic development is primarily under genetic control and glucocorticoid seems to act as a modulator and not the primary effector. The

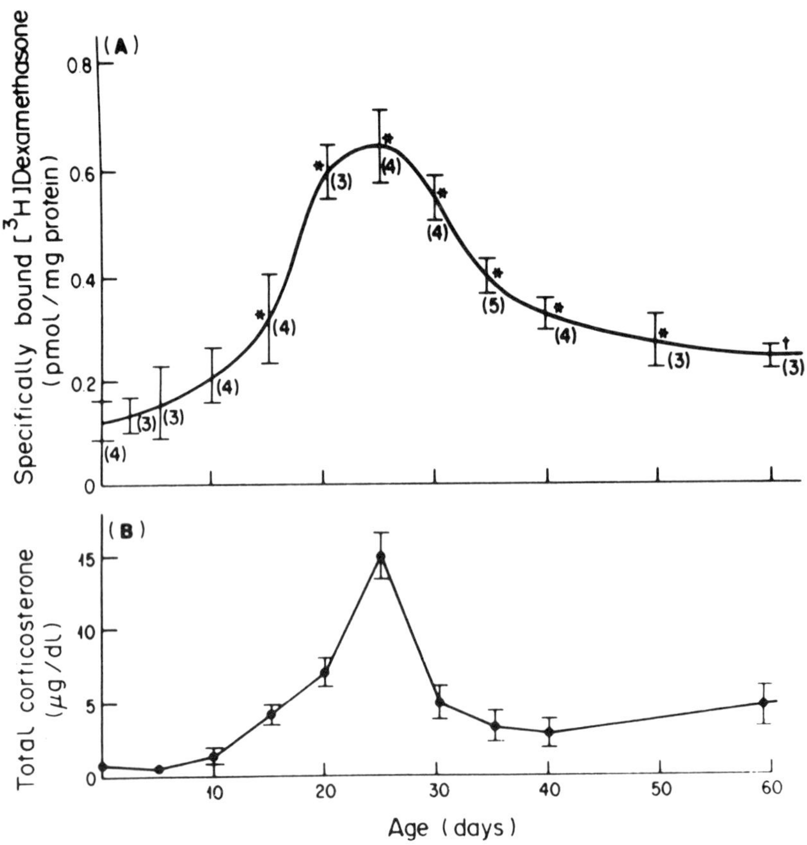

FIGURE 3. Developmental changes of glucocorticoid receptor binding activity in serum total corticosterone. (A) Ontogeny of glucocorticoid receptor binding activity in pancreatic cytosol. (B) Developmental patterns for total corticosterone in serum from rats. (From Lu, R. B., Lebenthal, E., and Lee, P. C., *J. Steroid Biochem.*, 26, 213, 1987. With permission.)

mechanism of the action of glucocorticoid at the molecular level is not known. Little is known about the complicated interaction of glucorticoids with other hormones.

Thyroxine is another important hormone in the growth and development of the pancreas. In rats, serum levels of thyroxine normally begin to increase from the second to the third postnatal week.[35,61-63] A peak of circulating thyroxine is detected at 14 to 16 d of age. Temporally, the rise in thyroxine level in the serum precedes that of the exocrine enzyme accumulation suggesting a possible cause and effect relationship. Thyroidectomy of suckling rats resulted in a delay in pancreatic amylase development. Moreover, administration of thyroxine to suckling rats caused precocious accumulation of pancreatic enzymes while treatment with propylthiouracil, an agent which suppresses thyroxine synthesis, suppressed enzyme accumulation.[59]

The mode of action of thyroxine on pancreatic enzyme development is not well understood. The relationship of the effect of thyroxine and that of corticosterone on the exocrine pancreas is possibly complex. Individually, each hormone has an effect on pancreatic development and they have been shown to have synergistic effects on lung development[64] and on growth hormone synthesis in pituitary cell lines.[65] Thyroxine has been shown to lead to an increase in plasma corticosterone level.[59,66-68] Therefore, it has been suggested that the action of thyroxine may be via a stimulation of adrenal function. This leads to an increase in circulating level of corticosterone which in turn is responsible for the developmental changes. However, in adrenalectomized rat pups which had a low level of pancreatic enzymes, injection of either corticosterone or thyroxine resulted in normal enzyme levels. Moreover, injection of both thyroxine and corticosterone to these pups resulted in an increase of pancreatic enzyme activities higher than those caused by either hormone alone.[59] Therefore, thyroxine has both a direct and indirect role in the modulation of the development of the exocrine pancreas.

Recent studies on steroid receptors revealed that analyses of these receptors are important for understanding molecular details of transcriptional control, as well as providing insight as to how an individual transacting factor contributes to cell identity and function. These studies have led to the identification of a superfamily of regulatory proteins that include receptors for thyroid hormone and the vertebrate morphogen retinoic acid. It is likely that these receptor-related molecules present in a wide range of species are responsible for the regulation of morphogenesis and homeostasis in these species.[69]

ii. Gastrointestinal Hormones[70]

Gastrointestinal hormones such as cholecystokinin (CCK), gastrin, secretin, and bombesin have been shown to have trophic effects on the adult pancreas *in vivo*[71-77] and *in vitro*.[78] The roles of these hormones in the development of the gastrointestinal tract have not been well defined. Gastrin and cholecystokinin administered in doses which could elicit a trophic response in adult animals had no action on the development of the pancreas in suckling animals.[72,79] The CCK analogue, caerulein, also exhibited different effects on pancreatic enzymes and DNA and RNA contents in suckling rats as compared to adults.[75] While caerulein led to both hyperplasia and hypertrophy when given to adult rats, only hyperplasia but no hypertrophy was seen when caerulein was given to suckling rats. Secretin, however, produced both hyperplasia and hypertrophy in the suckling as well as adult rats.

Bombesin has a trophic effect on the neonatal pancreas.[80] Administration of bombesin to 1-week-old rats for 6 d resulted in an increase in pancreatic weight due mainly to hypertrophy of acinar cells. Individual exocrine enzymes were affected differently: (1) chymotrypsinogen and trypsinogen increased, (2) lipase and colipase remained the same, and (3) amylase level decreased. The role of this peptide on development is not clear and further studies on its action as well as those of other gastrointestinal hormones should be performed in order to elucidate the role of these important regulatory substances.

Various other factors with growth promoting potentials have also been shown to play potential modulating roles in pancreatic development. Among these, EGF (epidermal growth factor), due to the attention it received in the study of tumorogenesis and its relation to the erb-B oncogene, has been extensively studied. EGF has been shown to result in a threefold increase in thymidine incorporation into cultured mouse pancreatic cells.[78] Moreover, mouse pancreatic acini have specific receptors for EGF[81] and receptors have been found to be present in fetal and neonatal rat pancreata.[82] Very little is known about the effect of EGF on the ontogeny of the pancreas. One study showed that both intraperitoneal and orogastric EGF increased pancreatic weight.[83] When given to suckling rats, orogastric EGF had no efffect on pancreatic amylase levels, while intraperitoneally administered EGF induced precocious maturation of pancreatic amylase. It seems that the effect of EGF on pancreatic development varies with the route of administration. Addition of EGF had no effects on the development of term-fetal pancreatic explants *in vitro*, while CCK and dexamethasone did, suggesting that in the perinatal stage, EGF does not have any direct effect on the pancreas.[84]

Insulin is also a very important growth promoting hormone. In humans, insulin deficiency is accompanied by abnormalities in the exocrine pancreas.[85-91] In adult rats, insulin deficiency induced by streptozotocin and alloxan treatment resulted in drastic decreases in pancreatic amylase concentrations[92-94] and reduced sensitivity to CCK stimulation of amylase release.[92] Administration of insulin completely normalized these changes.[95] Insulin also appears to be indispensable for the maintenance of rat pancreatic acinar cells or fetal pancreata in culture.[76,78] However, the role of insulin in exocrine pancreatic development has only been examined to a limited extent. In one study, it was shown that streptozotocin-induced diabetes in rat pups resulted in a transient lowering of amylase activity which was reversed by supplementation with insulin.[96] The results suggest that disturbances in glucose homeostasis in the early neonatal period have an effect on subsequent development of the exocrine pancreas. The requirement of a functioning insulin-acinar axis during pancreatic development has been further examined by using fetal pancreatic explants cultured in a serum-free medium.[84] Results indicated that although the addition of insulin to the defined medium appeared to have no effect on tissue amylase concentrations, initial exposure of pancreatic explants to streptozotozin did lead to a lowering of amylase concentration which could be prevented by exogenous supplementation with insulin. These results suggest that insulin has a role in prenatal amylase synthesis by the acinar cells.

Information regarding the potential role of these various factors in human pancreatic development is not available. Further studies using human explant materials are required to elucidate the relative roles of these factors and to confirm the findings in animals.

IV. GENE EXPRESSION DURING PANCREATIC DEVELOPMENT

Functional development of the exocrine pancreas depends on the availability of specific proteins which in turn is dependent on the regulation of the expression of the genes of various proteins. Enhancer domains related to genes for pancreatic enzymes have been demonstrated.[97,98] It is hoped that further identification of the interaction of the various endogenous and exogenous factors with the enhancer domains would result in an understanding of the molecular mechanisms involved.

V. SUMMARY

The development of exocrine pancreatic function as described in this chapter indicates a series of molecular, cellular, and physiologic processes which are well-coordinated by various endogenous and exogenous factors. The pancreas starts as an evagination of the foregut. Subsequent development involves the differentiation into endocrine and exocrine tissues early in fetal life. Morphogenesis and cytodifferentiation which occur simultaneously *in utero* result in a pancreas which resembles the adult pancreas histologically and ultrastructurally by term. Dramatic changes in enzyme patterns occur around weaning which may be due to changes in dietary composition and hormonal effects. Adult levels of secretory enzymes are acquired shortly after weaning. The various factors which play important roles in postnatal modulation of pancreatic exocrine enzymes include: glucocorticoids, thyroxine, secretin, CCK, gastrin, bombesin, EGF, insulin, and the pituitary hormones. The exact relationship of the various factors which play important roles in pancreatic development remains to be identified. Further studies using recently described techniques in biochemistry, physiology, and molecular biology would shed light on the mechanistic issues involved in the development of the pancreas.

REFERENCES

1. **Pictet, R. L., Clark, W. R., Williams, R. H., and Rutter, W.,** An ultrastructural analysis of the developing embryonic pancreas, *Dev. Biol.*, 29, 436, 1972.
2. **Pictet, R. L. and Rutter, W. J.,** Development of the embryonic endocrine pancreas, in *Handbook of Physiology, Section 7: Endocrinology*, Vol. 1, Steiner, D. F. and Freinke, N., Eds., American Physiological Society, Washington, D.C., 1980, 25.

3. **Moore, K. L.,** *The Developing Human,* W.B. Saunders, Philadelphia, PA,1982, 234.
4. **Laitio, M., Lev, R., and Orlic, D.,** The developing human fetal pancreas: an ultrastructural and histochemical study with special reference to exocrine cells, *J. Anat.,* 117, 619, 1974.
5. **Lebenthal, E., Lev, R., and Lee, P. C.,** Perinatal development of the exocrine pancreas, in *Textbook of Gastroenterology and Nutrition in Infancy*, Vol. 1, 1st ed., Lebenthal, E., Ed., Raven Press, New York, 1981, 149.
6. **Lebenthal, E. and Lee, P. C.,** Review article. Interactions of determinants in the ontogeny of the gastrointestinal tract: a unified concept, *Pediatr. Res.*, 102, 1, 1983.
7. **Van Nest, G. A., MacDonald, R. J., Raman, R. K., and Rutter, W. J.,** Protein synthesized and secreted during rat pancreatic development, *J. Cell Biol.,* 86, 784, 1980.
8. **Maylie-Pfenninger, M. F. and Jamieson, J. D.,** Distribution of cell surface saccharides on pancreatic cells. I and II, *J. Cell Biol.*, 80, 69, 1979.
9. **Maylie-Pfenninger, M. F., Doyle, C. M., and Jamieson, J. D.,** Differential lectin receptor appearance during the development of the pancreas, *J. Cell Biol.*, 75, 67a, 1977.
10. **Spooner, B. S., Cohne, H. I., and Faubion, J.,** Development of the mammalian embryonic pancreas: the relationship between morphogenesis and cytodifferentiation, *Dev. Biol.*, 61, 119, 1977.
11. **Parsa, I., Marsh, W. H., and Fitzgerald, P. J.,** Pancreas acinar cell differentiation. I. Morphologic and enzymatic comparisons of embryonic rat pancreas and pancreatic analage grown in organ culture, *Am. J. Pathol.*, 57, 457, 1969.
12. **Uchiyama, Y. and Watanabe, M.,** A morphometric study of developing pancreatic acinar cells of rats during prenatal life, *Cell Tissue Res.*, 237, 117, 1984.
13. **Parsa, I., Marsh, W. H., and Fitzgerald, P. J.,** Pancreas acinar cell differentiation. II. Comparison of DNA and protein synthesis of the embryonic pancreas and the pancreatic anlage grown in organ culture, *Am. J. Pathol.*, 57, 489, 1969.
14. **Golosow, N. and Grobstein, C.,** Epithelio-mesenchymal interaction in pancratic morphogenesis, *Dev. Biol.*, 4, 242, 1962.
15. **Ronzio, R. A. and Rutter, W. J.,** Effects of a partially purified factor from chick embryos on macromolecular synthesis of embryonic pancreatic epithelia, *Dev. Biol.*, 30, 307, 1973.
16. **Pictet, R. L., Rall, L., deGasparo, M., and Rutter, W. J.,** Regulation of differentiation of endocrine cells during pancreatic development *in vitro*, in *Early Diabetes in Early Life*, Camerini-Davalos, R. A. and Cole, H. S., Eds., Academic Press, New York, 1975, 25.
17. **Rutter, W. J., Kemp, J. D., Bradshaw, W. S., Clark, W. R., Ronzio, R. A., and Sanders, T. G.,** Regulation of specific protein synthesis in cytodifferentiation, *J. Cell Physiol.*, 72(Suppl. 1), 1, 1968.
18. **Sanders, T. G. and Rutter, W. J.,** The developmental regulation of amylolytic and proteolytic enzymes in the embryonic pancreas, *J. Biol. Chem.*, 249, 3500, 1974.
19. **Enesco, N. and Leblond, C. P.,** Increase in cell number as a factor in the growth of the organs and tissues of the young male rats, *J. Embryol. Exp. Morphol.*, 10, 530, 1962.
20. **Kachar, B., Taga, R., Kniebel, G. A., and Sessa, A.,** Morphometric evaluation of the number of exocrine pancreatic cells during early postnatal growth in the rat, *Acta Anat. (Basel)*, 103, 11, 1979.
21. **Borgstrom, B., Lindquist, B., and Lundh, G.,** Enzyme concentration and absorption of protein and glucose in duodenum of premature infants, *Am. J. Dis. Child.*, 99, 338, 1960.
22. **Lebenthal, E. and Lee, P. C.,** Development of functional response in human exocrine pancreas, *Pediatrics*, 66, 556, 1980.
23. **Delachaume-Salem, E. and Sarles, H.,** Evolution en fonction de l'age de la secretion pancreatique humaine normale, *Biol. Gastroenterol.*, 11, 135, 1970.
24. **Antonwicz, I. and Lebenthal, E.,** Developmental pattern of small intestinal enterokinase and disaccharidase activities in the human fetus, *Gastroenterology,* 72, 1299, 1977.
25. **Lester, R., Smallwood, R. A., Little, J. M., Brown, A. S., Piasecki, G. J., and Jackson, B. T.,** Fetal bile salt metabolism. The intestinal absorption of bile salts, *J. Clin. Invest.*, 59, 1009, 1977.
26. **Watkins, J. B.,** Mechanisms of fat absorption and the development of gastrointestinal function, *Pediatr. Clin. North Am.*, 22, 721, 1975.
27. **Ogita, S., Noma, H., Kawamura, T., Shimura, K., Ohnishi, M., Ishiko, O., and Sugawa, T.,** Differentiation of amylase isozyme in amniotic fluid with fetal age, *Am. J. Ob. Gyn.*, 131, 63, 1978.
28. **Lebenthal, E. and Lee, P. C.,** Glucoamylase and disaccharidase activities in normal subjects and in patients with mucosal injury of the small intestine, *J. Pediatr.*, 97, 389, 1980.
29. **Doyle, C. M. and Jamieson, J. D.,** Development of secretagogue response in rat pancreatic acinar cells, *Dev. Biol.*, 65, 11, 1978.
30. **Leung, Y. K., Lee, P. C., and Lebenthal, E.,** Maturation of cholecystokinin receptors in pancreatic acini of rats, *Am. J. Physiol.*, 250, G594, 1986.
31. **Lebenthal, E. and Lee, P. C.,** Development of functional response in human exocrine pancreas, *Pediatrics*, 66, 556, 1980.

32. **Williams, J. A. and Hootman, S. R.,** Stimulus-secretion coupling in pancreatic acinar-cells, in *The Exocrine Pancreas: Biology, Pathobiology and Diseases*, Go, V. L. W., Brooks, F. P., Dimagno, E. P., Gardner, J. D., Lebenthal, E., and Scheele, G., Eds., Raven Press, New York, 123, 1986.
33. **Leung, Y. K., Jirapinyo, P., Lebenthal, E., and Lee, P. C.,** Effect of hydrocortisone on the maturation of cholecystokinin binding and CCK stimulated amylase release in pancreatic acini of neonatal rats, *Pancreas*, 2, 73, 1987.
34. **Dumont, Y., Larose, L., Poirer, G. G., and Morisset, J.,** Modulation of pancreatic muscarinic receptors by weaning, *Life Sci.*, 30, 253, 1982.
35. **Dussault, J. H. and Labrie, F.,** Development of the hypothalamic-pituitary-thyroid axis in the neonatal rat, *Endocrinology*, 97, 1321, 1975.
36. **Chang, A. and Jamieson, J.,** Stimulus-secretion coupling in the developing exocrine pancreas: secretory responsiveness to cholecystokinin, *J. Cell Biol.*, 103, 2353, 1986.
37. **Shimizu, K., Lebenthal, E., and Lee, P. C.,** Immature stimulus-secretion coupling in the developing pancreas: ontogenic changes of protein kinase C, *Biochim. Biophys. Acta*, 968, 186, 1988.
38. **Gorelick, F. S., Chang, A., and Jamieson, D.,** Calcium-calmodulin stimulated protein kinase in developing pancreas, *Am. J. Physiol.*, 253, G469, 1987.
39. **Deschodt-Lanckman, M., Robberecht, P., Camus, J., and Christophe, J.,** Short term adaptation of pancreatic hydrolases to nutritional and physiological stimuli in adult rats, *Biochimie*, 53, 789, 1971.
40. **Deschodt-Lanckman, M., Robberecht, P., Camus, J., Baya, C., and Christophe, J.,** Hormonal and dietary adaptation of rat pancreatic hydrolases before and after weaning, *Am. J. Physiol.*, 226, 39, 1974.
41. **Snook, J. T.,** Dietary regulation of pancreatic enzyme synthesis. Secretion and inactivation in the rat, *J. Nutr.*, 87, 297, 1965.
42. **Snook, J. T.,** Effect of diet on development of exocrine pancreas of the neonatal rats, *Am. J. Physiol.*, 221, 1388, 1971.
43. **Dumont, Y., Larose, L., Poirer, G. G., and Morisset, J.,** Maturation of mascarinic agonist receptors in rat developing pancreas and its relation to maximal enzyme secretion, *Life Sci.*, 29, 2771, 1981.
44. **Lee, P. C., Kim, O., and Lebenthal, E.,** Effect of early weaning and prolonged nursing on development of the rat pancreases, *Pediatr. Res.*, 16, 470, 1982.
45. **Zoppi, G., Andreotti, G., Pajno-Ferrara, F., Njai, D. M., and Gaburro, D.,** Exocrine pancreas function in premature and full-term neonates, *Pediatr. Res.*, 6, 880, 1972.
46. **Lebenthal, E., Choi, T. S., and Lee, P. C.,** The development of pancreatic function in premature infants after milk-based and soy-based formulas, *Pediatr. Res.*, 15, 1240, 1981.
47. **Pitchumoni, C. S., Schelle, G., Lee, P. C., and Lebenthal, E.,** Effects of nutrition on the exocrine pancreas, in *The Exocrine Pancreas: Biology, Pathobiology and Diseases*, Go, V. W. L. et al, Eds., Raven Press, New York, 1986, 387.
48. **Klotz, A. P., Murdock, A. L., and Svoboda, D. J.,** The effect of protein deprivation on pancreatic function in young animals and in animals in utero, *Dig. Dis.*, 17, 399, 1972.
49. **Hatch, T. F., Lebenthal, E., Drasner, H., and Branski, D.,** Effect of postnatal malnutrition in pancreatic zymogen enzymes in the rat, *Am. J. Clin. Nutr.*, 32, 1224, 1979.
50. **Rossi, T. M., Lee, P. C., and Lebenthal, E.,** Effect of feeding regimens on the functional recovery of pancreatic enzymes in postnatally malnourished weanling rats, *Pediatr. Res.*, 17, 806, 1983.
51. **Young, C. M., Lee, P. C., and Lebenthal, E.,** Maternal dietary restriction during pregnancy and lactation: effect on digestive organ development in suckling rats, *Am. J. Clin. Nutr.*, 46, 36, 1987.
52. **Lebenthal, E. and Lee, P. C.,** Interaction of determinants in the ontogeny of the gastrointestinal tract: a unified concept, *Pediatr. Res.*, 17, 19, 1983.
53. **Lu, R. B., Lebenthal, E., and Lee, P. C.,** Developmental changes of glucocorticoids receptors in the rat pancreas, *J. Steroid Biochem.*, 26, 213, 1987.
54. **Takeuchi, T., Ogawa, M., Furihata, C., et al.,** Perinatal changes in amylase and serum corticosterone levels in rats, *Biochim. Biophys. Acta*, 497, 657, 1977.
55. **Lee, P. C. and Lebenthal, E.,** Role of corticoids independent of food intake in premature increase in pancreatic enzyme activities following early weaning in rats, *J. Nutr.*, 113, 1381, 1983.
56. **Morisset, J. and Jolicour, L.,** Effect of hydrocortisone on pancreatic growth in rats, *Am. J. Physiol.*, 239, G95, 1980.
57. **Werlin, S. L.,** Effects of hydrocortisone and cholecystokinin-octapeptide on neonatal rat pancreas, *J. Pediatr. Gastroenterol. Nutr.*, 1, 591, 1982.
58. **Takuma, T., Nakanishi, M., Takagi, Y., Tanemura, T., and Kumegawa, M.,** Precocious differentiation of mouse parotid glands and pancreas induced by hormones, *Biochim. Biophys. Acta*, 538, 376, 1978.
59. **Lu, R. B., Chaichanwatanakul, K., Lin, C. H., Lebenthal, E., and Lee, P. C.,** Thyroxine effect on exocrine pancreatic development in rats, *Am. J. Physiol.*, 254, G315, 1988.
60. **Lerner, A., Lee, P. C., and Lebenthal, E.,** Chemical adrenalectomy by aminoglutethimide and the pancreas in suckling rats, *Am. J. Physiol.*, 247, G346, 1984.

61. **Clos, J., Crepel, F., Legrand, C., et al.,** Thyroid physiology during the postnatal period in the rat. A study of the development of thyroid function and of the morphogenetic effects of thyroxine with special reference to cerebellar maturation, *Gen. Comp. Endocrinol.*, 23, 178, 1974.
62. **Partridge, N. C., Hoh, C. H., Weaver, P. K., and Oliver, I. T.,** Premature induction of glucokinase in the neonatal rat by thyroid hormone, *Eur. J. Biochem.*, 51, 49, 1975.
63. **Lee, P. C.,** Functional development of the exocrine pancreas, in *Textbook of Gastroenterology and Nutrition in Infancy*, 2nd ed., Lebenthal E., Ed., Raven Press, NY, 1989.
64. **Hitchcock, K. R.,** Hormones and the lung, *Anat. Res.*, 194, 15, 1979.
65. **Shapiro, L. E. and Sachchidananda, J.,** Regulation of proteins by thyroid hormone and glucocorticoid. The responses of hepatic 2-globulin and pituitary growth hormone differ in adult male hypothyroid rats, *Endocrinology*, 111, 653, 1982.
66. **D'Agostino, J. and Henning, S. J.,** Role of thyroxine in co-ordinate control of corticosterone and CBG in postnatal development, *Am. J. Physiol.*, 242, E33, 1982.
67. **Malinowska, K. W., Chan, W. S., Nathaniels, P. W., and Hardy, R. N.,** Plasma adrenocorticosteroid changes during thyroxine-induced accelerated maturation of neonatal rat intestine, *Experientia*, 30, 61, 1974.
68. **Tseng, C. C. and Johnson, J. R.,** Role of thyroxine in functional gastric development, *Am. J. Physiol.*, 251, G111, 1986.
69. **Evans, R. M.,** The steroid and thyroid hormone receptor superfamily, *Science*, 240, 889, 1988.
70. **Leung, Y. K. and Lebenthal, E.,** Gastrointestinal peptides — physiology, ontogeny and clinical significance, in *Textbook of Gastroenterology and Nutrition in Infancy*, 2nd ed., Lebenthal E., Ed., Raven Press, New York, 1989.
71. **Barrowman, J. A. and Mayston, P. D.,** The trophic influence of cholecystokinin on the rat pancreas, *J. Physiol. (London)*, 238, 73, 1974.
72. **Brants, E. and Morisset, J.,** Trophic effects of cholecystokinin-pancreozymin on pancreatic acinar cells from rats of different ages, *Proc. Soc. Exp. Biol. Med.*, 153, 523, 1976.
73. **Johnson, L. R.,** New aspects of the trophic action of gastrointestinal hormones, *Gastroenterology*, 72, 788, 1977.
74. **Lhoste, E., Aprahamian, M., Pousse, A., and Stock-Damge, C.,** Trophic effect of bombesin on the rat pancreas. Lack of inhibition by antrectomy or proglumide-treatment, in *Regulatory Peptides in Digestive, Nervous and Endocrine Systems*, Lewin, M. J. M. and Bonfils, S., Eds., Elsevier, Amsterdam, 1985, 417.
75. **Morisset, J.,** Stimulation of pancreatic growth by secretion and caerulein in suckling rats, *Biomed. Res.*, 1, 405, 1980.
76. **Parsa, I. and Marsh, W. H.,** Long-term organ culture of embryonic rat pancreas in a chemically defined medium, *Am. J. Pathol.*, 82, 119, 1976.
77. **Peterson, H., Solomon, T., and Grossman, M. I.,** Effects of chronic pentagastrin, cholecystokinin, and secretin on pancreas of rats, *Am. J. Physiol.*, 234, E286, 1978.
78. **Logsdon, C. D.,** Stimulation of pancreatic acinar cell growth by CCK, EGF and insulin *in vitro*, *Am. J. Physiol.*, 251, G487, 1976.
79. **Zahavi, I., Kelly, J., and Gall, G.,** Role of gastrin and cholecystokinin in the ontogenic development of the gastrointestinal tract, *Biol. Neonate*, 45, 95, 1984.
80. **Lehy, T., Puccio, F., Chariot, J., and Labeille, D.,** Stimulating effect of bombesin on the growth of gastrointestinal tract and pancreas in suckling rats, *Gastroenterology*, 90, 1942, 1986.
81. **Logsdon, C. D. and Williams, J.,** Epidermal growth factor binding and biological effects on mouse pancreatic acini, *Gastroenterology*, 85, 339, 1983.
82. **Leung, Y. K., Elitsur, Y., Lee, P. C., and Lebenthal, E.,** Binding of EGF to pancreatic acini of fetal and postnatal rats and possible transfer of EGF through the placenta, *Gastroenterology*, 90, 1518, 1986.
83. **O'Loughlin, E. V., Chung, M., Hollenberg, M., Hayden, J., Zahavi, I., and Gall, G.,** Effect of epidermal growth factor in ontogeny of the gastrointestinal tract, *Am. J. Physiol.*, 249, G674, 1985.
84. **Sarles, J., Lee, P. C., and Lebenthal, E.,** Organ culture of fetal rat pancreas. Effect of CCK on amylase activity, *Pediartr. Res.*, 23, 539, 1988.
85. **Adrian, T. E., Barnes, A. J., and Bloom, S. R.,** Hypotrypsinaemia in diabetes mellitus, *Clin. Chim. Acta*, 97, 213, 1979.
86. **Bazin, C. and Lavau, M.,** Diet composition and insulin effect on amylase to lipase ratio in pancreases of diabetic rats, *Digestion*, 19, 386, 1979.
87. **Chey, W. Y., Shay, H., and Shuman, C. R.,** External pancreatic secretion in diabetes mellitus, *Ann. Intern. Med.*, 59, 812, 1963.
88. **Dandona, P., Elias, E., and Beckett, A. G.,** Serum trypsin concentration in diabetes mellitus, *Br. Med. J.*, 2, 1125, 1978.
89. **Domschke, W., Tympner, F., Domschke, S., and Demling, L.,** Exocrine pancreatic function in juvenile diabetics, *Am. J. Dig. Dis.*, 20, 309, 1975.

90. **Frier, B. M., Saunders, J. H. B., Wormsley, K. G., and Bouchier, I. A. D.,** Exocrine pancreatic function in juvenile-onset diabetes mellitus, *Gut*, 17, 685, 1976.
91. **Moffat, A., Marks, V., and Gamble, D. R.,** Serum immunoreactive trypsin concentrations in diabetic children, *J. Clin. Pathol.*, 33, 871, 1980.
92. **Otsuki, M. and Williams, J. A.,** Effect of diabetes mellitus on the regulation of enzyme secretion by isolated rat pancreatic acini, *J. Clin. Invest.*, 70, 148, 1982.
93. **Palla, J. C., Ben Abdeljlil, A., and Desnuelle, P.,** Action de l'insuline sur la biosynthese de l'amylase et de quelques autres enzymes du pancreas de rat, *Biochim. Biophys. Acta*, 158, 25, 1965.
94. **Soling, H. D. and Unger, K. O.,** The role of insulin in the regulation of alpha-amylase synthesis in the rat pancreas, *Eur. J. Clin. Invest.*, 2, 199, 1972.
95. **Okabayashi, Y., Ohki, A., Sakamoto, C., and Otsuki, M.,** Relationship between the severity of diabetes mellitus and pancreatic hormone dysfunction in rats, *Diabetes Res. Clin. Pract.*, 1, 21, 1985.
96. **Lee, P. C., Brooks, S., and Lebenthal, E.,** Streptozotozin effect on the development of the exocrine pancreas in neonate rats, *Diabetes Res. Clin. Pract.*, 2, 269, 1986.
97. **Swift, G., Hammer, R., MacDonald, R., and Brinster, R.,** Tissue specific expression of the rat pancreatic elastase 1 gene in transgenic mice, *Cell*, 38, 639, 1984.
98. **Walker, M., Edlund, T., Bonlet, A., and Rutter, W.,** Cell specific expression controlled by the 5′ flanking region of insulin and chymotrypsin gene, *Nature*, 306, 557, 1983.

Chapter 5

PANCREATIC ADAPTATION TO DIET — REGULATION OF SPECIFIC ENZYME SYNTHESIS

J. C. Dagorn and D. Giorgi

TABLE OF CONTENTS

I. INTRODUCTION

The exocrine pancreas provides the intestine with a mixture of enzymes capable of digesting the three kinds of nutritional substrates generally found in a normal diet: protein, carbohydrate, and lipid. Several mechanisms regulating the rate of pancreatic secretion and its enzyme composition have developed during evolution. These mechanisms enable the organism to adjust quantitatively and qualitatively the enzyme output to changes in the pattern of food intake and to changes in food composition. Quantitative adjustments (control of enzyme output) are exerted by hormones and neurotransmitters released during feeding . These messengers are also involved in the regulation of pancreatic growth (see Chapter 1). Qualitative adjustments of enzyme levels in secretion have been described following changes in the proportions of dietary constituents. This phenomenon has been called pancreatic adaptation to diet, since the changes in digestive enzyme proportions were directed towards an optimization of digestion. The purpose of this chapter is to review our present knowledge on pancreatic adaptation. Contributions which led to the understanding that adaptation is not a matter of independent secretion but rather a gradual phenomenon involving changes in the relative rates of enzyme synthesis will be summarized in the first part. Recent evidence that this is actually due to differential regulation of pancreatic gene expression will be reported in the second part. The third part will deal with the important question of the messengers transmitting information on food composition from the intestinal lumen to the pancreas, an area in which our knowledge is still limited. Pancreatic response to nonphysiologic alterations of diet, such as protein deprivation, malnutrition, or the ingestion of toxic substances, will not be reviewed.

II. PANCREATIC RESPONSE TO DIETARY CHANGES

A. DEMONSTRATION OF PANCREATIC ADAPTATION

At the turn of the century, Pavlov observed in the dog that a diet rich in carbohydrates stimulated the secretion of amylase, and a protein-rich diet the secretion of proteolytic enzymes.[1] This is the first report of the adaptive properties of the pancreas. Few details were provided, however, on the experimental conditions used. A fully documented report on rat pancreatic adaptation was given in 1943 by Grossman et al.[2] They fed animals with diets of controlled composition, assayed amylase and trypsin in homogenates, by established methods, and were therefore able to conclude that the pancreatic levels of enzymes were regulated in direct proportions to nutritional substances. Since then, such an adaptive phenomenon has been demonstrated in other species such as dog,[3] mouse,[4] pig,[5] chicken,[6] and snake.[7] The situation is less clear in humans, since the few studies which have been conducted led to conflicting results. Studies in infants[8] concluded that amylase and trypsin adaptated to the dietary content of starch and protein, whereas in the adult, long-term intraduodenal infusion of different nutrients did not affect the enzyme pattern in an adaptive manner.[9] It would be surprising if the mechanisms of pancreatic adaptation are totally absent in man. They may be more subtle than in other species, however, and this, added to the ethical constraints limiting nutritional manipulations in human beings, could explain why so little information is presently available.

B. KINETICS OF ADAPTATION

Further studies in the dog by Pavlov[1] led to the hypothesis that adaptation was immediate, as the secretion of each class of enzymes in pancreatic juice appeared to be regulated by the amount of the nutrient substrate of each enzyme present in the duodenum. This proposition was very attractive and had important implications regarding pancreatic secretory mechanisms. For example, secretion of the different classes of enzymes from their storage compartments in the pancreas had to be independently regulated. Yet, in the dog, Grossman et al.[2] observed that the enzyme composition of juice was always similar to that of pancreatic tissue, a result later

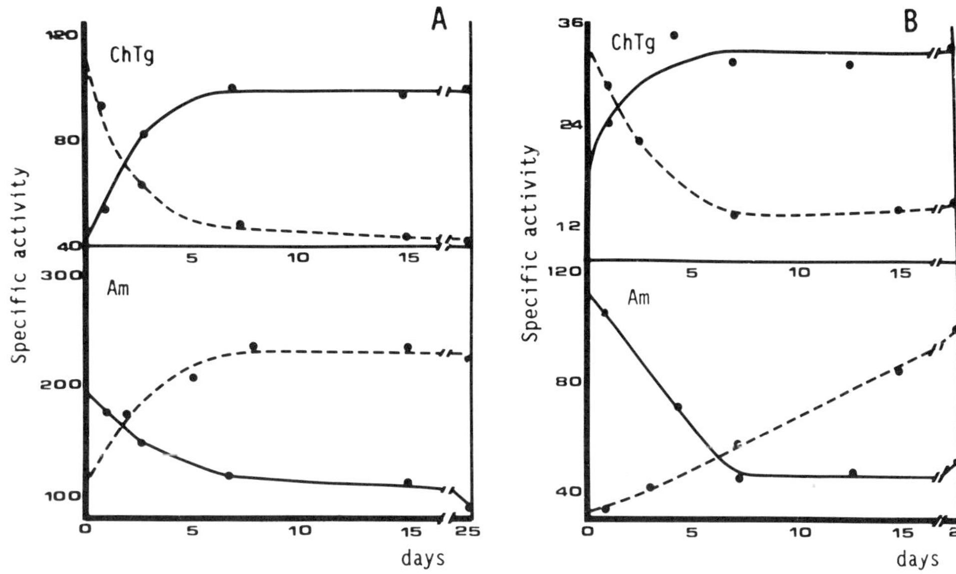

FIGURE 1. Kinetics of variations in amylase and chymotrypsinogen specific activities in rat juice (A) and pancreatic content (B), following a change from a carbohydrate rich (G: 75% starch, 25% casein) to a protein rich (P: 70% casein, 20% starch) diet (——) or from P to G (— —). (Adapted from Ben Abdeljlil, A. and Desnuelle, P., *Biochim. Biophys. Acta*, 181, 136, 1964. With permission.)

confirmed by many authors.[10-12] Information obtained from kinetic studies on adaptation was also difficult to reconcile with the hypothesis of immediate adaptation. Indeed, the groups of Desnuelle[13] and Christophe[14] demonstrated that adaptation of pancreatic enzyme content to a change in the protein to carbohydrate ratio in food took 5 to 10 d to be completed; they further observed that similar kinetics were obtained in secretion (Figure 1).

Meanwhile, many advances were being made in our understanding of the secretory process for enzymes and proteins, and numerous studies were using the pancreas as a model system for exocrine secretion. The intracellular pathway followed by newly synthesized exocrine enzymes was described by Palade and his colleagues.[15] It became widely accepted that all secretory proteins were segregated from the cytoplasm into membrane-bound vesicles from the moment of their synthesis until exocytosis. The molecular mechanisms of enzyme segregation were explored by Blobel, who demonstrated that secretory proteins, contrary to cytoplasmic proteins, were synthesized as precursors.[16] In the precursor molecule, an extra peptide, the signal peptide, is attached to the *N*-terminal end of the molecule and directs its cotranslational segregation into the cisternae of the rough endoplasmic reticulum.[17,18] Maturation of the precursor thus begins in the rough endoplasmic reticulum with the cleavage of the signal peptide and continues into the saccules of the Golgi network. Mature proteins are finally concentrated into the zymogen granules. Immunocytochemical methods have shown that all digestive enzymes were present in all zymogen granules,[19] albeit in different proportions.[20] Hence, the only possibility for independent secretion of enzymes would be that different areas of the pancreas store enzymes in very different proportions. This is apparently not the case, because the heterogeneities of enzyme distribution reported to date are either minor[21] or restricted to a very small region of the tissue.[22]

Evidence based on kinetic studies,[12] that pancreatic adaptation is progressive and not immediate, is consistent with our present understanding of the pancreatic secretory process. Intriguing results from Grendell et al. suggest, however, that short-term (within minutes) adaptive changes may be obtained when end products of digestion (such as certain amino acids, glucose, etc.) are injected into the celiac trunk of anesthetized rats.[23] For reasons mentioned above, the enzymes involved in such a rapid regulation must escape the general secretory pathway. It is conceivable that alternative pathways, like those first described by Kelly,[24] then by Arvan and Castle[25] in pancreas for constitutive secretion, take place in experimental conditions in which nonparallel secretion has also been demonstrated.[26] Their participation in the adaptative process appears nevertheless very limited, especially under physiological conditions.

Evidence that modifications in the relative rates of enzyme secretion were not involved in adaptation prompted investigators to look for mechanisms influencing enzyme production.

III. MECHANISMS INVOLVED IN PANCREATIC ADAPTATION

A. REGULATION OF ENZYME CONTENTS

Adaptive modifications observed in pancreatic juice reflect modifications in zymogen granule content (see above and Figure 1). Under physiological conditions, most enzymes follow the same secretory pathway, the last step of which is enzyme condensation in the zymogen granules. Hence, changes in the enzyme proportions of that compartment should reflect changes in the relative rates of enzyme synthesis.

Reboud et al.[27] tested in the rat whether adaptation to a protein-rich (P) or a carbohydrate-rich (C) diet induced variations in the rates of amylase and chymotrypsinogen biosynthesis after having labeled *in vivo* all pancreatic proteins with a radiolabeled amino acid. They purified the two enzymes mentioned above by column chromatography and measured the amount of radioactivity incorporated per minute in each polypeptide. From their data, they concluded that shifting from P to G diet induced an eightfold increase in amylase synthesis and a twofold decrease in chymotrypsinogen synthesis, strong evidence of an adaptative regulation of protein synthesis. The demonstration that changes in enzyme contents were the consequence of their regulation at the level of synthesis was provided by kinetic studies (Figure 2) where the animals were shifted from diet G to diet P, and subsequent changes in synthesis and content were monitored in parallel for 10 d.[28,29] Modifications in the relative rate of amylase and chymotrypsinogen synthesis were observed after 4 h, as 20% of the total change expected after full adaptation was already completed at that time. Therefore, adaptation of synthesis started almost immediately after the first meal of new composition was ingested. After 24 h, 60% of the full adaptation was attained and adaptation was maximal after 3 d. Modifications of enzyme ratios in pancreatic tissue followed a similar pattern, but at a slower pace, since 45% of total change was observed at 24 h with full adaptation reached only after 5 d. A similar slower adaptative process of tissue enzyme contents was observed when shifting from diet G to diet P. It is therefore highly probable that pancreatic adaptation occurs mostly by regulating the relative rates of synthesis of the different pancreatic enzymes. This has been confirmed by two other groups.[30,31]

B. NUTRITIONAL REGULATION OF PANCREATIC ENZYME SYNTHESIS

The synthesis of a specific protein is modulated either by the amount of its mRNA available for translation or by the rate at which ribosomes translate that mRNA. When studying the mechanisms regulating the synthesis of a protein, the first step is generally to look for changes in the concentrations of the corresponding mRNAs.

1. Estimation of mRNA Levels in Pancreas

Until recently, the only technique available to quantify a specific tissue mRNA was to purify total RNA and use it to direct protein synthesis in a cell-free translation system (rabbit reticulocyte lysate or wheat-germ extract) in the presence of radioactive amino acids.[32] The labeled polypeptides generated in that system were then resolved by SDS-polyacrylamide gel electrophoresis and their relative abundance measured by scanning the autoradiogram of the gel. Assuming that all mRNAs were translated with the same efficiency, the level of a given mRNA

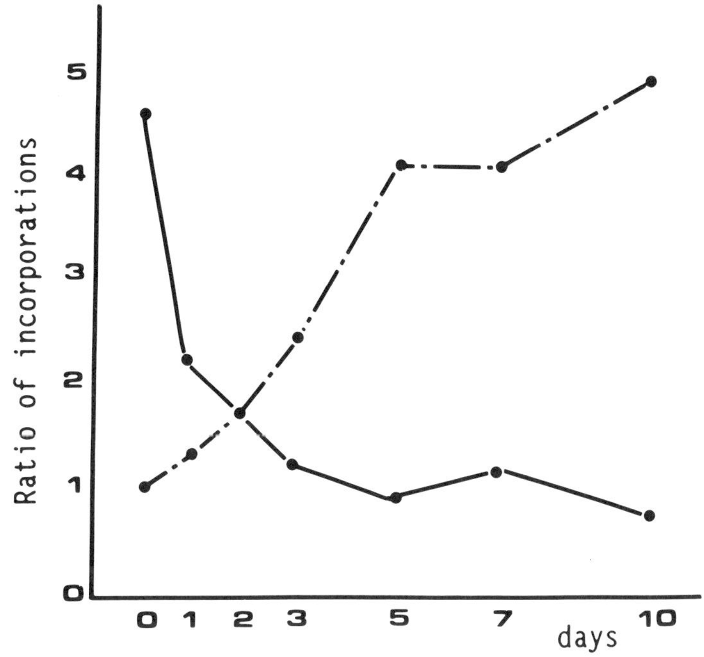

FIGURE 2. Kinetics of variations in the relative rate of amylase and chymotrypsinogen biosynthesis in rats, following a change from a G diet to a P diet (see legend of Figure 1) (——) or from P to G (— —). After protein labeling *in vivo* with ^{3}H-phenylalanine, pancreatic amylase and chymotrypsinogen were purified by isoelectric focusing on gels. ^{3}H-phenylalanine incorporation in the enzymes was measured by scintillation counting, and the ratio of those values was plotted against time. (From Lahaie, R. and Dagorn, J. C., *Biochim. Biophys. Acta*, 654, 119, 1981. With permission.)

was considered proportional to the amount of the related polypeptide synthetized in the cell-free system. That technique however presents several drawbacks, especially when applied to the pancreas. The RNA has to be totally free of degradation, since a single cut would generate a shortened polypeptide migrating in a different position on the gel. Good preparations of pancreatic RNA are sometimes difficult to obtain, due to the high content of RNAase in the gland.[33] Also, when several mRNAs encoding proteins of similar sizes are present in a tissue, it may be difficult to quantitate them independently, since the corresponding polypeptides synthesized *in vitro* would migrate together on polyacrylamide gel. This is the case for rat serine proteases which migrate in the 25 kDa region.[34]

Recent developments in recombinant DNA technology make possible more accurate estimations of mRNA concentrations by selective hybridization to cloned cDNA probes. In "dot-blot" assays, known amounts of total RNA are blotted onto nitrocellulose and incubated with an excess of a radiolabeled cDNA homologous to the mRNA to be quantitated. Following stringent washes, the amount of probe bound to the filter, which is proportional to the amount of the homologous mRNA in the blotted sample, is measured by quantitative autoradiography. The mRNAs of numerous pancreatic proteins from different species have been cloned in the past decade,[35-37] and their cDNAs have been used as probes in nutritional studies.

2. Nutritional Regulation of mRNA Levels

Three diets enriched in protein, carbohydrate, or lipids, defined by Desnuelle to study rat pancreatic enzyme adaptation in the juice and tissue,[12] and later to monitor adaptive variations

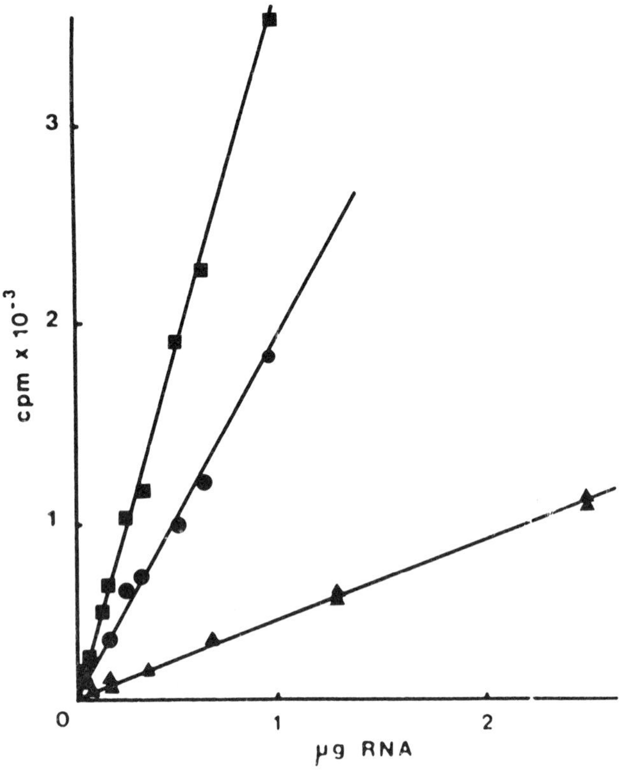

FIGURE 3. Amylase mRNA concentrations in the pancreas of rats adapted to diets containing respectively 11% (▲), 25% (●) and 75% (■) carbohydrate (w/w). mRNA concentrations were estimated by hybridization to an excess of ^{32}P-labeled amylase recombinant cDNA clone; equal amounts of total RNA prepared from the pancreas of the three groups of rats were "dot blotted" in sequential dilution onto nitrocellulose. After hybridization to the cDNA probe, the radioactivity borne by each spot was measured by scintillation counting. Regression lines for each diet were calculated by the least squares method. The slopes of the lines correspond to the amounts of labeled amylase probe hybridized per µg of total mRNA, i.e., to the concentration of amylase mRNA in the tissue. The results presented here show that amylase mRNA concentration was nine times higher in the pancreas of rats fed the 75% carbohydrate diet than in the 11% diet. (From Giorgi, D., Bernard, J. P., Lapointe, R., and Dagorn, J. C., *EMBO J.*, 3, 1521, 1984. With permission.)

in enzyme synthesis,[28,29] were also chosen to study further the mechanism of adaptation at the RNA level.[38,39] Hybridization of total RNA to an amylase cDNA probe indicated that the tissue concentration of amylase mRNA was nine times higher in the 75% starch diet than in its 11% counterpart (Figure 3). These data strongly support an adaptation at the mRNA level. Under such conditions, amylase synthesis was increased only 4.5-fold; this suggests that regulation of mRNA concentrations can account for changes observed in the rates of synthesis of specific enzymes and also that not all the mRNA present in the cells is actually engaged in protein synthesis. A fraction of it might be stored in an inactive form as ribonucleoproteins[40] and be available for use in case of urgent need. These experiments strongly suggested that in fully adapted animals, enzyme proportions are regulated at a pretranslational level. In contrast to carbohydrates, proteins are digested by a family of endo- and exoproteases. Concentrations of mRNAs encoding three serine proteases, trypsinogen 1, chymotrypsinogen B, and elastase I, and the exometalloprotease procarboxypeptidase A were also determined in similar diet-controlled experiments. The mRNA levels of the serine proteases increased in a coordinate fashion, in direct proportion to the amount of protein in the diet. The amplitude of the changes differed for each enzyme mRNA. That of chymotrypsinogen was increased the most, while no change occurred for procarboxypeptidase.[39] Similar observations were made with diets of different carbohydrate to protein ratio in experiments where relative rates of enzyme synthesis were measured *in vitro* and changes in mRNA concentrations estimated by cell-free translation of total RNA.[41] These findings indicate that nutritional regulation of the tissue levels of pancreatic enzymes is primarily mediated by changes in the content of active cytoplasmic mRNAs. The relative contribution to those changes of transcriptional regulation and modifications in the turnover rate of mRNAs is still unknown.

IV. EFFECTORS OF ADAPTATION IN RESPONSE TO ALTERED FOOD COMPOSITION

Evidence that the expression of genes encoding specific pancreatic proteins in the adaptation process is modulated by changes in food composition raises two questions. First, what are the effectors released from the intestine that carry the information on the nature of the ingested nutrients? Second, how can they or their secondary messengers eventually get into the pancreatic nuclei to interact with the genes and ensure proper regulation of their expression? Among potential effectors are the products of food digestion and the gastrointestinal hormones released into the circulation after a meal. A possible role in adaptation of the neurotransmitters involved in pancreatic secretory stimulation is less likely since complete adaptation to diet occurred in vagotomized animals.[42] End products of digestion are at first sight among the most plausible candidates, since they could conceivably regulate the production of their catalytically related enzymes by "simple" feed-back mechanisms.

A. END PRODUCTS OF FOOD DIGESTION AS PRIMARY EFFECTORS OF ADAPTATION

Most enzymes secreted by the pancreas are involved in the hydrolysis of oligosaccharides, lipids, or proteins. Oligosaccharides are degraded to absorbable monosaccharides by the complementary action of amylase and brush-border enzymes.[43] Triglycerides are hydrolyzed by the lipase-colipase complex to fatty acids and monoglycerides, absorbed as such, and later resynthesized by esterification.[44] Proteins are transformed by the combined action of endo- and exoproteases into amino acids and small peptides. After absorption, the remaining peptides are degraded by intestinal intracellular enzymes so that dietary protein enters portal blood in the form of free amino acids.[45]

1. Oligo- and Monosaccharides

Several studies have investigated the occurrence of amylase adaptation to carbohydrate-rich diets in which starch was replaced by monosaccharides.[2,12,14,46] Glucose, sucrose, and to a lesser extent fructose were potent effectors, but lactose and galactose were without effect. Other evidence that products of starch digestion, but not starch itself, nor effectors released by starch, are involved in amylase adaptation was provided by Fölsch et al.,[47] who fed rats a diet supplemented with an amylase inhibitor. In such conditions of total and constant inhibition of amylase present in the intestine, pancreatic amylase content decreased by 67% in 20 d. To test whether the presence of monosaccharides in the intestine was necessary to trigger adaptation, glucose was injected into the peritoneal cavity[48] or given by parenteral infusion (Table 1).[49] In both cases, adaptation occurred. Hence, circulating monosaccharides derived from dietary carbohydrates are effectors of pancreatic amylase adaptation.

TABLE 1
Influence of Intravenous Infusion of Products of Digestion on Pancreatic Enzyme Contents

Diets	Amylase	Lipase	Tg	ChTg
High fat				
Infusion of NaCl	35	122	5.0	23
Infusion of glucose	106*	63*	3.9	18
High starch				
Infusion of NaCl	187	65	5.4	21
Infusion of lipids	100*	92*	4.0	22
High starch				
Infusion of NaCl	190	55	4.3	22
Infusion of amino acids	170	54	5.2	26

Note: The effects of intravenous infusion of 30% glucose (1.2 ml/h), lipids (intralipids, 0.7 ml/h) and 20% amino acids (1 ml/h) on the pancreatic levels of amylase, lipase, and trypsinogen were studied in rats. Enzyme specific activities are expressed as units per 1 g proteins. Asterisks represent values significantly different from those in control animals (infusion of NaCl).

Adapted from Lavau, M., Bazin, R., and Herzog, J., *J. Nutr.*, 104, 1432, 1974.

2. Lipids

In the early reports on lipase adaptation to diet, Grossman et al.[2] and Reboud et al.[50] could not observe an increase in lipase in the pancreas of rats fed a high-fat diet. Later, however, Bucko and Kopec,[51] Snook,[46] and others[14,52] clearly demonstrated that pancreatic lipase content was dependent on the amount of lipids present in food, showing thereby that the adaptive process involved lipolytic enzymes as well. Two reasons may explain why the phenomenon was not immediately recognized: (1) the rather complicated relationship that exists between the amount and type of fat in the diet and lipase adaptation and (2) the unsuspected influence of colipase on lipase activity. Apparently conflicting results were reported on the effect of the type of dietary fat ingested and pancreatic lipase contents. Indeed, unsaturated fats were found to be better inducers than saturated fats;[53] others denied such an effect of saturation,[54] while emphasizing the importance of chain length, with long-chain triglycerides increasing pancreatic lipase contents more than medium-chain fatty acids.

Sabb et al.[55] recently provided clarification of this problem with the important finding that a threshold amount of dietary fat (47% kcal) was needed to induce lipase adaptation. Above that threshold amount, a small increase in fat (7% kcal) induced full lipase adaptation without any influence of chain length or saturation. Under that threshold, only highly unsaturated triglycerides could induce limited adaptation. That report is also one of the few dealing with lipase adaptation in which "true" lipase activity was monitored by excluding the influence of colipase. Colipase is a small peptide[56] that binds to lipase and restores its activity while inhibited by physiological amounts of bile salts. It also has a protective effect against lipase denaturation.[57] When measured by standard methods, lipase activity is maximal only when the colipase to lipase ratio is above one. Then, if in dietary manipulations intended to study lipase adaptation, lipase and colipase are not regulated in parallel, standard determination of lipase activity will not reflect actual changes in lipase concentration. Emergence of that problem prompted several groups to study dietary-induced variations in colipase activity. Adaptation to dietary lipid content was at first unclear since it could not be established by Vandermeers-Piret et al. in rat,[58] but was later reported by others in rat[59] and pig.[60] It turned out, however, that colipase expression in the pancreas was only partially related to the fat content of the diet, but depended mostly on its protein content.[54] These results emphasize that in studies dealing with lipase adaptation, lipase activity has to be assayed in the presence of an excess of exogenous colipase. Thus, care has to be taken when analyzing earlier reports on that matter and results from studies in which colipase concentration was controlled (see above) give a better view of the dietary stimuli of lipase adaptation.

There is nevertheless a consensus on the fact that a high-fat diet induces lipase accumulation in the pancreas. As done with carbohydrates, the next strategy was to bypass the intestine and see whether intravenous lipids would influence lipase contents. Lavau et al.[49] demonstrated a significant lipase increase in the pancreas of rats infused with soybean oil emulsified with lecithins and concluded that blood lipids or their metabolites were potent signals for lipase adaptation (Table 1). On the other hand, they also noticed that lipase increased in diabetic rats fed a balanced diet. Diabetes and a lipid-rich diet share the common feature of favoring lipid catabolism at the expense of glucose utilization.[61,62] In fact, acetoacetate and β-hydroxybutyrate are significantly augmented in blood of diabetic rats as in animals fed high-fat diets.[49] Bazin et al.[63] reproduced those conditions by infusing β-hydroxybutyrate in control animals and observed a subsequent increase in pancreatic lipase. It was then proposed that ketone bodies rather than blood lipids might be the major effectors of lipase adaptation. This might explain why long chain fatty acids are better inducers of lipase adaptation;[54] indeed, they will generate on a molar basis more β-hydroxybutyrate than medium-chain fatty acids.

3. Peptides and Amino Acids

With the exception of Snook,[64] all investigators who used either an amino acid mixture or caseine hydrolysate instead of dietary protein failed to demonstrate any adaptation of the pancreatic proteolytic enzymes in rats.[66,67] It therefore came as no surprise that giving amino acids either by intraperitoneal injection[48] or by intravenous infusion[49] did not influence pancreatic chymotrypsinogen content. Hence, opposite to carbohydrates and lipids, proteins have to be present in the intestinal lumen, either intact or partly hydrolyzed, to promote adaptation of proteases (Table 1). Therefore, messengers other than degradation products of proteins have to be considered, but they should be released from the gut by dietary proteins or their hydrolysis products.

B. HORMONES INVOLVED IN THE ADAPTATIVE PROCESS

Grossman et al.[65] already suspected in 1944 that hormones like insulin and the recently discovered pancreozymin could play a role in pancreatic enzyme adaptation. This suspicion was based on the observations that dextrose could induce amylase adaptation, implying that glycemia had something to do with the process and that amino acids could not induce protease adaptation. This suggested that a pancreatotropic factor had to be released from the gut. These premises have remained the same 45 years later, and our knowledge on the role of these hormones has improved but still remains incomplete.

1. Insulin

The expectation that insulin would be the unique modulator of amylase expression quickly disappeared when Grossman et al.,[65] then Palla et al.,[68] demonstrated that even if insulin could reverse the decrease in pancreatic amylase concentration seen in diabetes, it was unable to increase amylase content in normal animals. Similar findings were obtained by quantitating amylase mRNA in normal and diabetic tissue.[69] Yet, intravenous glucose, the level of which remains under insulin control, is a potent effector of amylase adaptation. Hence, insulin should interact in some way with the adaptative regulation of amylase. To test this hypothesis, one needs conditions in which insulinemia and glycemia can be experimentally dissociated. Such an approach was achieved by Dranginis et al.[70] who studied in mice the effects of long-term fasting

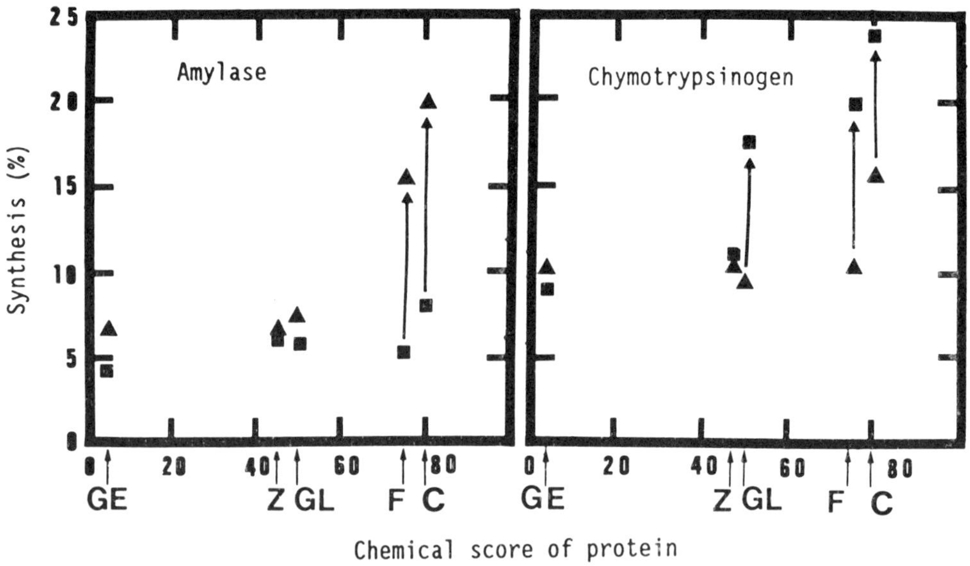

FIGURE 4. Effect of quality of dietary protein on synthesis of pancreatic enzymes. The squares represent average values of rats fed for 7 d diets containing 64% sucrose (19% protein). The triangles represent average values of rats fed for 7 d diets containing 64% protein. The protein used was either gelatin (GE), zein (Z), gluten (GL), fish protein (F), or casein (C). The regimen are ordered on the abscissa according to the chemical score of their protein moiety. Arrows between squares and triangles indicate significant difference in values. (Adapted from Johnson, A., Hurwitz, R., and Kretchmer, N., *J. Nutr.*, 107, 87, 1977. With permission.)

on pancreatic amylase content. Fasting animals have very low blood levels of insulin, comparable to that of diabetic animals, but they are hypoglycemic compared to hyperglycemic diabetic animals. Under these conditions, amylase contents were equally low in both normal and diabetic animals, demonstrating that amylase expression does not depend on glucose availability in the absence of adequate insulin supply. Bazin and Lavau[62] proposed that insulin acts indirectly by regulating pancreatic glucose metabolism. Diabetic rats were therefore given insulin in conjunction with low- or high-fat diets, and then compared to normal rats. Feeding the low-fat diet increased amylase content to control values while the high-fat diet, known to impair insulin action on glucose metabolism, kept amylase content at a very low level. Further work by these same authors using pancreatic acini *in vitro*[71] established that high-fat diet impaired the rate of glucose transport into acinar cells. These results, and those of Korc et al.[69] and Williams et al.,[72] demonstrate that insulin increases glucose transport into pancreatic acini, thereby suggesting that amylase expression may at least be in part controlled by intracellular levels of glucose or its metabolites. The observation that obese Zucker rats, that manifest insulin resistance and decreased glucose utilization, exhibit low pancreatic amylase content supports this hypothesis.[73] In these studies, increases in lipase contents were also reported, comparable to those found in diabetic rats[62] and in animals fed a high-fat diet.[52]

Such data led to speculations that modifications in the balance between glucose metabolism and lipid metabolism dictated by the physiological status of an animal may influence the amylase to lipase ratio in pancreas. This hypothesis could also be consistent with reports that decreased pancreatic amylase content following adrenalectomy can be reversed by glucocorticoids,[74] if one presumes that these hormones act by regulating blood glucose and stimulating insulin secretion. On the other hand, demonstration that potential glucocorticoid-responsive sequences are present in the promoter region of the 5′ flanking sequence of the amylase gene[75,76] and that dexamethasone increases content, synthesis, and mRNA level of amylase in the rat pancreatic acinar cell line AR4-2J[77] do not necessarily indicate a direct influence of the hormone, as

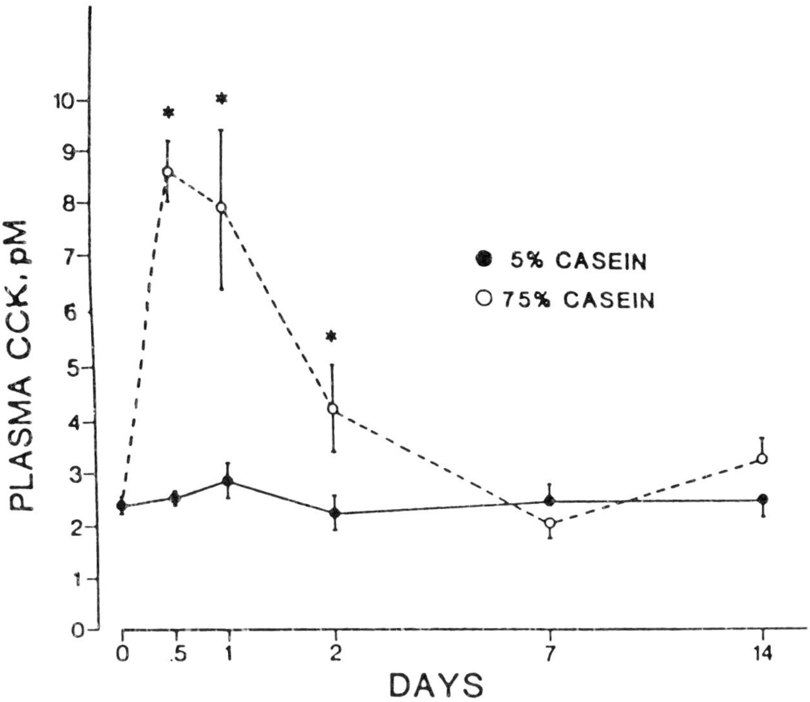

FIGURE 5. Changes in plasma cholecystokinin during adaptation of rats to a high-protein diet. Plasma CCK concentrations were measured by bioassay (ability to stimulate amylase release from pancreatic acini). Plasma CCK was expressed as CCK-PZ equivalents (pmol/l). Results are means ± SE; asterisks denote significant difference between 5% and 75% casein-fed rats at each period. (From Green, G. M., Levan, V. H., and Liddle, R. A., *Am. J. Physiol.*, 251, G70, 1986. With permission.)

suggested at first, since these effects occur relatively slowly and require induction of an unidentified protein.[78]

2. CCK and Related Peptides

As discussed earlier, the first indication that substances released by the gut play a role in the mechanism of adaptation came from the observation that intraluminal protein could not be replaced by intravenous amino acids in triggering adaptation to a protein-rich diet.[49] It has also been shown that amylase adaptation to a carbohydrate-rich diet requires a minimum supply of dietary protein to take place.[14] Johnson et al.[79] further investigated the influence of the protein moiety of carbohydrate-rich diets on stimulation of amylase synthesis (Figure 4). They observed that if rats, adapted to a protein-rich diet, were fed a carbohydrate-rich diet (64% sucrose and 19% protein) in which protein had a low chemical index (C.I. < 50, e.g., gelatin, zein, or gluten), no adaptation occurred. However, with fish protein (C.I. = 75) or casein (C.I. = 80), amylase synthesis increased as expected. The nature of the substance released from the gut by casein and not by proteins with low C.I. is not known. Evidence points to CCK as a possible candidate; Green et al.[80] monitored plasma CCK in the rat following a change from a low-protein to a high-protein (casein) diet. They found that plasma CCK increased by approximately threefold in the first 24 h, which corresponds to the period of maximal adaptative induction, and returned to control values by day 7 when rats were fully adapted to their new diet (Figure 5). On the other hand, Liddle et al.[81] established in the rat that CCK secretion was stimulated by intraluminal casein but not by amino acids, carbohydrates, or lipids. Furthermore, the effect of casein on CCK-PZ release was not shared by bovine serum albumin or lactalbumin, because of their poor

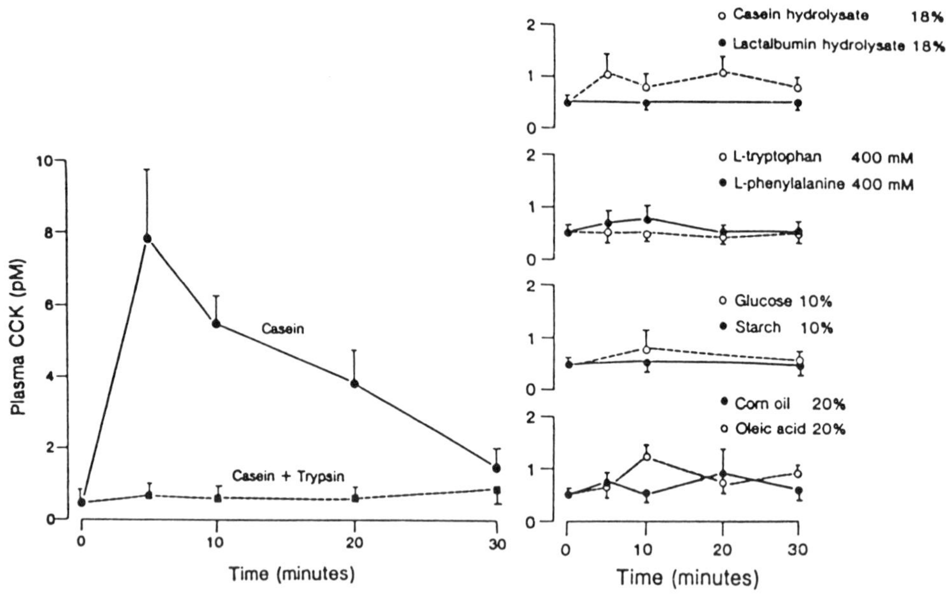

FIGURE 6. Plasma CCK response to various dietary components. Rats were fed by orogastric tube 5 ml of either: 18% casein, 18% casein + 20% trypsin (left panel), 18% casein hydrolysate or lactalbumin hydrolysate, 400 m*M* L-tryptophan or L-phenylalanine, 10% glucose or starch or 20% corn oil or oleic acid (right panel). CCK concentrations were estimated as described in Figure 5. The only significant increase in plasma CCK was obtained with casein. (Adapted from Liddle, R. A., Green, G. M., Conrad, C. K., and Williams, J. A., *Am. J. Physiol.*, 251, G243, 1986. With permission.)

affinity for trypsin. On the contrary, soybean trypsin inhibitor was a very strong CCK releaser (Figure 6). The authors concluded that, in the rat, CCK release was regulated through a feed-back mechanism involving trypsin inhibition by dietary protein. No data are available on the trypsin inhibitory activity of gelatin, zein, and gluten. Those three proteins, however, are characterized by a very low content in basic amino acids (the targets of trypsin hydrolysis) and a peculiar secondary structure due to their high proline content[82,83] suggesting limited interaction with trypsin. As a result, it is tempting to speculate that failure to adapt to diets containing those proteins was due to insufficent CCK release. This would support the hypothesis that CCK plays a permissive role in pancreatic adaptation, at least for carbohydrate-rich and lipid-rich diets.[14] The situation may be more complicated for protein-rich diets since CCK alone is unable to mimick that adaptation. Chronic administration of the hormone or its analog caerulein resulted in an increased synthesis of proteolytic enzymes and decreased amylase synthesis,[84,85] but the corresponding mRNA levels were not modified to the same extent,[86] contrary to what was observed in adaptation.[38,39] Thus, CCK seems to act primarily at the translational level,[87] and other effectors, equivalent to glucose or ketone bodies, may be involved in the transcriptional regulation occurring during adaptation to protein-rich diets.

REFERENCES

1. **Pavlov, J. P.,** *Die Arbeit der Verdauungsdrüsen,* Bergmann, Wiesbaden, 1898, 48.
2. **Grossman, M. J., Greengard, H., and Ivy, A. C.,** The effect of dietary composition on pancreatic enzymes, *Am. J. Physiol.*, 138, 676, 1942.

3. **Magee, D. F. and Hong, S. S.,** Daily output of pancreatic juice and some dietary factors which influence it, *Am. J. Physiol.*, 197, 27, 1959.
4. **Roy, D. M. and Schneeman, B. O.,** Effect of soy protein, casein and trypsin inhibitors on cholesterol, bile acids and pancreatic enzymes in mice, *J. Nutr.*, 111, 878, 1981.
5. **Corring, T., Lebas, F., and Courtot, D.,** Sécrétion pancréatique sur porc fistulé : adaptation à la teneur en protéines du régime, *Ann. Biol. Anim. Biochem. Biophys.*, 12, 233, 1972.
6. **Salman, A. J., Dal Borgo, G., Pubols, M. A., and McGinnis, J.,** Changes in pancreatic enzymes as a function of diet in the chick, *Proc. Soc. Exp. Biol. (N.Y.)*, 126, 694, 1967.
7. **Zhalka, M. and Bdolah, A.,** Dietary regulation of digestive enzyme levels in the water snake, *Natrix tessellata*, *J. Exp. Zool.*, 243, 9, 1987.
8. **Zoppi, G., Andreotti, G., Pajno-Ferrara, F., Njai, D. M., and Gaburro, D.,** Exocrine pancreatic function in premature and full term neonates, *Pediatr. Res.*, 6, 880, 1972.
9. **Emde, C., Liehr, R. M., Gregor, M., Pleul, O., Riecken, E. O., and Menge, H.,** Lack of adaptative changes in human pancreatic amylase and lipase secretion in response to high carbohydrate, low fat diet applied by a 10-day continuous intraduodenal infusion, *Dig. Dis. Sci.*, 30, 204, 1985.
10. **Keller, P. J. and Cohen, E.,** Enzymic composition of some cell fractions of bovine pancreas, *J. Biol. Chem.*, 236, 1407, 1961.
11. **Green, L. J., Hirs, C. H. W., and Palade, G. E.,** On the protein composition of bovine pancreatic zymogen granules, *J. Biol. Chem.*, 231, 2054, 1963.
12. **Ben Abdeljlil, A. and Desnuelle, P.,** Sur l'adaptation des enzymes exocrines du pancréas à la composition du régime, *Biochim. Biophys. Acta*, 81, 136, 1964.
13. **Desnuelle, P., Reboud, J. P., and Ben Abdeljlil, A.,** Influence of the composition of the diet on the enzyme content of rat pancreas, in *Ciba Foundation Symposium on the Exocrine Pancreas: Normal and Abnormal Function*, De Reuck, A. V. S. and Cameron, M. P., Eds., Churchill, London, 1962, 90.
14. **Christophe, J., Camus, J., Deschodt-Lankman, M., Rathe, J., Robberecht, P., Vandermeers-Piret, M. C., and Vandermeers, A.,** Factors regulating biosynthesis, intracellular transport and secretion of amylase and lipase in the rat exocrine pancreas, *Horm. Metab. Res.*, 3, 393, 1971.
15. **Palade, G. E.,** Intracellular aspects of the process of protein synthesis, *Science*, 189, 347, 1975.
16. **Blobel, G. and Doberstein, B.,** Transfer of proteins across membranes. Presence of proteolytically processed and unprocessed nascent immunoglobulin light chains on membrane bound ribosomes of murine myeloma, *J. Cell. Biol.*, 67, 835, 1975.
17. **Devillers-Thiery, A., Kindt, T., Scheele, G., and Blobel, G.,** Homology in amino-terminal sequences of precursors to pancreatic secretory proteins, *Proc. Natl. Acad. Sci. U.S.A.*, 72, 5016, 1975.
18. **Walter, P., Gilmore, R., and Blobel, G.,** Protein translocation across the endoplasmic reticulum, *Cell*, 38, 5, 1984.
19. **Kraehenbuhl, J. P., Racine, L., and Jamieson, J.,** Immunocytochemical localization of secretory proteins in bovine pancreatic exocrine cells, *J. Cell. Biol.*, 72, 406, 1977.
20. **Mroz, E. A. and Lechene, C.,** Pancreatic zymogen granules differ markedly in protein composition, *Science*, 232, 871, 1986.
21. **Bruzzone, R., Trimble, E. R., Gjinovci, A., and Renold, A. E.,** Difference in pancreatic enzyme release from ventral and dorsal areas of the pancreas, *Am. J. Physiol.*, 251, G56, 1986.
22. **Malaisse-Lagae, F., Ravazzola, M., Robberecht, P., Vandermeers, A., Malaisse, W., and Orci, L.,** Exocrine pancreas: evidence for topographic partition of secretory function, *Science*, 190, 795, 1975.
23. **Grendell, J. H., Tseng, A. C., and Rothman, S. S.,** Regulation of digestion. I. Effects of glucose and lysine on pancreatic secretion, *Am. J. Physiol.*, 246, G445, 1984.
24. **Kelly, R. B.,** Pathways of protein secretion in eukaryotes, *Science*, 230, 25, 1985.
25. **Arvan, P. and Castle, J. D.,** Phasic release of newly synthetized secretory proteins in the unstimulated rat exocrine pancreas, *J. Cell. Biol.*, 104, 243, 1987.
26. **Dagorn, J. C.,** Non-parallel secretion from rat pancreas: in vivo studies, *J. Physiol. London*, 280, 435, 1978.
27. **Reboud, J. P., Marchis-Mouren, G., Pasero, L., Cozzone, A., and Desnuelle, P.,** Adaptation de la vitesse de biosynthèse de l'amylase pancréatique et du chymotrypsinogène à des régimes riches en amidon ou en protéines, *Biochim. Biophys. Acta*, 117, 351, 1966.
28. **Dagorn, J. C. and Lahaie, R.,** Dietary regulation of pancreatic protein synthesis. I. Rapid and specific modulation of enzyme synthesis by changes in dietary composition, *Biochim. Biophys. Acta*, 654, 111, 1981.
29. **Lahaie, R. and Dagorn, J. C.,** Dietary regulation of pancreatic protein synthesis. II. Kinetics of adaptation of protein synthesis and its effects on enzyme content, *Biochim. Biophys. Acta*, 654, 119, 1981.
30. **Schick, J., Verspohl, R., Kern, H., and Scheele, G.,** Two distinct adaptive responses in the synthesis of exocrine pancreatic enzymes to inverse changes in protein and carbohydrate in the diet, *Am. J. Physiol.*, 247, G611, 1984.
31. **Wicker, C. and Puigserver, A.,** Effects of inverse changes in dietary lipids and carbohydrate on the synthesis of some pancreatic secretory proteins, *Eur. J. Biochem.*, 162, 25, 1987.

32. **Pelham, R. B. and Jackson, R. J.,** An efficient mRNA dependent translation system from reticulocyte lysates, *Eur. J. Biochem.,* 67, 247, 1976.
33. **Beintema, J. J., Campagne, R. N., and Gruber, M.,** Rat pancreatic ribonuclease: isolation and properties, *Biochim. Biophys. Acta,* 310, 148, 1973.
34. **Scheele, G. A.,** Analysis of the secretory process in the exocrine pancreas by two-dimensional isoelectric focusing SDS gel electrophoresis, *Methods Cell. Biol.,* 23, 345, 1981.
35. **MacDonald, R. J., Crerar, M. M., Swain, W. F., Pictet, R. L., Thomas, G., and Rutter, W. J.,** Structure of a family of rat amylase genes, *Nature,* 287, 117, 1980.
36. **MacDonald, R. J., Stary, J. S., and Swift, G. H.,** Two similar but non-allelic rat pancreatic trypsinogens, *J. Biol. Chem.,* 257, 9724, 1982.
37. **Bell, G. I., Quinto, C., Quiroga, M., Valenzuela, P., Craik, C. S., and Rutter, W. J.,** Isolation and sequence of a rat chymotrypsinogen B gene, *J. Biol. Chem.,* 259, 14265, 1984.
38. **Giorgi, D., Bernard, J. P., Lapointe, R., and Dagorn, J. C.,** Regulation of amylase messenger RNA concentration in rat pancreas by food content, *EMBO J.,* 3, 1521, 1984.
39. **Giorgi, D., Renaud, W., Bernard, J. P., and Dagorn, J. C.,** Regulation of proteolytic enzymes activities and mRNA concentration in rat pancreas by food content, *Biochem. Biophys. Res. Commun.,* 127, 937, 1985.
40. **Spirin, A. D.,** Informosomes, *Eur. J. Biochem.,* 10, 20, 1969.
41. **Wicker, C., Puigserver, A., and Scheele, G.,** Dietary regulation of levels of active mRNA coding for amylase and serine protease zymogens in the rat pancreas, *Eur. J. Biochem.,* 139, 381, 1984.
42. **Morisset, J. and Dunnigan, J.,** Exocrine pancreas adaptation to diet in vagotomized rats, *Rev. Can. Biol.,* 26, 11, 1967.
43. **Gray, G. M.,** Intraluminal and surface membrane digestion of dietary carbohydrates and protein, in *The Exocrine Pancreas: Biology, Pathobiology and Diseases,* Go, V. L. W., Gardner, J. D., Brooks, F. P., Lebenthal, E., Dimand, E. P., and Scheele, G. A., Eds., Raven Press, New York, 1986, 375.
44. **Borgstrom, B.,** Luminal digestion of fat, in *The Exocrine Pancreas: Biology, Pathobiology and Diseases,* Go, V. L. W., Gardner, J. D., Brooks, F. P., Lebenthal, E., Dimand, E. P., and Scheele, G. A., Eds., Raven Press, New York, 1986, 361.
45. **Matthews, D. M.,** Intestinal absorption of peptides, *Physiol. Rev.,* 55, 537, 1975.
46. **Snook, J. T.,** Dietary regulation of pancreatic enzymes with special emphasis on carbohydrates, *Am. J. Physiol.,* 221, 1383, 1971.
47. **Folsch, V. R, Grieb, N., Caspary, W. F., and Creutzfeldt, W.,** Influence of short and long-term feeding of an alpha amylase inhibitor (BAY 4609) on the exocrine pancreas of the rat, *Digestion,* 21, 74, 1981.
48. **Morisset, J. and Dunnigan, J.,** Effects of glucose, amino-acids and insulin on adaptation of exocrine pancreas to diet, *Proc. Soc. Exp. Biol. Med.,* 136, 231, 1971.
49. **Lavau, M., Bazin, R., and Herzog, J.,** Comparative effect of oral and parental feeding on pancreatic enzymes in the rat, *J. Nutr.,* 104, 1432, 1974.
50. **Reboud, J. P., Ben Abdeljlil, A., and Desnuelle, P.,** Variations de la teneur en enzymes du pancréas de rat en fonction de la composition des régimes, *Biochim. Biophys. Acta,* 58, 326, 1962.
51. **Bicko, A. and Kopec, Z.,** Adaptation of enzyme activity of the rat pancreas on altered food intake, *Nutr. Dieta,* 10, 276, 1968.
52. **Gidez, L. I.,** Effect of dietary fat on pancreatic lipase levels in rat, *J. Lipid Res.,* 14, 169, 1973.
53. **Deschodt-Lankman, M., Robberecht, P., Camus, J., and Christophe, J.,** Short-term adaptation of pancreatic hydrolases to nutritional and physiological stimuli in adult rats, *Biochimie,* 53, 789, 1971.
54. **Saraux, B., Girard-Globa, A., Ouagued, M., and Vacher, D.,** Response of the exocrine pancreas to quantitative and qualitative variations in dietary lipids, *Am. J. Physiol.,* 243, G10, 1982.
55. **Sabb, J. E., Godfrey, P. M., and Brannon, P.,** Adaptative response of rat pancreatic lipase to dietary fat: effect of amount and type of fat, *J. Nutr.,* 116, 892, 1986.
56. **Maylie, M. F., Charles, M., Gache, C., and Desnuelle, P.,** Isolation and partial characterization of pancreatic colipase, *Biochim. Biophys. Acta,* 229, 286, 1971.
57. **Borgström, B.,** Pancreatic colipase: chemistry and physiology, *J. Lipid Res.,* 20, 805, 1979.
58. **Vandermeers-Piret, M. C., Vandermeers, A., Wijns, W., Rathe, J., and Christophe, J.,** Lack of adaptation of pancreatic colipase in rats and mice, *Am. J. Physiol.,* 232, E131, 1977.
59. **Girard-Globa, A., Bourdel, G., and Lardeux, B.,** Regulation of protein synthesis and enzyme accumulation in the rat pancreas by amount and timing of dietary protein, *J. Nutr.,* 110, 1380, 1980.
60. **Mourot, J. and Corring, T.,** Adaptation of the lipase-colipase system to dietary lipid content in pig pancreatic tissue, *Ann. Biol. Anim. Biochim. Biophys.,* 19, 119, 1979.
61. **Christophe, J. and Mayer, Z.,** Infuence of diet on utilization of glucose and incorporation of acetate $1^{14}C$ into liver fatty acids and cholesterol in rats, *Am. J. Physiol.,* 197, 55, 1959.
62. **Bazin, R. and Lavau, M.,** Diet composition and insulin effect on amylase to lipase ratio in pancreas of diabetic rats, *Digestion,* 19, 386, 1979.
63. **Bazin, R., Lavau, M., and Herzog, J.,** Ketogenic conditions and pancreatic lipase, *Biomedicine,* 28, 160, 1978.

64. **Snook, J. T.,** Dietary regulation of pancreatic enzyme synthesis, secretion and inactivation in the rat, *J. Nutr.*, 87, 297, 1965.
65. **Grossman, M. J., Greengard, H., and Ivy, A. C.,** On the mechanism of the adaptation of pancreatic enzymes to dietary composition, *Am. J. Physiol.*, 141, 38, 1944.
66. **Howard, F. and Yudkin, J.,** Effect of dietary change upon the amylase and trypsin activities of the rat pancreas, *Br. J. Nutr.*, 17, 281, 1963.
67. **Hong, S. S. and Magee, D. F.,** Influence of dietary amino acids on pancreatic enzymes, *Am. J. Physiol.*, 197, 71, 1957.
68. **Palla, J. C., Ben Abdeljlil, A. and Desnuelle, P.,** Action de l'insuline sur la biosynthèse de l'amylase et de quelques autres enzymes du pancréas de rat, *Biochim. Biophys. Acta*, 158, 25, 1968.
69. **Korc, M., Owerbach, D., Quinto, C., and Rutter, W. J.,** Pancreatic islet-acinar cell interaction: amylase mRNA levels are determined by insulin, *Science*, 213, 351, 1981.
70. **Dranginis, A., Morley, M., Nesbitt, M., Rosenblum, B. B., and Meisler, M. H.,** Independent regulation of nonallelic pancreatic amylase genes in diabetic mice, *J. Biol. Chem.*, 259, 12216, 1984.
71. **Bazin, R. and Lavau, M.,** Effect of high-fat diet on glucose metabolism in isolated pancreatic acini of rats, *Am. J. Physiol.*, 243, G448, 1982.
72. **Williams, J. A., Bailey, A. C., Preissler, M., and Goldfine, I.,** Insulin regulation of sugar transport in isolated pancreatic acini from diabetic mice, *Diabetes*, 31, 674, 1982.
73. **Bruzzone, R., Trimble, E. R., Gjinovci, A., and Renold, A. E.,** Changes in content and secretion of pancreatic enzymes in the obese Zucker rat, *Biochem. J.*, 219, 333, 1984.
74. **Logsdon, C. D., Akana, S. F., Meyer, C., Dallman, M., and Williams, J. A.,** Pancreatic acinar cell amylase gene expression: selective effects of adrenalectomy and corticosterone replacement, *Endocrinology*, 121, 1242, 1987.
75. **Darlington, G. J., Tsai, C. C., Samuelson, L. C., Gumucio, D. L., and Meisler, M. H.,** Simultaneous expression of salivary and pancreatic amylase genes in cultured mouse hepatoma cells, *Mol. Cell. Biol.*, 6, 969, 1986.
76. **Osborn, L., Rosenberg, M. P., Keller, S. A., and Meisler, M. H.,** Tissue specific and insulin-dependent expression of the amylase gene in transgenic mice, *Mol. Cell. Biol.*, 7, 326, 1987.
77. **Logsdon, C. D., Moessner, J., Williams, J. A., and Goldfine, J.,** Glucocorticoids increase amylase mRNA levels, secretory organelles and secretion in pancreatic acinar AR4-2J cells, *J. Cell. Biol.*, 100, 1200, 1985.
78. **Logsdon, L. D., Perot, K. J., and McDonald, A. R.,** Mechanism of glucocorticoid-induced increase in pancreatic amylase gene transcription, *J. Biol. Chem.*, 262, 15765, 1987.
79. **Johnson, A., Hurwitz, R., and Kretchmer, N.,** Adaptation of rat pancreatic amylase and chymotrypsinogen to changes in diet, *J. Nutr.*, 107, 87, 1977.
80. **Green, G. M., Levan, V. H., and Liddle, R. A.,** Plasma cholecystokinin and pancreatic growth during adaptation to dietary protein, *Am. J. Physiol.*, 251, G70, 1986.
81. **Liddle, R. A., Green, G. M., Conrad, C. K., and Williams, J. A.,** Proteins but not amino acids carbohydrates or fats stimulate cholecystokinin secretion in the rat, *Am. J. Physiol.*, 251, G243, 1986.
82. **Kasarda, D. D., Okita, T. W., Bernardin, J. E., Baecker, P. A., Nimmo, C. C., Lew, E. J. L., Dietler, M. D., and Greene, F. C.,** Nucleic acid (cDNA) and amino acid sequences of alpha-type gliadins from wheat (*Triticum aestiorum*), *Proc. Natl. Acad. Sci. U.S.A.*, 81, 4712, 1984.
83. **Argos, P., Pedersen, K., Marks, D. M., and Larkins, B.A.,** A structural model for maize zein proteins, *J. Biol. Chem.*, 257, 9984, 1982.
84. **Dagorn, J. C. and Mongeau, R.,** Different action of hormonal stimulation on the biosynthesis of three pancreatic enzymes, *Biochim. Biophys. Acta*, 498, 76, 1977.
85. **Schick, J., Kern, H., and Scheele, G.,** Hormonal stimulation in the exocrine pancreas results in coordinate and anticoordinate regulation of protein synthesis, *J. Cell. Biol.*, 99, 1569, 1984.
86. **Renaud, W., Giorgi, D., Iovanna, J., and Dagorn, J. C.,** Regulation of concentration of mRNA for amylase, trypsinogen and chymotrypsinogen in the rat pancreas by secretagogues, *Biochem. J.*, 235, 305, 1986.
87. **Lahaie, R. G.,** Translational control of protein synthesis in isolated pancreatic acini: role of CCK, carbachol and insulin, *Pancreas*, 1, 403, 1986.

Chapter 6

PANCREATIC REGENERATION

H. P. Elsässer and H. F. Kern

TABLE OF CONTENTS

I. GENERAL CONSIDERATIONS

Regeneration is a common phenomenon in biology and has been investigated for 250 years.[1] Studies on regeneration are always experimental and therefore depend on model systems. In the analysis of pancreatic regeneration various model systems have been used in a variety of animal species.[2-7] It has to be kept in mind when interpreting the data from regeneration studies that common phenomena of pancreatic regeneration have to be distinguished from those which are species- or even model-specific. Furthermore, the transfer of findings from animal experiments to the human situation and drawing conclusions with clinical relevance should be done with great care.

In the following we will mainly summarize data from regeneration experiments performed in the rat pancreas, since this species has been used most extensively. Three major models of injury have been used in regard to regeneration: (1) injection of ethionine together with feeding a protein deficient diet,[4] (2) surgical resection of large parts of the pancreas,[6-7] and (3) acute pancreatic destruction due to supramaximal hormonal stimulation.[3] In all three models regeneration is causally linked to injury and therefore complex temporal and spatial events overlap. These include tissue inflammation, replication of specific parenchymal cells, and pattern formation of newly formed pancreatic tissues. Each of these aspects has its own regulation and at the same time exerts a regulatory function in the whole process of organ regeneration.

A. INFLAMMATION

It has become clear that during the inflammatory reaction infiltrating mononuclear cells not only remove injured tissue and cellular debris by phagocytosis, but also release a variety of secretory products, including enzymes, complement components, various binding proteins, oxygen metabolites, arachidonate metabolites, and a variety of growth and differentiation factors.[8,9] It is beyond the scope of this article to discuss all these aspects of the inflammatory reaction. For the context of pancreatic regeneration it is important to note that it has been established that monokines from activated mononuclear cells, especially from monoctyes/macrophages, play an important role in modulating the biological activity of various target cells. Especially the modulation of fibroblast function by interleukin 1 has been well established.[10] Thus, the inflammatory reaction is tightly linked to the production and deposition of collagens in the injured organ.

B. CELL REPLICATION

In general cells can be divided into three categories according to their state in the cell cycle.[11]

1. Continuously Cycling Cells

This type of cell has a G_1-phase of 6 to 10 h and continuously undergoes cell division. Examples are cells of the bone marrow or the krypt cells in the small intestine.

2. G_0-Arrested Cells

These cells do not replicate normally but can be stimulated to enter the cell cycle. The time between the stimulation and the beginning of the S-phase is usually 18 to 24 h which is in contrast to the normal duration of a G_1-phase in continuously cycling cells.

3. Terminally Differentiated Cells

For these cells there are no conditions known under which the cell cycle is entered.

4. The Situation in the Pancreas

Studies using autoradiography after injection of ^{3}H-thymidine have shown that the labeling index of pancreatic parenchymal cells (acinar cells, duct cells, and islet cells) is less than 0.5%

in control animals. It therefore seems unlikely that the pancreas contains continuously cycling cells. Then the question remains whether pancreatic parenchymal cells are terminally differentiated or G_0-arrested cells. In the first case a G_0-arrested stem cell has to be assumed. Stem cells are originally defined in the hematopoietic system as cells with two main properties of pluripotentiality and self maintenance.[12] Even in the hematopoietic system it was difficult to demonstrate cells with these properties and this is true for other systems to which the stem cell concept has been applied like the intestinal epithelium, [12] the epidermis,[13] and the testes.[14] In the pancreas there is no functional or structural evidence for a stem cell according to this definition. The available data show that the pancreatic parenchymal cells are G_0-arrested cells, which will be discussed in detail below.

C. PATTERN FORMATION

Except for the embryonic development of the pancreas, where the influence of extracellular components for regular development has been demonstrated,[15] there are no studies of pattern formation in this organ during regeneration. In other experimental systems three major mechanisms are discussed which are responsible for the regulation of pattern formation.

1. Humoral Factors

Since the introduction of "physiological gradients" by C. M. Child[16] and the development of a mathematical theory of how such gradients can control morphogenesis,[17] much work has been done about the significance of humoral factors in pattern formation, which were designated as "morphogens". Although the theoretical models had been very powerful in simulating morphogenetic processes, no substance could be established with the features of a morphogen. Only recently retinoic acid was shown to act as a morphogen in the development of the bird limb.[18]

2. Cell-Cell Interactions

Although the idea of cell-cell interaction during tissue development has long been discussed,[19-20] it was substantiated only during the last years by the isolation and characterization of several membrane proteins which became known as cell-adhesion molecules (CAMs).[21] Beside the CAMs, membrane glycolipids also might play a role in cell-cell interaction.[22] For some CAMs it has been demonstrated that they exhibit a complex temporal and spatial pattern during tissue development.[23] In liver regeneration after partial hepatectomy the involvement of the membrane glycoprotein cell-CAM-105 was verified.[24]

3. Cell-Matrix Interactions

As mentioned above, the importance of extracellular matrix components for the development of the pancreas was first shown about 30 years ago.[15] Meanwhile, considerable information has accumulated about the structure and function of extracellular matrix molecules, including the structure and function of the basement membrane.[25-28] It is now clear that the extracellular matrix plays a general important role for cell movement and differentiation during embryonic development and regeneration.[29-31]

In the case of pancreatic regeneration experimental work about pattern formations is at the very beginning. Only a few recent studies suggest that the extracellular matrix might have a key role in this aspect of regeneration.[32-34]

II. MODELS OF PANCREATIC REGENERATION

A. INJECTION OF ETHIONINE AND FEEDING A PROTEIN-FREE DIET

Basic insights into pancreatic regeneration were obtained by Fitzgerald and his co-workers using the model of ethionine/protein-free diet for pancreatic injury in the rat.[4] In this model, the standard period of injury induction is 10 d, where rats receive a daily intraperitoneal injection of ethionine combined with a protein-free diet. After this induction period the animals are refed with a normal standard diet without further ethionine application. One to two hours after the first ethionine injection, changes in the ultrastructure of acinar cells can be observed mainly manifested at the rough endoplasmic reticulum (RER): detachment of ribosomes, enlargement of RER-lumina, and development of autophagosomes containing RER-membranes.[35,36] The degeneration of the RER is progressive until day 10, while other cytoplasmic organelles like mitochondria, Golgi-complex, and zymogen granules are only slightly damaged until day 5. In the second half of the induction period autophagocytosis of the latter organelles also became evident.

Furthermore, vacuoles containing lysosomal enzymes like β-glucuronidase and acid phosphatase are apparent in most acinar cells,[37] which is in agreement with an increase of these two enzymes in pancreatic homogenates. This increase has its maximum at day 10, where β-glucuronidase is elevated 25-fold and acid phosphatase 3.5-fold.[36]

The enzymatic activity of chymotrypsin, lipase, and amylase in the gland was slightly increased in the first 6 d which might reflect a decreased secretion of these products. Thereafter the content of these enzymes dropped very rapidly to 5 to 20 % of control values on day 12. For day 10 (end of the induction period) a loss of 80 to 90% of acinar cell mass was calculated.[35]

These changes were accompanied by an inflammatory reaction with slight edema and infiltration of the pancreas with polymorphonuclear leukocytes and macrophages. Other tissue components like islets, ducts, or vessels have been found damaged only slightly, mainly as a result of the inflammatory reaction. All investigated parameters returned to control about 30 d after the end of the induction period, except for pancreatic weight which remained reduced by 30% in regenerated pancreata.

B. PANCREATITIS FOLLOWING SUPRAMAXIMAL HORMONAL STIMULATION

In 1977 the model of hormone-induced acute pancreatitis was introduced,[39] which offered a number of advantages in the study of both acute pancreatitis and pancreatic regeneration. In this model the synthetic CCK analog caerulein[40] is infused in conscious animals at a dose of 5 to 10 $\mu g\ kg^{-1}\ h^{-1}$ for periods ranging from 3 to 12 h. The short induction period and the absence of side effects in other organs which are observed in the ethionine/protein-free diet model (including a 50% reduction of body weight) make this model a powerful and reliable tool in studies of pancreatic regeneration.

The initial alteration in the acinar cells in hormone-induced pancreatitis is a disturbance of protein sorting and membrane fusion in the normal secretory pathway. Within 3 h fusion of zymogen granules with each other and with the basolateral plasmamembrane occurs, whereas in normal pancreatic acinar cells zymogen granules exclusively fuse with the apical plasmamembrane. This leads to a release of pancreatic enzymes into the interstitial space of the gland.[41] Furthermore, a disturbance of the sorting of lysosomal enzymes from pancreatic hydrolases is observed in the Golgi complex, leading to condensing vacuoles containing both types of enzymes.[42] In addition, lysosomal activation is supposed, confirmed by an increase of cathepsin B (2-fold) and β-galactosidase (2- to 3-fold) in pancreatic homogenates after a 12 h infusion. This leads to autophagic destruction of cytoplasmic compartments and possibly to cell death.[43] In addition, the polarity of acinar cells is destroyed as shown by the disruption of tight junctions and a subsequent uniform distribution of membrane particles in freeze-fracture replicas.[41] At the end of the 12-h induction period the acinar cells are vacuolated and a massive interstitial edema has formed. Cellular infiltration with macrophages, polymorphonuclear granulocytes, and lymphocytes is evident in the inter- and intralobular space and increases during the following 24 to 48 h. The interstitial edema is absorbed during the same time period.

The content of amylase, chymotrypsin, and lipase in the pancreas decreases rapidly after the end of the infusion and reaches a minimum 48 h later. At this point only 24% of chymotrypsin, 4.9% of amylase, and 18% of lipase, compared to control pancreata, are found. The protein synthesis has dropped to 30 to 40% of control values measuring *in vitro* ^{3}H-leucine incorporation into proteins in isolated lobules.[44]

Already after a 3-h infusion period acinar cell necrosis is frequently seen and results in a reduction of pancreatic weight to 50% on day 3 to 4 after a 12-h infusion.[33] Keeping in mind the cellular infiltration of the gland, at the same time period, acinar cell destruction in this model is greater than 50%, while the islets and ducts seem to be unaffected. Although light and electron microscopic studies have not presented evidence for damage of pancreatic vessels so far, it was recently shown by scanning electron microscopy of caerulein-infused dogs that within a 2-h infusion the capillaries of the pancreas were markedly reduced and the remaining vessels showed ectasia and abrupt termination.[45] The fast establishment of these alterations indicates that they are part of primary rather than of secondary inflammatory events. All parameters investigated returned to control value on day 14.

C. SURGICAL RESECTION OF PARTS OF THE PANCREAS

Regeneration of the pancreas after partial pancreatectomy seems to be very low. This is discussed most extensively in the paper of Lehv et al.[7] where pancreatic regeneration was studied after a 55% resection of this organ in rats. After a period of 9 to 12 months the organ weight reached only 68% of pancreas from control animals of the same age. Histological features of the remaining segment, which weighed 50% more than a comparable normal segment, were not distinguishable from control pancreata.

It is noteworthy that in the resection model only a local inflammatory reaction is induced, mainly at the site of resection and in peripancreatic areas. The inflammation which changes in severity shows a slight edema and infiltration of macrophages and polymorphonuclear leukocytes. This is observed until the end of the first postoperative week. One might assume that a less pronounced and more localized inflammatory reaction might be the reason for a decreased regeneration. In the case of the liver, however, a two-third resection is restored within 2 to 3 weeks[46] indicating that beside inflammatory cells and their mediators, there are other factors regulating regeneration in these two different organs. Some hints in this respect are given by other authors who showed that pancreatic regeneration after partial pancreatectomy is a function of the age of the animals or the diet they received.[4] In other model systems the importance of sex, mass of resected organ, various hormones, and other effector substances like folic acid has been stressed.[47-49]

While in the three models described, mainly regeneration of exocrine pancreatic tissue was studied, the regeneration of the islets of Langerhans was investigated in alloxan or streptozotocin diabetic rats or mice. For a detailed discussion of islet cell regeneration the reader is referred to Logothetopulos.[50]

III. PANCREATIC REGENERATION AND CELL REPLICATION

Since the beginning of pancreatic regeneration research, the cellular source from which the lost parenchyma is restored was discussed with controversy. In analogy to the embryonic development of the pancreas[51] it was thought for a long time that certain duct cells could function as precursor or stem cells.[52] This would imply, in regard to the stem cell definition give above, that these duct cells are G_0-arrested, tripotential stem cells with the developmental potency for duct, acinar, and islet cells, which are then terminally differentiated. That this concept is not true was unequivocally proven in the models described and also under conditions with enhanced pancreatic growth without injury.[63-65] In light microscopical autoradiography after ^{3}H-thymidine injection 1 h before sacrifice, all three cell types are labeled,[50,53] indicating that these cells are in the S-phase of the cell cycle.

Furthermore, mitotic figures were observed in all pancreatic parenchyma cells[53] providing their ability to pass the whole cell cycle (Figure 1). Electron microscopical investigation showed

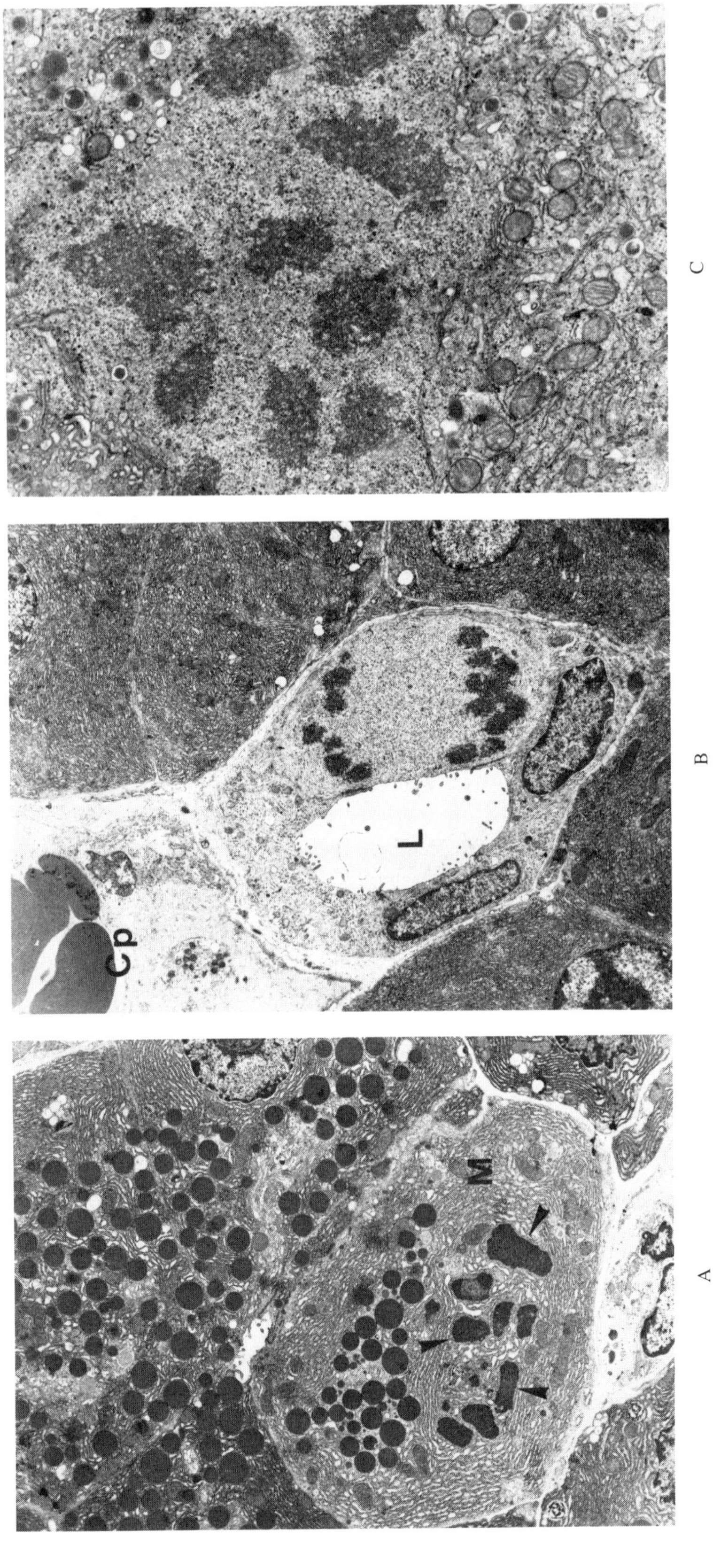

FIGURE 1. Pancreatic parenchymal cells in mitosis. (A) Acinar cell (M). (B) Intralobular duct cell. (C) Islet B-cell. (L) Luminal space and (Cp) capillary. Arrows indicate condensed chromosomes.

TABLE 1
Labeling Indices of Rat Pancreatic Parenchymal Cells Under Control Conditions and During Regeneration in Different Models of Pancreatic Injury

	Labeling index					
	Acinar cells		Duct cells		Centroacinar cells	
Model	Control	Regeneration	Control	Regeneration	Control	Regeneration
Ethionine/ protein-free diet[61]	0.46 (0.10—1.03)	10.32 (6.34—14.83) day 5	0.77 (0.00—2.22)	1.64 (0.71—2.48) day 4	n.d.	n.d.
Caerulein[57]	0.38 (0.08—1.05)	11.21 (7.31—15.78) day 4	0.41 (0.16—0.65)	0.50 (0.28—0.73)	0.62 (0.00—1.55)	19.4 (14.10—27.81) day 1
Rejection[7]	0.56 (0.05—1.02)	3.36 (2.69—4.03) day 1.5	n.d.	n.d.	n.d.	n.d.

Note: For regenerating pancreata the mean maximal values are given together with the time point after injury. The range of observed labeling indices is given in parentheses.

that structurally fully differentiated duct, acinar, and islet cells are observed undergoing mitosis. Taking these data together a terminal differentiated state of these cell types cannot be assumed. Meanwhile, it is accepted that each parenchymal cell has its own capability for cell renewal and is a G_0-arrested rather than a terminal differentiated cell. In the case of acinar and duct cells it was shown, after stimulation of their proliferation by low doses of caerulein without pancreatic damage[53,55] and by feeding serine-protease inhibitors,[54,56] that an increase in labeling index after ^{3}H-thymidine injection is first observed after 18 to 36 h during stimulation. This also fits well with the definition of G_0-arrested cells. While in the models of pancreatic growth enhancement without pancreatic injury (hormonal stimulation) all cell types enter the cell cycle simultaneously, the time course of an increase in labeling indices in the regeneration models is more complex. Both in the ethionine and caerulein model the duct cells first react with proliferation showing a maximum 24 h after the end of induction period in the caerulein model and 72 h after the end of induction period in the ethionine model. It seems to be important that in the ethionine model all duct cells were comprised, whereas in the caerulein model the labeling index of centroacinar cells was separately quantitated from the labeling index of intra- and interlobular duct cells.[57] The labeling index was only elevated in centroacinar cells. The labeling index of acinar cells showed a maximum at 24 h after the duct cell maximum in both models. The maximum values are summarized in Table 1. While the duct cells show an increase of labeling index only for 48 h (caerulein) and 72 h (ethionine), the acinar cell labeling index was 12.4-fold higher as compared to control values until day 7 after the end of induction in the caerulein model (data of further time points are not available) and normalized in the ethionine model on day 26 after the end of ethionine treatment.

Unfortunately, in the resection model, the time course of labeling index was only documented for acinar cells, with a maximum between day 1 and 2 after pancreatectomy.[7] From day 3 on the labeling index in this cell population was not different from control. Hence, early histological observations of mitotic figures in the different cell types of the pancreas after a 30% resection in dogs and guinea pigs also showed that entrance of duct cells into the cell cycle is earlier than that of acinar cells,[6] indicating that the time course of labeling index is uniform in pancreatic regeneration independent of the kind of injury, but that it differs in pancreatic growth enhancement without injury.

The fact that the duct cell proliferation precedes the acinar cell proliferation raises the

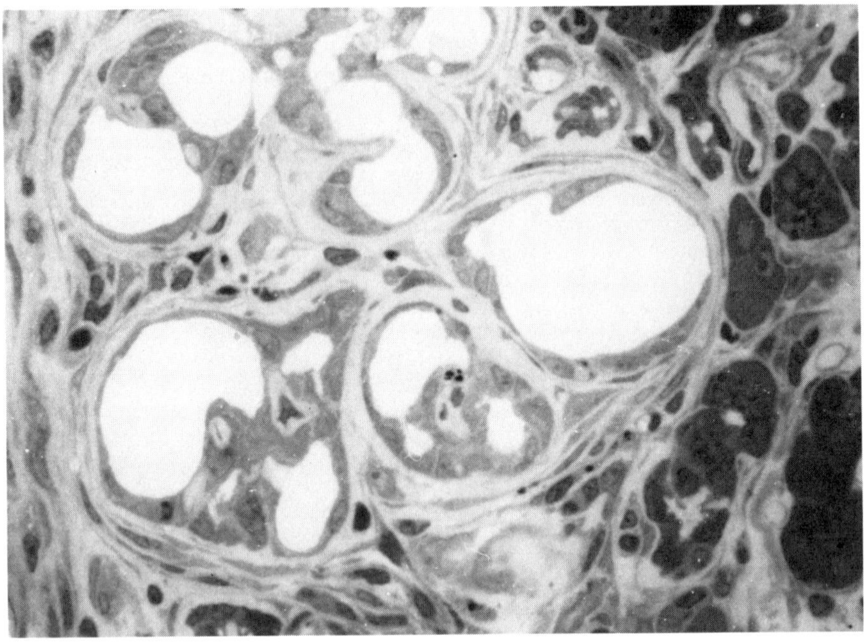

FIGURE 2. Tubular complex in the rat pancreas during regeneration from hormone-induced pancreatitis.

possibility that in regeneration, some duct cells first divide and then undergo differentiation into acinar cells which divide further. This opens again the discussion of a precursor cell. This assumption could not be verified in a study using the caerulein model where some animals received a ^{3}H-thymidine injection on the first day after hormone infusion (maximum of duct cell labeling index and no increased labeling index of acinar cells) while the pancreas was investigated by light microscopic autoradiography 14 d later.[57] No increase in acinar cell labeling index could be found, making a transdifferentiation from duct cells to acinar cells unlikely.

Most pancreatic lesions in animal models and human pathological specimens lead to common structures known as tubular complexes which are now discussed as being derived from acini (Figure 2).[58] Recently it was shown in pancreatic regeneration after trypsin injection into the pancreatic duct that these structures might have regenerating capacities in terms of cell replication, because 1.84% of the epithelial cells lining these structures showed mitotic figures.[59] Another study, however, using light microscopic autoradiography in the caerulein model could not confirm this observation.[58] A reason for this discrepancy might be the dynamics of tubular complexes and hence the state of development of these structures at which the investigation was performed. While, early after an injury, many acini show widened lumina and cuboidal rather than isoprismatic form, fully developed tubular complexes have an irregular delineation of their lumen and flat rather than cuboidal epithelial cells. Furthermore, the whole structure is separated from the neighboring tissue by fibroblasts, inflammatory cells, and fibers of connective tissue. It could be that not all early acinar lesions give rise to late tubular complexes but return to a normal acinar structure. This would mean that some acinar cells not only enter the cell cycle from a G_0-arrested state, but also pass through complex steps of dedifferentiation and redifferentiation, similar to the mode of amphibian limb regeneration.[1]

IV. PANCREATIC REGENERATION AND PATTERN FORMATION

During embryonic development it was clearly demonstrated that mesenchymal constituents

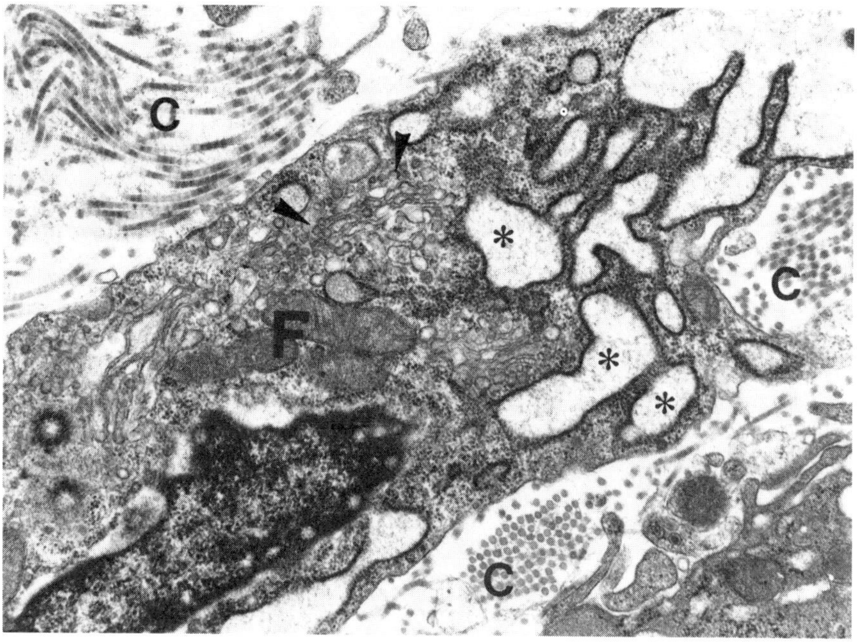

FIGURE 3. Activated fibroblast (F) from rat pancreas during regeneration from hormone-induced pancreatitis. Arrows indicate the Golgi complex. Asterisks indicate the rough endoplasmic reticulum.

are necessary for the formation of an ordered structure of this gland,[15] although the source of mesenchym has not been shown to be as specific as demonstrated for organs with branching morphogenesis.[31] Some recent studies also indicate that in pancreatic regeneration extracellular matrix components are involved and might have regular functions in an ordered reconstruction of the injured gland.[32-34] In the caerulein model it was shown, using ^{3}H-proline as an *in vivo* precursor and quantifying ^{3}H hydroxyproline, that the biosynthesis of collagen was maximally enhanced (15-fold) at day 1.5 to 2.5 after the end of induction and remains elevated 4-fold until day 10.5. This biosynthetic change was preceded by an increase of labeling index after ^{3}H-thymidine injection in interstitial cells which were shown to be mainly fibroblasts by electron microscopical autoradiography.

Furthermore, many fibroblasts had a fine structure of biosynthetically stimulated cells (enlarged RER-cisternae, prominent Golgi apparatus, and numerus secretory vesicles with small filaments; Figure 3). Additionally it was found that most of the newly synthesized collagen was degraded between day 2.5 and 5.5.[33] In a modified ethionine model (induction time 4 d with normal diet), collagenase activity in pancreatic homogenates was increased between day 5 to 7 after the start of ethionine application and the overall hydroxyproline content per milligram dry tissue was maximal on day 7, whereas it was normal again on day 9 after ethionine application.[34] In a third study using trypsin injection into the pancreatic duct to injure the pancreas, the hydroxyproline content of the gland was maximally increased on day 8 after trypsin injection and was of control value again at day 18.[32] These data give a first insight into the whole collagen metabolism during pancreatic regeneration: enhanced collagen biosynthesis occurs very early after injury followed by deposition of a transient matrix and rapid degradation of this matrix when regeneration proceeds. It is assumed that collagen is only a connective tissue marker molecule of the transient matrix which might be constituted by many other matrix components (glycoproteins, proteoglycans). In light of pancreatic embryology one might suppose that the transient matrix is involved in pattern formation during pancreatic regeneration.

To our knowledge no information about the involvement of humoral factors or cell surface molecules in pattern formation during pancreatic regeneration is yet available. While again in pancreatic development a change of pattern in cell surface molecules was demonstrated using lectins as tools,[60] such studies have not been performed during regeneration. This lack of information in the field of pattern formation, which includes the termination of regeneration process, leaves us in a situation where most regulatory features of pancreatic regeneration remain obscure.

REFERENCES

1. **Goss, R. J.**, *Principles of Regeneration*, Academic Press, New York, 1969.
2. **Di Mattei, E.**, Degli effeti della irritazione sugli elementi glandulari del pancreas, *Giornale Dell Reale Acad. Med. Torino Ser.*, 3, 476, 1885.
3. **Elsässer, H. P., Lütcke, H., and Kern, H. F.**, Acinar and duct cell replication and regeneration, in *The Exocrine Pancreas: Biology, Pathobiology and Diseases*, Go, V. L. W., Gardner, J. D., Brooks, F. P., Lebenthal, E., DiMagno, E. P., and Scheele, G. A., Eds., Raven Press, New York, 1985, 45.
4. **Fitzgerald, P. J.**, Pancreatic acinar cell regeneration: the sequence of major morphological and functional changes, in *Regulation of Organ and Tissue Growth*, Goss, R. J., Ed., Academic Press, New York, 1972, 233.
5. **Grauer, T. P.**, Regeneration in the pancreas of the rabbit, *Am. J. Anat.*, 38, 233, 1926.
6. **Kyrle, J.**, Über die Regenerationsvorgänge im tierischen Pankreas, *Arch. Mikrosk. Anat. Entwicklungsgesch.*, 72, 141, 1908.
7. **Lehv, M. and Fitzgerald, P. J.**, Pancreatic acinar cell regeneration. IV. Regeneration after surgical resection, *Am. J. Pathol.*, 53, 513, 1968.
8. **Nathan, C. F.**, Secretory products of macrophages, *J. Clin. Invest.*, 79, 3319, 1987.
9. **Neveu, P. J.**, The mononuclear phagocyte system, *Bull. Inst. Pasteur*, 84, 23, 1986.
10. **Postlethwaite, A. E., Raghow, R., Stricklin, G. P., Poppleton, H., Seyer, J. M., and Kang, A. H.**, Modulation of fibroblast functions by interleukin I: increased steady-state accumulation of type I procollagen messenger RNAs and stimulation of other functions but not chemotaxis by human recombinant interleukin 1 and β, *J. Cell. Biol.*, 106, 311, 1988.
11. **Baserga, R.**, *The Biology of Cell Reproduction*, Harvard University Press, Cambridge, MA, 1985.
12. **Potten, C. S., and Lajtha, L. G.**, Stem cells versus stem lines, in *Annals of the New York Academy of Sciences*, Vol. 397, Baserga, R., Ed., The New York Academy of Sciences, New York, 1982, 49.
13. **Potten, C. S.**, The epidermal proliferative unit: the possible role of the central basal cell, *Cell Tissue Kinetics*, 7, 77, 1974.
14. **Potten, C. S., Schofield, R., and Lajtha, L. G.**, A comparison of cell replacement in bone marow, testis and three regions of surface epithelium, *Biochim. Biophys. Acta*, 560, 281, 1979.
15. **Grobstein, C.**, Morphogenetic interaction between embryonic mouse tissues separated by a membrane filter, *Nature*, 172, 869, 1953.
16. **Child, C. M.**, Physiological dominance and physiological isolation in development and reconstitution, *W. Roux' Arch. Entwicklungsnech*, 117, 21, 1929.
17. **Turing, A. M.**, The chemical basis of morphogenesis, *Philos. Trans. R. Soc.*, B237, 37, 1952.
18. **Thalier, C. and Elchele, G.**, Identification and spatial distribution of retinoids in the developing chick limb bud, *Nature*, 327, 625, 1987.
19. **Wilson, H. V.**, On some phenomena of coalescence and regeneration in sponges, *J. Exp. Zool.*, 5, 245, 1907.
20. **Holtfreter, J.**, Gewebeaffinität, ein Mittel der embryonalen Formbildung, *Arch. Exp. Zellforsch. Gewebezüchtung*, 23, 169, 1939.
21. **Edelman, G. M.**, Cell adhesion molecules in the regulation of animal form and tissue pattern, *Ann. Rev. Biochem.*, 50, 733, 1981.
22. **Hakomori, S.**, Glycosphingolipids in cellular interaction, differentiation and oncogenesis, *Annu. Rev. Biochem.*, 50, 733, 1981.
23. **Chuong, C. M.**, Expression of cell-adhesion molecules in embryonic induction. II. Morphogenesis of adult feathers, *J. Cell. Biol.*, 101, 1027, 1985.
24. **Odin, P. and Öbrink, B.**, Dynamic expression of the cell adhesion molecule cell-CAM 105 in fetal and regenerating rat liver, *Exp. Cell Res.*, 164, 103, 1986.
25. **Hay, E. D.**, *Cell Biology of the Extracellular Matrix*, Plenum Press, New York, 1981.
26. **Hukins, D. W. L.**, *Connective Tissue Matrix*, Verlag Chernie, Basel, 1986.
27. **Kühn, K. and Krieg, T.**, *Connective Tissue: Biological and Clinical Aspects*, Karger, New York, 1986.

28. **Timpl, R. and Dziadek, M.**, Structure, development and molecular pathology of basement membranes, *Int. Rev. Exp. Pathol.*, 29, 1, 1986.
29. **Ekblom, P., Vestweber, D., and Kemler, R.**, Cell-matrix interactions and cell adhesion during development, *Ann. Rev. Cell Biol.*, 2, 27, 1986.
30. **Saxen, L., Sariola, H., and Lethonen, E.**, Sequential cell and tissue interactions governing organogenesis of the kidney, *Anat. Embryol.*, 175, 1, 1986.
31. **Spooner, B. S. and Thompson-Pletscher, H. A.**, Matrix accumulation and the development of form, in *Regulation of Matrix Accumulation*, Mecham, R. P., Ed., Academic Press, New York, 1986, 399.
32. **Uscanga, L., Kennedy, R. H., Choux, R., Druguet, M., Grimaud, J. A., and Sarles, H.**, Sequential connective matrix changes in experimental acute pancreatitis. An immunohistochemical and biochemical assessment in the rat, *Int. J. Pancreatol.*, 2, 33, 1987.
33. **Elsässer, H. P., Adler, G., and Kern, H. F.**, Fibroblast structure and function during regeneration from hormone-induced acute pancreatitis in the rat, *Pancreas*, 4, 169, 1989.
34. **Funatomi, H., Hatta, J., Taguchi, S., Shin, S., Omachi, H., Nakano, I., and Nilkawa, J.**, Studies of interrelation between pancreatic cell regeneration and collagen metabolism in rat acute pancreatitis, *Dig. Dis. Sci.*, 32, 1165, 1987.
35. **Fitzgerald, P. J., and Herman, L.**, Pancreatic acinar cell regeneration. I. Cytologic, cytochemical and pancreatic weight changes, *Am. J. Pathol.*, 52, 963, 1968.
36. **Herman, L. and Fitzgerald, P. J.**, Restitution of pancreatic acinar cells following ethionine, *J. Cell Biol.*, 12, 297, 1962.
37. **Horie, A., Takino, T., Herman, L., and Fitzgerald, P. J.**, Pancreatic acinar cell regeneration. VIII. Relationship of acid phosphatase and β-glucuronidase to intracellular organelles and ethionine lesions, *Am. J. Pathol.*, 63, 299, 1971.
38. **Marsh, W. H., Goldsmith, S., Crocco, J., and Fitzgerald, P. J.**, Pancreatic acinar cell regeneration. II. Enzymatic, nucleic acid and protein changes, *Am. J. Pathol.*, 52, 1013, 1968.
39. **Lampel, M. and Kern, H. F.**, Acute interstitial pancreatitis in the rat induced by excessive doses of a pancreatic secretagogue, *Virchows Arch. A: Pathol. Anat. Histol.*, 373, 97, 1977.
40. **Anastasi, A., Erspamer, V., and Endean, R.**, Isolation and structure of caerulein, an active decapeptide from the skin of *Hyla caerulea*, *Experientia*, 23, 699, 1967.
41. **Adler, G., Rohr, G., and Kern, H. F.**, Alteration of membrane fusion as a cause of acute pancreatitis in the rat, *Dig. Dis. Sci.*, 27, 933, 1982.
42. **Steer, M. L. and Meldolesi, J.**, The cell biology of experimental pancreatitis, *N. Engl. J. Med.*, 316, 144, 1987.
43. **Adler, G., Hahn, C., Kern, H. F., and Rao, K. N.**, Cerulein-induced pancreatitis in rats: increase lysosomal enzyme activity and autophagocytosis, *Digestion*, 32, 10, 1985.
44. **Adler, G., Hupp, T., and Kern, H. F.**, Course and spontaneous regression of acute pancreatitis in the rat, *Virchows Arch. A: Pathol. Anat. Histol.*, 382, 31, 1979.
45. **McEntee, G. P., Leahy, A., Catell, D., Dervan, P., McGeeney, K. F., and Fitzpatrick, J. M.**, Three-dimensional study of the microvasculature in caerulein-induced pancreatitis, *Digestion*, 38, 39, 1987.
46. **Bucher, N. L. R.**, Regeneration of mammalian liver, *Int. Rev. Cytol.*, 15, 245, 1963.
47. **Bucher, N. L. R., and Malt, R. A.**, *Regeneration of Liver and Kidney*, Little, Brown and Co., Boston, 1971.
48. **Brownie, A. C., Bernardis, L. L., Niwa, T., Kamura, S., and Skelton, F. R.**, The influence of age and sex on the development of adrenal regeneration hypertension, *Lab. Invest.*, 15, 1342, 1966.
49. **Threlfall, G., Taylor, D. M., and Buck, A. T.**, Studies of changes of growth and DNA synthesis in rat kidney during experimentally induced renal hypertrophy, *Am. J. Pathol.*, 50, 1, 1967.
50. **Logothetopulos, J.**, Islet cell regeneration and neogenesis, in *Handbook of Physiology, Section 7*, Vol. 2, Freinkel, N. and Renold, A. E., Eds., American Physiological Society, Washington, DC, 1972, 67.
51. **Githens, S.**, Differentiation and development of the exocrine pancreas in animals, in *The Exocrine Pancreas: Biology, Pathobiology and Diseases*, Go., V. L. W., Gardner, J. D., Brooks, F. P., Lebenthal, E., DiMagno, E. P., and Scheele, G. A., Eds., Raven Press, New York, 1985, 21.
52. **Fitzgerald, P. J.**, The problem of the precursor cell of regenerating pancreatic acinar epithelium, *Lab. Invest.*, 9, 67, 1960.
53. **Lütcke, H., Scheele, G. A., and Kern, H. F.**, Time course and cellular site of mitotic activity in the exocrine pancreas of the rat during sustained hormone stimulation, *Cell Tissue Res.*, 247, 385, 1987.
54. **Oates, P. S., and Morgan, R. G. H.**, Pancreatic growth and cell turn-over in the rat fed raw soya flour, *Am. J. Pathol.*, 108, 217, 1982.
55. **Solomon, T. E., Vanier, M., and Morisset, J.**, Cell site and time course of DNA synthesis in the pancreas after caerulein and secretin, *Am. J. Physiol.*, 245, G99, 1983.
56. **Puplat, D., Elsässer, H. P., Adler, G., and Kern, H. F.**, Stimulation of pancreatic secretory process in the rat by low molecular weight proteinase inhibitor. III. Stimulation of DNA synthesis and cell replication, *Cell Tissue Res.*, in press, 1990.

57. **Elässer, H. P., Adler, G., and Kern, H. F.,** Time course and cellular source of pancreatic regeneration following acute pancreatitis in the rat, *Pancreas*, 1, 421, 1986.
58. **Willemer, S., Elsässer, H. P., Kern, H. F., and Adler, G.,** Tubular complexes in cerulein- and oleic acid-induced pancreatitis in the rats: glycoconjugate pattern, immunocytochemical and ultrastructural findings, *Pancreas*, 2, 669, 1987.
59. **Lechenne de la Port, P., Iovanna, J., Berger, Z., Odaira, C., Choux, R., and Sarles, H.,** The cells of tubular complexes participate in the regeneration after acute pancreatitis, *Digestion*, 38, 35, 1987.
60. **Maylie-Pfenninger, M. F. and Jamieson, J. D.,** Development of cell surface saccharides on embryonic pancreatic cells, *J. Cell. Biol.*, 86, 96, 1980.
61. **Fitzgerald, P. J., Vinijchalkul, K., Carol, B., and Rosenstock, L.,** Pancreatic acinar cell regeneration. III. DNA synthesis of pancreas nuclei as indicated by thymidine-H^3 autoradiography, *Am. J. Pathol.*, 52, 1039, 1968.

Stomach

Chapter 7

GROWTH AND MATURATION OF THE GASTRIC MUCOSA

Adhip P. N. Majumdar

TABLE OF CONTENTS

In this chapter, the pattern of structural and functional development of the gastric mucosa will be reviewed. In addition, nutritional and hormonal regulation of gastric mucosal growth and function during the early stages of life will also be reviewed. Since most of the information on these topics has been gathered from experiments performed on rats and mice, only the results of these studies will be discussed.

I. PATTERN OF GROWTH AND MATURATION

A. GROWTH

Organ growth may be due to either hyperplasia, hypertrophy, or a combination of both. Hyperplasia, defined as an increased number of cells, can be readily assessed by measuring DNA content of the tissue, since DNA is constant in a diploid cell. Thus, an increase in DNA content would reflect hyperplastic growth. In hypertrophy, the cellular mass or volume increases without affecting DNA content. This will result in an increment in weight of the tissue, RNA, or protein content.

We applied these criteria to evaluate the growth pattern of the gastric mucosa during the first 4 weeks of postnatal life in rats.[1,2] We have observed that gastric mucosal DNA content rises slowly up to about 15 d of age and then increases very rapidly over the next 7 d.[1] Such an observation suggests a rapid rise in gastric mucosal cell populations during the third postnatal week in rats. Gastric mucosal ribosomal RNA content during the first 4 weeks of postnatal life also revealed a pattern similar to that of DNA (Figure 1A). When RNA/DNA ratios for each age group were calculated, we observed a dramatic rise in RNA/DNA ratios between 18 and 30 d of age (Figure 1B). This, together with the observation that gastric mucosal DNA content also increases throughout the first 4 weeks of life,[1] suggests that mucosal growth during this period consists of both hyperplasia and hypertrophy. A similar but less pronounced phenomenon occurs between 7 and 15 d of age (Figure 1B).

Although gastric mucosal DNA content rises throughout the first 4 to 5 weeks of postnatal life,[1] the rate of DNA synthesis has been found to remain elevated only for the first 2 weeks of age and then to decrease dramatically over the next 2 to 3 weeks.[1] The rate of gastric mucosal protein synthesis also shows a similar phenomenon.[2] However, the fact that a steady rise in mucosal DNA content occurs throughout the developmental period indicates an increase in cell number with growth of the organ. Thus, when the total ^{3}H-thymidine incorporation into mucosal DNA of the whole stomach was determined, we observed a progressive increase (in incorporation) with age.[1] Taken together, our results show that, in spite of the fall in the rate of mucosal DNA synthesis, expressed per unit of DNA that begins at 15 d of age, the total number of cells undergoing division rises throughout the first 4 weeks of life.[1] During this period stomach weight increases almost linearly as does body weight. However, none of the cell types grow in a linear fashion. Jacobs and Ackerman[3] have measured the rate of growth of different gastric mucosal cells between 15 and 100 d of age in rats. They have observed that whereas the relative number of surface epithelial cells decreases, chief and base parietal cell populations increase during this period. In contrast, the relative number of neck parietal cells remains constant and surface epithelial cells decreases.[3]

B. MATURATION

In rats and mice, the functional properties of various tissues of the gastrointestinal (GI) tract, including that of the stomach, undergo dramatic changes around the third week of postnatal life, which coincides with weaning. Thus, the sudden spurt in gastric mucosal growth that occurs with weaning is also associated with functional maturation of the tissue.

To assess the functional maturation of gastric mucosa numerous investigators have studied the pattern of development of acid and pepsin secretion during the first 4 to 5 weeks of postnatal life in rats *in vivo*[4-9] and *in vitro*.[10-12] This topic has been the subject of a recent review[13] and consequently will be discussed here only briefly.

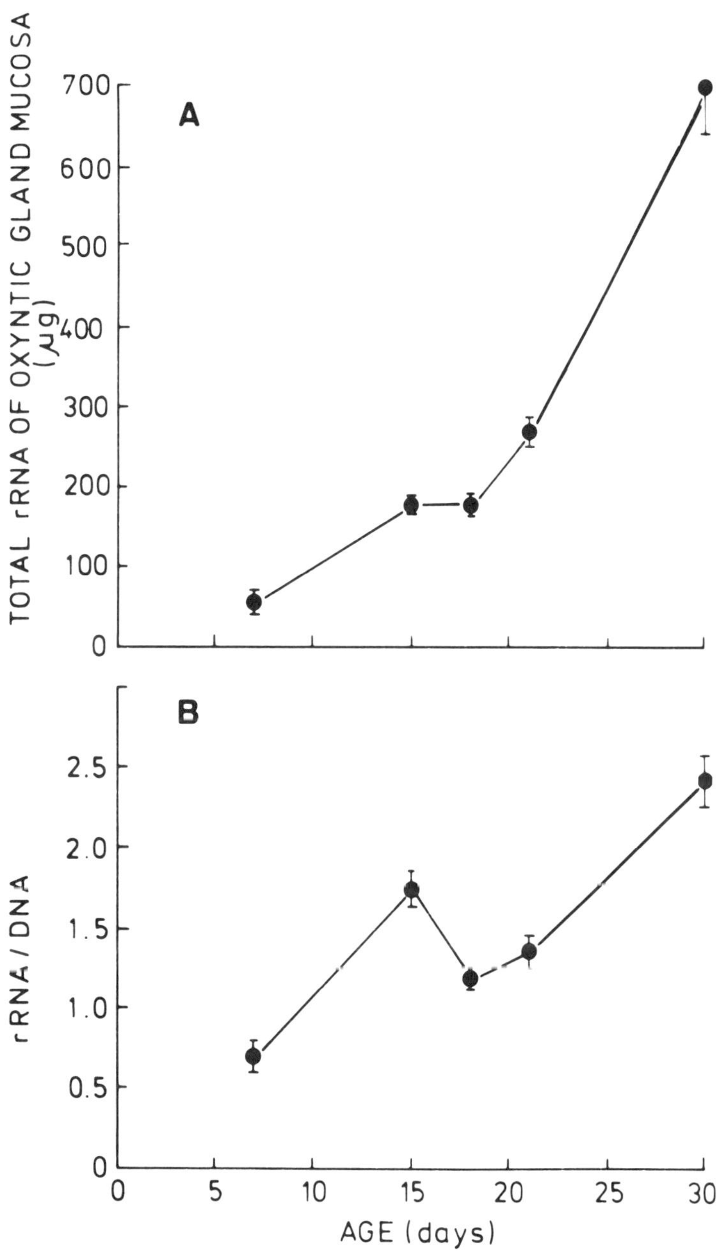

FIGURE 1. Changes in (A) ribosomal RNA (rRNA) and (B) rRNA/DNA ratio in the oxyntic gland mucosa at various stages of development. rRNA represents about 80% of total cellular RNA. Each value represents the mean ± SE of four to six observations. (From Majumdar, A. P. N., *J. Pediatr. Gastroenterol. Nutr.*, 3, 123, 1984. With permission.)

Although most species including the rat are capable of secreting acid shortly after birth,[6,11] the amount of acid secreted increases postnatally with a sharp rise occurring around the third postnatal week,[8,11,14] attaining 25% of the adult level.[13,15] Over the next 2 weeks the capacity of the gastric mucosa to secrete acid increases further and by about 40 d of age the rate of acid secretion reaches the adult level.[13,15]

As has been observed for acid secretion, gastric mucosa is also capable of secreting pepsinogen early in life, though at a very low rate.[8,9] Helander[16] has demonstrated pepsin activity

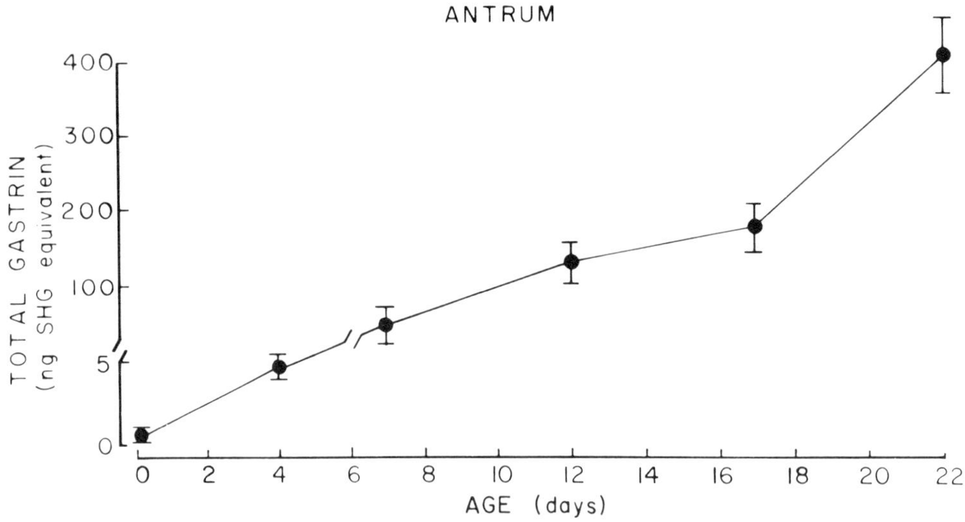

FIGURE 2. Changes in antral gastrin levels during development. Each value represents the mean ± SE (n = 8 to 10 in each age group). (Adapted from Majumdar, A. P. N., *Scand. J. Gastroenterol.*, 19 (Suppl. 101), 13, 1984.)

in the rat gastric mucosa as early as the 19th day of gestational age. Furihata et al.[17] have reported that in rats, pepsinogen levels in the gastric mucosa remain essentially unchanged for the first 17 d of age, then increase to adult levels by 30 d of age. A similar observation has also been made by Kumegawa et al.[18] Ikezaki and Johnson[8] have also measured pepsinogen secretion in 5- to 40-d-old rats following pylorus ligation. They have observed basal pepsinogen secretion to be very low until 18 d of age, whereafter it increased sharply. In a subsequent study, Seidel and Johnson[9] have found a steady rise in basal pepsinogen output during the first 4 to 5 weeks of postnatal life. The reason for this discrepancy is not apparent.

The stomach is also a source for a number of hormones, including gastrin and somatostatin.[19,20] Whereas the antropyloric mucosa of the stomach constitutes the main source of gastrin,[19,21,22] somatostatin cells are present throughout the gastric mucosa.[19,23] The pattern of development of gastrin and somatostatin in the gastric mucosa has been investigated. In rats, gastrin cells in the antrum appear at day 19 to 20 of fetal life, but do not show any significant rise for up to 2 d postnatally, whereafter they increase steadily with age, attaining the adult level around 2 ½ months of age.[24] We have examined the changes in antral gastrin levels in rats from the time of birth to 20 to 22 d of age.[25] The results depicted in Figure 2 demonstrate that antral gastrin content increases slowly for the first 17 d of postnatal life, followed by an abrupt rise between 17 and 22 d of age. This observation is slightly different from what has been reported by Lichtenberger and Johnson[26] and Takeuchi et al.[27] They have reported that in rats, antral gastrin levels remain very low and essentially unaltered for up to the age of about 18 d, whereafter they increase dramatically.

As opposed to what has been observed for gastrin cells, somatostatin cells in the gastric mucosa of rats do not appear until about 4 d of age and then increase steadily with age.[28] Essentially the same phenomenon was observed when gastric mucosal somatostatin levels were measured during the first 4 to 5 weeks of life in rats.[23]

II. REGULATION OF GROWTH AND MATURATION

The structural and functional properties of the GI tract undergo dramatic changes before the adult pattern is attained. A variety of dietary and hormonal factors have been found to be involved in regulating GI mucosal growth and function. Most of these have been examined in

TABLE 1
Effect of Undernutrition and Subsequent Nutritional Rehabilitation on Weight and Protein, DNA, and RNA Content of the Oxyntic Gland Area

Treatment	Final age (d)	Tissue weight (mg)	DNA (mg)	RNA (mg)	Protein (mg)
Well-nourished (control)	14	116 ± 3	0.59 ± 0.03	0.44 ± 0.03	7.03 ± 0.39
Undernourished	14	93 ± 6[a] (–20)	0.55 ± 0.04 (–7)	0.33 ± 0.04[b] (–25)	6.47 ± 0.31 (–8)
Well-nourished (control)	27	385 ± 23	1.21 ± 0.06	1.82 ± 0.09	29.73 ± 0.93
Nutritional rehabilitation (weaned)	27 (5)	406 ± 25	1.22 ± 0.05 (12)	2.03 ± 0.10 (10)	32.57 ± 0.93
Undernourished (nonunweaned)	27	213 ± 10[c] (–45)	0.86 ± 0.06[d] (–29)	1.37 ± 0.10[c] (–25)	17.80 ± 0.61[e] (–40)

Note: Each value represents the mean ± SEM. Figures in parentheses represent percent increase or decrease, compared with the corresponding well-nourished controls. Number of animals in each group was six.

[a] $p < 0.05$, as compared with the 14-d-old well-nourished controls.
[b] $p < 0.01$, as compared with the 14-d-old well-nourished controls.
[c] $p < 0.025$, as compared with the 27-d-old well-nourished controls.
[d] $p < 0.001$, as compared with the 27-d-old well-nourished controls.
[e] $p < 0.005$, as compared with the 27-d-old well-nourished controls.

From Majumdar, A. P. N., *Nutr. Rep. Int.*, 33, 187, 1986. With permission.

great detail in relation to small intestinal development. Little attention has, however, been paid to other tissues of the GI tract.

A. DIET

Infancy is a period of rapid growth, when immature organs grow at an accelerating rate.[29] However, the rate of growth of different organs during the early developmental period does not follow the same pattern.[29] Whereas some organs, such as the brain and lungs, show rapid growth during the early postnatal life, others grow steadily throughout this period of development.[29] In rats and mice, growth and maturation of various parts of the GI tract, including that of the stomach, show a dramatic change with the initiation of weaning. Thus, deprivation of weaning is likely to affect gastric mucosal growth and maturation. Indeed, Takeuchi et al.[27] have demonstrated that rats deprived of solid food between 18 and 25 d of postnatal life show marked reduction in antral gastrin levels and gastrin receptor populations in the gastric mucosa. We have also observed that in rats prevention of weaning results in diminution of gastric mucosal growth, as evidenced by decreased weight of the stomach and DNA, RNA, and protein content of the oxyntic gland mucosa.[30] Such an observation is similar to what has been reported earlier for the small intestine.[26] In weanling rats, prolonged intake of low-protein diet or restricted consumption of normal diet has also been shown to affect the growth and maturation of the small intestine.[31,32] To determine whether the changes in structural and functional properties of the gastric mucosa brought about by dietary changes at an early period of life could be normalized by nutritional rehabilitation, we have examined the effects of undernutrition and subsequent nutritional rehabilitation on growth and maturation of the gastric mucosa in rats.[33,34] We have observed that undernutrition from birth to 14 d of age and subsequent prevention from weaning of the 14-d undernourished rats for the next 13 d result in diminution in body weight and growth of the stomach (Table 1) as well as mucosal pepsinogen content.[33] Each of these parameters could be brought back to the respective well-nourished control levels when the 14-d-old undernourished rats were fed ad libitum solid food (nutritional rehabilitation) for the next 13 d (Table 1). In contrast, decreased antral gastrin levels resulting from undernutrition from birth to 14 d of age could not be normalized by nutritional rehabilitation.[34] This observation suggests that dietary insult at an early postnatal age affects the subsequent development of antral gastrin cells.

Okahata et al.[23] have studied the influence of diet on the development of gastrin and somatostatin in the gastric mucosa. They observed that when rats are suckled for up to 20 d of age and then switched to laboratory chow, a 3- to 5-fold rise in the levels of antral gastrin and somatostatin occurs when compared with rats that received only milk or were weaned normally. However, the somatostatin level in the fundus was not affected by the laboratory chow.[23] Thus, although weaning is associated with growth and maturation of the stomach, different cells in the gastric mucosa respond differently to solid food.

B. HORMONES

Besides food, various hormones are thought to be involved in the regulation of GI mucosal growth and function during development. In recent years a large body of evidence has appeared to indicate that gastrin, epidermal growth factor (EGF), and corticosterone may play important roles in growth and maturation of the gastric mucosa during the early stages of life.[35] Responsiveness of the gastric mucosa to gastrin, EGF, and corticosterone, however, changes during development.

Despite the fact that circulating gastrin levels in neonatal dogs, rats, and humans are relatively high,[27,36-38] gastrin does not appear to exert any of its known physiological actions at very early stages of life. Malloy et al.[38] reported that in dogs pentagastrin does not produce any change in either gastric secretion or motility until the animals are about 2 weeks old. Takeuchi et al.[27] and Ikezaki and Johnson[8] demonstrated that in rats gastrin begins to stimulate acid secretion around the age of 20 d. Gastrin-mediated stimulation of gastric mucosal histidine decarboxylase activity in rats also begins around the age of 20 d.[8] Since one of the physiological actions of gastrin is to stimulate gastric mucosal growth,[35,39] we studied the effect of pentagastrin on gastric mucosal proliferative activity at various stages of development.[1] Cellular proliferative activities were assessed by measuring the rate of mucosal DNA synthesis and thymidine kinase, an enzyme of the DNA-salvage pathway, whose activity correlates extremely well with DNA synthesis.[40-43] We observed that in 20- and 28-d-old rats, cellular proliferative activities are significantly increased following multiple injections of pentagastrin.[1] In 10-d-old suckling rats, however, the same treatment with pentagastrin had no effects on proliferation.[1] This, together with the observation that pentagastrin begins to stimulate acid secretion around the 20th day of life in rats,[8,27] suggests that sensitivity of the gastric mucosa to various actions of gastrin including its trophic action becomes apparent around the third postnatal week. This could in part be attributed to the appearance of gastrin receptors in the gastric mucosa.[27] Takeuchi et al.[27] demonstrated that receptors for gastrin do not appear in the gastric mucosa until around the 20th day of life in rats. A relationship between gastrin receptor populations and gastric mucosal sensitivity to gastrin has also been observed during the later part of life. We recently demonstrated that in rats, aging is associated with progressive loss of sensitivity of the gastric mucosa to both trophic and acid-secretory actions of gastrin.[44,45] This could be due, in part, to loss of functional receptors for gastrin. Singh et al.[46] have observed that the number of gastrin binding sites in the gastric mucosa of 24-month-old rats is lower than in their 3- and 6-month-old counterparts.

As stated above, the structural and functional properties of the gastric mucosa in rats and mice undergo dramatic changes during the third postnatal week, which coincides with weaning. Our finding of a steep rise in gastric mucosal DNA and RNA between 15 and 30 d of age indicates a rapid growth of the tissue during this period.[1,2] This rapid growth is also associated with functional maturation of the organ since gastric mucosal acid and pepsinogen secretion as well as gastrin receptor populations and antral gastrin levels increase dramatically during this period.[4-9,25-27] The role of glucocorticoids in this matter has been the subject of numerous investigations, since the steroids have been shown to hasten functional maturation of the small intestine.[47-54] As has been observed for the small intestine, glucocorticoids also appear to hasten the functional maturation of the gastric mucosa during the postnatal developmental period. In suckling rats, administration of corticosterone or hydrocortisone has been found to cause a precocious differentiation of the chief cells[55] and to increase mucosal pepsinogen levels[8,17,56] and basal acid output.[56] Tseng et al.[57] recently demonstrated that administration of corticosterone in the first 2 weeks of postnatal life in rats increases relative numbers of chief cells in the fundic mucosa without affecting the absolute number of total cells in each gland. The chief cells displayed more rough endoplasmic reticulum, zymogen granules, and Golgi saccules in corticosterone-treated rats than in control rats.[57] Glucocorticoids have also been found to induce premature appearance of gastrin receptors[58] and to increase antral gastrin levels.[25,58,59] However, in neonatal rats, glucocorticoids do not appear to affect the growth of the gastric mucosa[56,60] and inhibit colonic mucosal growth.[61]

Glucocorticoids also affect fetal gut development. Celano et al.[62] have reported that in rats, daily administration of cortisone during the third week of pregnancy results in precocious appearance of jejunal sucrase in the fetuses. Recently, we studied the effect of glucocorticoids on fetal gastric and small intestinal growth and maturation.[63] We observed that daily administration of hydrocortisone (50 mg/kg) to rats during the last 10 d of pregnancy not only causes precocious appearance of disaccharidases in the small intestine but also stimulates growth of the organ as evidenced by increased weight, DNA, RNA, and protein content at birth.[63] In addition, we have observed that newborns from hydrocortisone-treated mothers have significantly higher antral gastrin levels when compared with controls.[63] These observations suggest that glucocorticoids play a role in the regulation of growth and maturation of the fetal gut. Taken together, the results suggest that glucocorticoids induce maturation of both endocrine and exocrine cells of the gastric mucosa.

Most of the studies assessing the role of glucocorticoids in the regulation of gastric mucosal growth and maturation during the early stages of life have utilized intact animals. To further determine the role of glucocorticoids and also to assess the importance of gastrin in the regulation of gastric mucosal growth and function in the absence of the steroid, we have studied the effects of adrenalectomy and subsequent administration of hydrocortisone or pentagastrin on the structural and functional properties of the stomach in 20-d-old adrenalectomized rats.[64] The reason for choosing 20-d-old adrenalectomized rats is that around this time serum corticosterone rises markedly,[65] which has been thought to be responsible for the ensuing changes in the structural and functional properties of the GI mucosa.

As shown in Figure 3, 11 d after bilateral adrenalectomy (when rats attained 31 d of age), RNA and protein, but not DNA, content in the oxyntic gland mucosa were significantly reduced compared to the sham-operated control. Gastric mucosal pepsinogen content was also decreased by adrenalectomy.[64] Administration of either hydrocortisone or pentagastrin to adrenalectomized rats reversed these changes. In adrenalectomized rats, pentagastrin but not hydrocortisone increased DNA content (Figure 3) and thymidine kinase activity.[64] Our observation of increased DNA, RNA, and protein content, together with stimulation of thymidine kinase activity in the oxyntic gland area of the stomach of adrenalectomized rats by pentagastrin, suggests that the hormone-mediated stimulation of growth or hyperplasia is also associated with hypertrophy.[64] Hydrocortisone augmented RNA and protein content in the oxyntic gland area without affecting DNA, suggesting that growth is primarily due to hypertrophy.[64] These observations led us to conclude that (1) both glucocorticoid and gastrin are important for normal growth and maturation of the gastric mucosa and (2) gastrin exerts trophic action on the gastric mucosa in the absence of adrenal glucocorticoids.

In recent years, several lines of evidence have appeared to indicate that EGF, a polypeptide (M_r = 6045) which is structurally and functionally similar to human urogastrone,[66] plays an

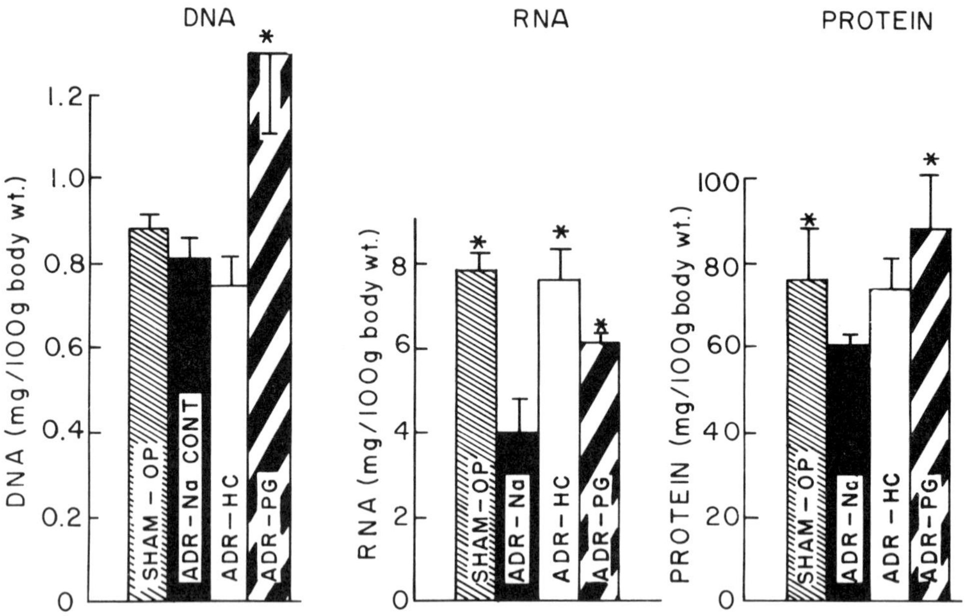

FIGURE 3. Effect of bilateral adrenalectomy and subsequent injections of hydrocortisone or pentagastrin on DNA, RNA, and protein content of the oxyntic gland area of the stomach. Groups of 20-d-old rats were bilaterally adrenalectomized or sham-operated and injected (i.m.) with 0.9% NaCl (control), pentagastrin (500 μg/kg) twice a day at 12-h intervals for 10 d. The animals were killed on the 11th morning, 12 to 14 h after the last injection. DNA, RNA, and protein content were measured and expressed as mg/100 g body weight. Each value represents the mean ± SE (n = 8 in each group). So = sham-operated; Adr = adrenalectomized controls; Adr-HC = hydrocortisone-treated adrenalectomized; Adr-PG = pentagastrin-treated adrenalectomized. *$p < 0.05$ or lower, when compared with the Adr-Na group. (Adapted from Majumdar, A. P. N., *Scand. J. Gastroenterol.*, 16, 151, 1981.)

important role in regulating growth and maturation of the digestive tract.[35] In humans, urogastrone has been localized in submandibular salivary glands, duodenal Brunner's glands, and in several other tissues.[67-69] Although in the mouse EGF concentration is found to be the highest in the submaxillary glands, considerable amounts are detected in the kidney, stomach, parotids, and pancreas.[70] In addition, EGF/urogastrone immunoreactivity has been detected in a number of body fluids, including urine, plasma, sweat, and milk.[71-73] Despite the widespread distribution of EGF/urogastrone and its receptors in the body, the physiological role of this polypeptide remains to be fully elucidated.

However, the fact that EGF stimulates proliferation and keratinization of epidermal tissues[74,75] prompted many investigators to evaluate its growth-promoting action in various tissues. The first evidence that EGF might stimulate gastrointestinal mucosal growth came from Feldman et al.[76] who demonstrated that administration of EGF to suckling mice stimulates gastric and duodenal mucosal ornithine decarboxylase activity. Subsequently, Malo and Menard[77] have demonstrated that EGF not only stimulates growth of the proximal small intestine but also the distal small intestine and proximal colon. Similar observations have also been made by others.[78] Using an organ culture system, we have demonstrated that EGF stimulates protein and DNA synthesis in the small intestine of both pre- and neonatal rats.[79] Such an observation not only indicates a direct effect of EGF on growth-related processes of the small intestine but also suggests that this polypeptide may have a role in regulating small intestinal mucosal growth in fetal and early neonatal life in rats. Like the small intestine, the stomach also responds to the growth-promoting action of EGF. Dembinski and Johnson[80] reported that in 10-d suckling rats, repeated injections of EGF for 5 d resulted in a significant increase in weight of the whole stomach and DNA, RNA, and protein content of the oxyntic gland mucosa. We have observed that in unweaned rats daily injection of EGF (20 μg/kg) for 7 d not only stimulates gastric and small intestinal mucosal protein and nucleic acid content, but also increases body weight of the animals.[30] Such a phenomenon is analogous to what has been observed after administration of growth hormone.[81] It appears, therefore, that EGF, as has been observed for growth hormone, promotes general body and tissue growth. This observation is, however, different from what has been noted with gastrin. Administration of pentagastrin to young unweaned or weaned adrenalectomized rats induces growth of the GI tissues without affecting the body weight.[26,64] These findings, together with the observations that milk possesses EGF-immunoreactivity[72] and that both EGF and milk stimulate DNA synthesis in cultured fibroblast,[82] suggest further that EGF may play an important role in regulating growth during the early stages of life.

There are also reports to indicate that EGF is involved in regulating functional maturation of the GI tract. Malo and Menard[77] have observed premature appearance of sucrase and increased activities of several other brush border enzymes in the small intestine of neonatal mice after administration of EGF. EGF has also been shown to stimulate maturation of fetal mouse duodenal absorptive cells both in organ culture[83] and *in utero*.[84] As opposed to what has been observed in the small intestine, none of the functional properties of the gastric mucosa were affected by EGF. In suckling rats, repeated injections of EGF produced no apparent change in antral or serum gastrin levels,[30,80] mucosal pepsinogen content, or the sensitivity of parietal cells to gastrin.[80] Taken together, the results suggest that although EGF stimulates gastric mucosal growth during the early stages of life, it does not induce premature functional development of the tissue.

III. SUMMARY

The structural and functional properties of the gastric mucosa in rats and mice undergo changes throughout the first 4 to 5 weeks of postnatal life to attain the adult level. The major changes, however, occur around the third postnatal week, which corresponds to the weaning period. This period is associated with a rapid increase in gastric mucosal growth, as evidenced by a steep rise in DNA and RNA content in the tissue. This rapid growth is also accompanied by functional maturation of the gastric mucosa since basal acid and pepsin output, mucosal pepsinogen levels, and antral gastrin content increase markedly. At this time, receptors for gastrin appear and the gastric mucosa becomes responsive to gastrin and secretes acid in response to the hormone. Gastrin also begins to exert its trophic action on the gastric mucosa around the third postnatal week. Glucocorticoids induce premature maturation of the gastric mucosa. This is evidenced by precocious appearance of mucosal pepsinogen and gastrin receptors and increased acid and pepsin output following administration of corticosterone or hydrocortisone to neonatal rats. Glucocorticoids also increase antral gastrin levels in neonatal rats. Bilateral adrenalectomy results in diminution of gastric mucosal growth and maturation of the gastric mucosa which could be restored to normal levels by hydrocortisone or pentagastrin. Glucocosteroids also appear to be important for fetal gut development, since administration of hydrocortisone during the last 10 d of pregnancy has been found to stimulate antral gastrin levels and growth and maturation of the small intestine.

Gastric mucosal growth but not maturation has also been found to be affected by EGF. Administration of this polypeptide to suckling rats stimulates gastric mucosal growth, but does not induce premature functional development of the tissue.

ACKNOWLEDGMENT

The work was in part supported by the Veterans Administration Medical Research Service.

REFERENCES

1. **Majumdar, A. P. N. and Johnson L. R.,** Gastric mucosal cell proliferation during development and effects of pentagastrin, *Am. J. Physiol.*, 242, G135, 1982.
2. **Majumdar, A. P. N.,** Protein synthesis by gastric mucosal ribosomes during development in rats, *J. Pediatr. Gastroenterol. Nutr.*, 3, 123, 1984.
3. **Jacobs, D. E. and Ackerman, S. H.,** Differential growth rate of rat gastric mucosal cells during postnatal ontogeny, *Am. J. Physiol.*, 247, G645, 1984.
4. **Ackerman, S. H.,** Ontogeny of gastric acid secretion in the rat: evidence for multiple response system, *Science*, 217, 75, 1982.
5. **Ducroc, R., Desjeux, J. F., Barzon, B., and Onolfo, J. P.,** Acid secretion in fetal rat stomach in vitro, *Am. J. Physiol.*, 240, G206, 1981.
6. **Garzon, B., Ducroc, R., and Geloso, J. P.,** Ontogenesis of gastric acid secretion in fetal rat, *Pediatr. Res.*, 15, 921, 1981.
7. **Gibson, R. B., Hirschowitz, I., and Huchison, G.,** Actions of metiamide and H_2-histamine receptor antagonist on gastrin H^+ and pepsin secretion in dogs, *Gastroenterology*, 67, 93, 1977.
8. **Ikezaki, M. and Johnson, L. R.,** Development of sensitivity to different secretagogues in the rat stomach, *Am. J. Physiol.*, 244, G165, 1983.
9. **Seidel, E. R. and Johnson, L. R.,** Ontogeny of gastric mucosal muscarinic receptor and sensitivity to carbachol, *Am. J. Physiol.*, 246, G550, 1984.
10. **Ducroc, R., Desjeux, J. F., Garzon, B., Onolfo, J. P., and Geloso, J. P.,** Acid secretion in fetal rat stomach *in vitro*, *Am. J. Physiol.*, 240, G206, 1981.
11. **Garzon, B., Ducroc, R., Onolfo, J. P., Desjeux, J. F., and Geloso, J. P.,** Biphasic development of pentagastrin sensitivity in rat stomach, *Am. J. Physiol.*, 242, G111, 1982.
12. **Yahav, J., Lee, P. G., and Lebenthal, E.,** Ontogeny of pepsin secretory response to secretagogues in isolated rat gastric glands, *Am. J. Physiol.*, 250, G200, 1986.
13. **Johnson, L. R.,** Functional development of the stomach, *Ann. Rev. Physiol.*, 47, 199, 1985.
14. **Ackerman, S. H. and Shindledecker, R. D.,** Maturational increases and decreases in acid secretion in the rat, *Am. J. Physiol.*, 247, G638, 1984.
15. **Christie, D. L.,** Development of gastric function during the first month of life, in *Textbook of Gastroenterology and Nutrition in Infancy*, Lebenthal, E., Ed., Raven Press, New York, 1981, 109.
16. **Helander, H. F.,** Ultrastructure and function of gastric mucoid and zymogen cells in the rat during development, *Gastroenterology*, 56, 53, 1969.
17. **Furihata, C., Kawachi, T., and Sugimura, T.,** Premature induction of pepsinogen in developing rat gastric mucosa by hormones, *Biochem. Biophys. Res. Commun.*, 47, 705, 1972.
18. **Kumegawa, M., Takuma, T., Hosoda, S., Kunii, S., and Kanda, Y.,** Precocious induction of pepsinogen in the stomach of suckling mice by hormones, *Biochim. Biophys. Acta*, 543, 243, 1978.
19. **Larsson, L. I.,** Pathology of gastrointestinal endocrine cells, *Scand. J. Gastroenterol.*, 14 (Suppl. 53), 1, 1979.
20. **Rehfeld, J. F.,** Heterogeneity of gastrointestinal hormones, in *Gastrointestinal Hormones*, Glass, G. B. J., Ed., Raven Press, New York, 1980, 433.
21. **Majumdar, A. P. N.,** Effects of fasting and refeeding on antral, duodenal and serum gastrin levels and on colonic thymidine kinase activity in rats, *Hormone Res.*, 19, 127, 1984.
22. **Larsson, L. I. and Rehfeld, J. F.,** Distribution of gastrin and CCK cells in the rat gastrointestinal tract. Evidence for the occurrence of three distinct cell types storing COOH-terminal gastrin immunoreactivity, *Histochemistry*, 58, 23, 1978.
23. **Okahata, H., Nishi, Y., Muraki, K., Kumamoto, T., Sumioka, M., Sumii, K., Hide, M., Tanaka, K., Migachi, Y., and Usui, T.,** Influence of diet on the development of gastrin and somatostatin-like immunoreactivity in the stomach of rats, *J. Pediatr. Gastroenterol. Nutr.*, 4, 463, 1985.
24. **Lehy, T., Gres, L., and Ferreira de Castro, E.,** Quantitation of gastrin and somatostatin cell populations in the antral mucosa of the rat, *Cell Tissue Res.*, 198, 325, 1979.
25. **Majumdar, A. P. N.,** Gastrin ontogeny and gastric mucosal growth, *Scand. J. Gastroenterol.*, 19 (Suppl 101), 13, 1984.
26. **Lichtenberger, L. and Johnson, L. R.,** Gastrin in the ontogenic development of the small intestine, *Am. J. Physiol.*, 227, 390, 1974.
27. **Takeuchi, K., Peitsch, W., and Johnson, L. R.,** Mucosal gastrin receptor. V. Development in newborn rats, *Am. J. Physiol.*, 240, G163, 1981.
28. **Onolfo, J. P. and Lehy, T.,** Comparative development of gastrin and somatostatin cell populations in the pancreas, stomach and duodenum of the rat during the perinatal period, *Anat. Rec.*, 288, 416, 1987.
29. **Winick, M. and Noble, A.,** Quantitative changes in DNA, RNA and protein during prenatal and postnatal growth in the rat, *Dev. Biol.*, 12, 451, 1965.

30. **Majumdar, A. P. N.,** Postnatal undernutrition: effect of epidermal growth factor on growth and function of the gastrointestinal tract in rats, *J. Pediatr. Gastroenterol. Nutr.,* 3, 618, 1984.
31. **Adams, J. and Leicher, J.,** Effects of protein-deficient diets with various amounts of carbohydrate on intestinal disaccharidase activities in the rats, *J. Nutr.,* 103, 1716, 1973.
32. **Solimano, G., Burgess, E. A., and Levin, B.,** Protein-calorie malnutrition: effect of deficient diets on enzyme levels of jejunal mucosa of rats, *Br. J. Nutr.,* 21, 55, 1967.
33. **Majumdar, A. P. N.,** Effect of undernutrition and subsequent nutritional rehabilitation or hydrocortisone administration on growth and function of the gastrointestinal tract in rats, *Nutr. Rep. Intr.,* 33, 187, 1986.
34. **Majumdar, A. P. N.,** Postnatal undernutrition: effect on antral gastrin levels at a later age, *Experientia,* 40, 751, 1984.
35. **Johnson, L. R.,** Regulation of gastrointestinal growth, in *Physiology of the Gastrointestinal Tract,* Johnson, L. R., Ed., Raven Press, New York, 1987, 307.
36. **Euler, A. R., Byrne, W. J., Cousins, L. M., Ament, M. E., and Walsh, J. H.,** Increased serum gastrin concentration and gastric acid hypersecretion in the immediate newborn period, *Gastroenterology,* 72, 1271, 1977.
37. **Euler, A. R., Byrne, P. J., Meis, R. D., Leake, R. D., and Ament, M. E.,** Basal and pentagastrin-stimulated acid secretion in newborn human infants, *Pediatr. Res.,* 13, 36 1979.
38. **Malloy, M. H., Morriss, F. H., Denson, S. E., Weisbrodt, N. W., Lichtenberger, L. M., and Adock, E. W.,** Neonatal gastric motility in dogs: maturation and response to pentagastrin, *Am. J. Physiol.,* 236, E562, 1979.
39. **Johnson, L. R.,** The trophic action of gastrointestinal hormones, *Gastroenterology,* 70, 278, 1976.
40. **Majumdar, A. P. N.,** Role of tryptophan in maintaining the structural and functional properties of the gastrointestinal mucosa, in *Advances in Diet and Nutrition,* Horwitz, C., Ed., John Libbey, London, 1984, 144.
41. **Majumdar, A. P. N.,** Pentagastrin stimulation of DNA synthesis and thymidine kinase activity in gastric mucosa of rats, *Hormone Res.,* 15, 37, 1981.
42. **Majumdar, A. P. N.,** Regulation of gastric mucosal DNA synthesis during fasting and refeeding in rats, *Digestion,* 27, 36, 1983.
43. **Fitzpatrick, L. R., Wang, P., and Johnson, L. R.,** Effect of epidermal growth factor on polyamine synthesizing enzymes in rat enterocytes, *Am. J. Physiol.,* 252, G209, 1987.
44. **Majumdar, A. P. N., Edgerton, E. A., Dayal, Y., and Murthy, S. N. S.,** Gastrin: levels and trophic action during advancing age, *Am. J. Physiol.,* 254, G538, 1988.
45. **Maitra, R. S., Edgerton, E. A., and Majumdar, A. P. N.,** Gastric secretion during aging in pyloric-ligated rats and effects of pentagastrin, *Exp. Gerontol.,* 23, 463, 1988.
46. **Singh, P., Rae-Venter, B., Townsend, C. M., Jr, Khalil, T., and Thompson, J. C.,** Gastrin receptors in normal and malignant gastrointestinal mucosa: age-associated changes, *Am. J. Physiol.,* 249, G761, 1985.
47. **Doel, R. G. and Kretchmer, N.,** Intestinal invertase; precocious development of activity after injection of hydrocortisone, *Science,* 143, 42, 1964.
48. **Doel, R. G., Rosen, G., and Kretchmer, N.,** Immunochemical studies of intestinal disaccharidase during normal and precocious development, *Proc. Natl. Acad. Sci. U.S.A.,* 54, 1268, 1965.
49. **Galand, G. and Forstner, G. G.,** Isolation of microvillus plasma membrane from suckling rat intestine. The influence of premature induction of digestive enzymes by injection of cortisol acetate, *Biochem. J.,* 144, 293, 1974.
50. **Henning, S. J., Helman, T. A., and Kretchmer, N.,** Studies on normal and precocious appearance of jejunal sucrase in suckling rats, *Biol. Neonate.,* 26, 249, 1975.
51. **Koldovsky, O., Herbst, J. J., Burke, J., and Sunshine, P.,** RNA and DNA in intestinal mucosa during development of normal and cortisol-treated rats, *Growth,* 34, 359, 1970.
52. **Raul, F., Simon, P. M., Kedinger, M., Grenier, J. F., and Haffen, K.,** Separation and characterization of intestinal brush border enzymes in adult rats under normal conditions and after hydrocortisone injections, *Enzymes,* 23, 89, 1978.
53. **Henning, S. J. and Sims, J. M.,** Delineation of the glucocorticoid-sensitive period of intestinal development in the rat, *Endocrinology,* 104, 1158, 1979.
54. **Majumdar, A. P. N.,** Bilateral adrenalectomy: effects of hydrocortisone and pentagastrin on the activity of intestinal disaccharidases and alkaline phosphatase in weanling rats, *Scand. J. Gastroenterol.,* 16, 177, 1981.
55. **Tatrematsu, M., Takahashi, M., Tsuda, H., Hirose, M., Furihata, C., and Sugimura, T.,** Precocious differentiation of immature chief cells in fundic mucosa of infant rats induced by hydrocortisone, *Cell Differentiation,* 4, 285, 1975.
56. **Tseng, C. C. and Johnson, L. R.,** Does corticosterone affect gastric mucosal growth during development?, *Am. J. Physiol.,* 250, G633, 1986.
57. **Tseng, C. C., Schmidt, K. L., and Johnson, L. R.,** Hormonal effects on development of the secretory apparatus of chief cells, *Am. J. Physiol.,* 253, G274, 1987.
58. **Peitsch, W., Takeuchi, K., and Johnson, L. R.,** Mucosal gastrin receptors. VI. Induction by hydrocortisone in newborn rats, *Am. J. Physiol.,* 240, G442 1981.

59. **Majumdar, A. P. N. and Rehfeld, J. F.,** Effects of hydrocortisone on gastrin cell functions in various tissues of suckling rats, *Digestion*, 27, 36, 1983.
60. **Arsenault, P. and Menard, D.,** Stimulatory effects of epidermal growth factor on deoxyribonucleic acid synthesis in the gastrointestinal tract of the suckling mouse, *Comp. Biochem. Physiol.*, 86B, 123, 1987.
61. **Buts, J. P., DeMeyer, R., and Kolanowski, J.,** Ontogeny of cell proliferation and DNA synthesis in rat colon: role of glucocorticoids, *Am. J. Physiol.*, 244, G469, 1983.
62. **Celano, P., Jumawan, J., Horowitz, C., Lace, H., and Koldovsky, O.,** Prenatal induction of sucrase activity in rat jejunum, *Biochem. J.*, 162, 469, 1977.
63. **Majumdar, A. P. N. and Nielsen, H.,** Influence of glucocorticoids on prenatal development of the gut and pancreas in rats, *Scand. J. Gastroenterol.*, 20, 65, 1985.
64. **Majumdar, A. P. N.,** Bilateral adrenalectomy: effects of hydrocortisone and pentagastrin on some structural and functional properties of the stomach of weanling rats, *Scand. J. Gastroenterol.*, 16, 151, 1981.
65. **Henning, S. J.,** Plasma concentrations of total and free corticosterone during development in the rat, *Am. J. Physiol.*, 235, E451, 1978.
66. **Gregory, H.,** Isolation and structure of urogastrone and its relationship to epidermal growth factor, *Nature*, 257, 325, 1975.
67. **Elder, J. B., Williams, G., Lacey, F., and Gregory, H.,** Cellular localization of human urogastrone/epidermal growth factor, *Nature (London)*, 271, 465, 1978.
68. **Heitz, P. H., Kasper, U., Van Noorden, S., Polak, J. M., Gregory, H., and Pearse, A. G. E.,** Immunohistochemical localization of urogastrone to human duodenal and submandibular glands, *Gut*, 19, 408, 1978.
69. **Kasselberg, A. G., Orth, D. N., Gray, M. E., and Stahlman, M. T.,** Immunohistochemical localization of human epidermal growth factor/urogastrone in several human tissues, *J. Histochem. Cytochem.*, 33, 315, 1985.
70. **Byyny, R. L., Orth, D. N., and Cohen, S.,** Radioimmunoassay of epidermal growth factor, *Endocrinology*, 90, 1261, 1972.
71. **Ances, I. G.,** Serum concentrations of epidermal growth factor in human pregnancy, *Am. J. Obstet. Gynecol.*, 115, 357, 1973.
72. **Hirata, Y. and Orth, D. N.,** Concentrations of epidermal growth factor, nerve growth factor, and submandibular gland renin in male and female mouse tissue fluids, *Endocrinology*, 105, 1382, 1979.
73. **Gregory, H., Walsh, S., and Hopkins, C. R.,** The identification of urogastrone in serum, saliva and gastric juice, *Gastroenterology*, 77, 313, 1979.
74. **Cohen, S. and Elliott, G. A.,** The stimulation of epidermal keritinization by a protein isolated from the submaxillary gland of the mouse, *J. Invest. Dermatol.*, 40, 1, 1963.
75. **Savage, C. R., Jr. and Cohen, S.,** Proliferation of corneal epithelium induced by epidermal growth factor, *Exp. Eye Res.*, 15, 361, 1973.
76. **Feldman, E. J., Aures, D., and Grossman, M.,** Epidermal growth factor stimulates ornithine decarboxylase activity in the digestive tract of mouse, *Proc. Soc. Exp. Biol. Med.*, 159, 400, 1978.
77. **Malo, C. and Menard, D.,** Influence of epidermal growth factor on the development of suckling mouse intestinal mucosa, *Gastroenterology*, 83, 28, 1982.
78. **O'Loughlin, E. V., Chung, M., Hollenberg, M., Hayden, J., Zahavi, I., and Gall, D. G.,** Effect of epidermal growth factor on ontogeny of the gastrointestinal tract, *Am. J. Physiol.*, 249, G674, 1985.
79. **Conteas, C. N., DeMorrow, J. M., and Majumdar, A. P. N.,** Effect of epidermal growth factor on growth and maturation of fetal and neonatal rat small intestine in organ culture, *Experientia*, 42, 950, 1986.
80. **Dembinski, A. B. and Johnson, L. R.,** Effect of epidermal growth factor on the development of rat gastric mucosa, *Endocrinology*, 116, 90, 1985.
81. **Bauman, D. E., Eisenmann, J. H., and Curie, W. B.,** Hormonal effects on partitioning of nutrients for tissue growth: role of growth hormone and prolactin, *Fed. Proc.*, 41, 2538, 1982.
82. **Klagsbrun, M.,** Human milk stimulates DNA synthesis and cellular proliferation in cultured fibroblasts, *Proc. Natl. Acad. Sci. U.S.A.*, 75, 5057, 1978.
83. **Beaulieu, J. F. and Calvert, R.,** The effect of epidermal growth factor (EGF) on the differentiation of the rough endoplasmic reticulum in fetal mouse small intestine in organ culture, *J. Histochem. Cytochem.*, 29, 765, 1987.
84. **Calvert, R., Beaulieu, J. F., and Menard, D.,** Epidermal growth factor (EGF) accelerates the maturation of fetal mouse intestinal mucosa *in utero*, *Experientia*, 38, 1096, 1982.

Chapter 8

NEUROHORMONAL REGULATION OF GASTRIC MUCOSAL GROWTH

G. Willems

TABLE OF CONTENTS

I. INTRODUCTION

The epithelial cells in the mucosa of the stomach are continuously renewed. Just as in the rest of the gastrointestinal tract, there is continuous proliferative activity of progenitor cells in the gastric glands which compensates for cell losses from the mucosal surface and maintains a morphological steady state. Self replication of well differentiated cells or differentiation from progenitor cells are alternative pathways by which more specialized cells in the deeper segments of the gastric glands are replaced. Increased mitotic activity or differentiation, producing more young cells, as well as prolonged survival of existing elements, are likely to produce hyperplasia of specific cell populations. Conversely, a negative imbalance between cell production and cell loss may result in mucosal ulcerations or atrophy.

An increasing number of factors which may influence the renewal rate of several epithelial cell types in the stomach have been described. Knowing the physiological significance may be important to interpretation of cell kinetic abnormalities and to our understanding of a number of human gastric diseases. Factors controlling gastric cell population kinetics may also be implicated in the development as well as in the treatment of abnormal mucosal growth.

II. NORMAL CELL TURNOVER IN THE GASTRIC GLANDS

The glands of the gastric fundus contain at least five different types of epithelial cells: mucous, mucous neck, zymogen, oxyntic, and immature neck cells. The two predominant endocrine cell types in the fundus are entero-chromaffin-like (ECL) and somatostatin (D) cells. In the antrum, mucous cells, pyloric type cuboid mucous cells, and immature cells are observed, whereas gastrin (G) and D cells are the predominant hormone producing cells.

The immature neck cells have generally been thought to represent the unique and multipotential precursor to all other cell types in the gastric glands. Convincing data in favor of this unitarian theory have been reported, namely the existence of intermediate cell types between most gastric glandular cells.[1] However, several other observations are incompatible with this hypothesis.[2]

The mucous cells are renewed through mitotic activity of immature neck cells which differentiate into mucous cells while migrating upwards to the surface. During their migration the young mucous cells retain their ability to divide until they leave the so-called progenitor area. Once the more superficial area is reached, the cells lose their mitotic capacity. The entire gastric surface epithelium is renewed within 3 to 6 d in most adult mammals.

Oxyntic (or parietal) cells are unable to synthesize DNA or to divide.[2,3] They differentiate in the progenitor area of the glands from the immature neck cells.[2] Thereafter they undergo a slow downward migration. It takes approximately 3 months for parietal cells to reach the bottom of the glandular tube in rodents. The mechanisms by which these cells die and are removed from this area are still unknown.

Experimental data on developing[4] and regenerating[5,6] gastric mucosa have suggested that zymogen cells (chief cells, pepsinogen cells), like parietal cells, may differentiate from immature cells in the neck area. In the normal adult however, mature zymogen cells reproduce themselves through mitosis.[2]

The consequence of such a kinetic autonomy of a cell type in the gastric glands is that regulation of its renewal may be different from that in the common progenitor cell zone of the same glands. For instance, acceleration of mitotic activity of progenitor cells by pentagastrin increases the number of mucous and oxyntic cells, but not of peptic cells.[7] Conversely, vagotomy may selectively decrease the number of peptic cells but not of oxyntic cells.[8]

Combination of specific G-cell staining with antigastrin antibodies and autoradiography after ^{3}H thymidine has allowed the study of antral G cell population kinetics in rodents.[9,10]

These studies have shown that G cells may be renewed by self replication. Whether G cells are also produced through an alternative mechanism, namely through differentiation from immature cells,[10] is still a matter of controversy. Using similar autoradiographic methods with antisomatostatin antibodies, the uptake DNA precursors by D cells as well as their ability to divide has been observed in the antral mucosa of rodents.[11] The slow renewal rate of D cells appears to be similar to that of G cells.

The kinetics of the ECL cells in the gastric fundus have remained unclear for a long time, perhaps because no important clinical problems involving the renewal of these cells were known. Moreover, the hormone released by these cells, which represent the predominant endocrine cell type in the fundus, is still unknown. An additional difficulty has been that specific staining of ECL cells was not possible in combination with autoradiography. Recent observations of ECL cell hyperplasia and carcinoids in several human diseases that are accompanied by hypergastrinemia[12,13] have focused interest on this cell type. The recent observation that ECL cells store histamine and that they can be stained selectively using histamine antibodies[14] makes kinetic studies of these cells possible.

The latter investigations have shown that the ECL cell population in the oxyntic glands is renewed entirely through mitotic division of mature ECL cells.[15] The evidence available thus indicates that G cells, D cells, and ECL cells are able to proliferate through mitotic division. This kinetic autonomy is another challenge to the unitarian theory of a common gastric cell lineage. Such a remarkable ability to divide despite a high degree of differentiation is not a normal property of all endocrine cells in the gut. For instance, secretin cells in the duodenum[16] or a series of endocrine cells in the small intestine[17] are unable to undergo mitosis and appear to be renewed through differentiation of immature precursor cells.

The fact that proliferation of several types of endocrine cells may be independent from that of the gland progenitor cells opens the possibility of selectively manipulating the size of these cell populations. Hyperplasia or atrophy of a given endocrine cell population may have considerable consequences on gastric function. A general principle of tumor induction is that specific mitogenic influences on a given cell population can occur when the cells are able to divide, particularly when their mitotic activity is stimulated.

III. NONSPECIFIC INFLUENCES ON CELL RENEWAL IN THE STOMACH

Cell renewal in the stomach can be modified by normal or pathological nonspecific influences and also by tissue or cell population specific neural or hormonal control factors. It is probable that a network of several individual factors, some of which are still unknown, are involved in the former.

A. FASTING AND FEEDING

Since feeding represents the most physiological stimulant to the stomach, the effect of a meal on gastric mucosal cell growth has been investigated for many years.[18]

Using *in vitro* labeling of oxyntic mucosal biopsies with ^{3}H-thymidine and autoradiography, a burst of DNA synthetic activity was observed in the progenitor area of the glands between 8 and 20 h after refeeding of fasted dogs.[19] A peak of mitotic activity followed the peak of DNA synthesis. At 24 h after the meal, the production of young cells began to restore the gastric pits that had been desquamated by feeding (Figure 1). DNA synthesis is preceded by activation of several enzymes, namely DNA polymerase and thimidine kinase, during the so-called "lag period".[20] Because the delayed increase in DNA synthesis activity may reach 300% of control values and higher values are seen as late as 20 h after feeding, studies on cell renewal in the stomach should take this source of variation into account. Unfortunately, most human studies have been carried out after fasting periods that are too short to avoid interference from previous meals. This may explain questionable proliferative index values published in human patients

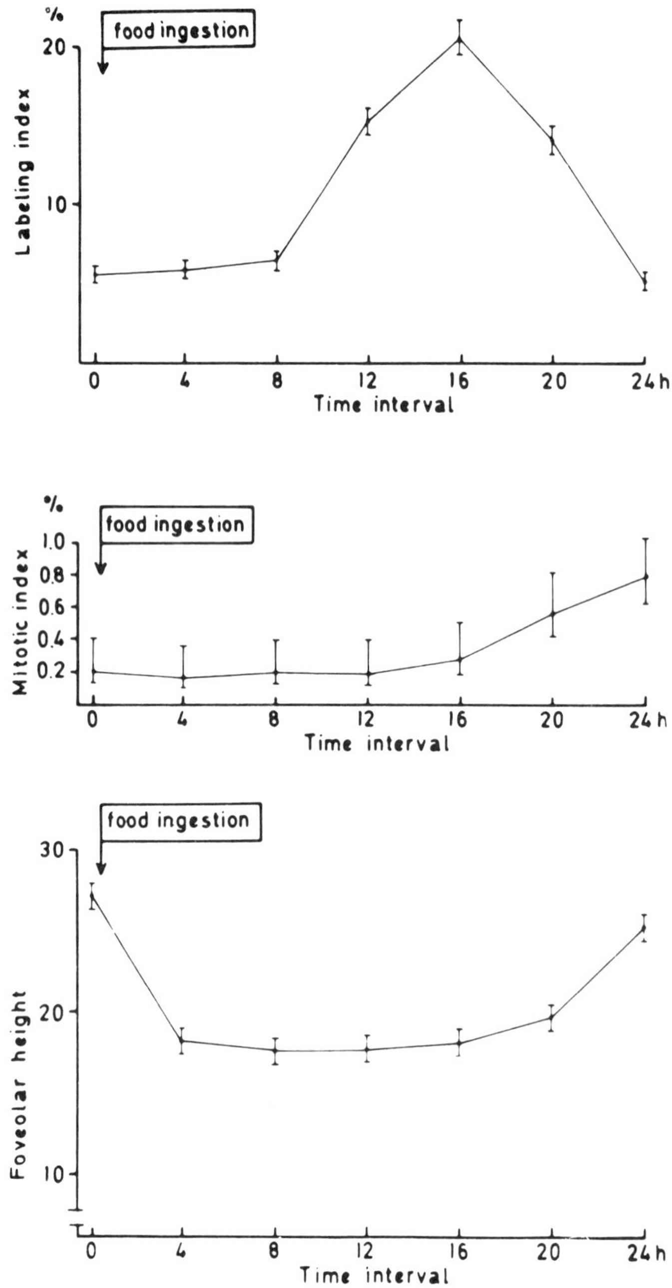

FIGURE 1. Effect of a test meal on labeling (DNA synthesis) index, mitotic index, and mucosal thickness. Mean values in the fundic mucosa of four dogs. Note the burst of proliferative activity starting between 8 and 12 h after feeding. (From Willems, G., Vansteenkiste, Y., and Smets, P., Effects of food ingestion on the cell proliferation kinetics in the canine fundic mucosa, *Gastroenterology,* 61, 323, 1971. © by Williams & Wilkins, 1971. With permission.)

because the minimum 24-h fasting period required before biopsy was not always performed. Moreover, this nonspecific effect of the meal could be prolonged further by retarded gastric emptying in some pathological conditions.

The mechanisms by which fasting and refeeding modify the rate of epithelial cell renewal in the stomach are probably multiple and are still poorly understood. The hypothesis of increased workload inducing mitosis is a simplistic but frequently used explanation which may actually disguise our ignorance about the true factors involved. The effect of luminal nutrients on cell proliferation has never been investigated in the stomach. Two main regulators are generally considered to be involved in the variation of gastric epithelial cell renewal after feeding. The release of a negative feedback system from the surface epithelium which could involve a chalone effect on the progenitor cell zone is one of them. On the other hand, neurohormonal factors released by the meal clearly play an important role.

B. CELL LOSS AND GASTRIC CHALONES

Feeding induces a massive loss of epithelial cells from the gastric surface, sometimes removing the entire mucosa down to the lamina propria. Recent studies have shown that within 30 min after desquamation, an immediate epithelial salvage mechanism is initiated which brings cells through replication and migration from the inside of the crypts toward the surface.[21] Thus, the continuity of the epithelial surface is repaired very rapidly. Actual restoration of the initial mucosal cell population occurs only between 20 and 24 h later, after the wave of proliferative activity described earlier.[19] As in other epithelia, the latter mitotic response could be a tissue specific reaction to epithelial cell loss. Indeed, in most tissues, mitotic control depends on a negative feedback mechanism, involving tissue specific antimitotic messengers called "chalones". This type of control ensures that one cell is produced for each cell that is lost, because the local concentration of chalones decreases in parallel with the number of cells in this tissue. When, for instance, epidermal tissue is damaged by trauma, a DNA synthesis peak is observed in the basal layer around the lesion within 9 h of injury. Such latent periods are similar to those elapsing between feeding and the DNA synthetic wave in the stomach. Philpott[22] has observed inhibition of DNA synthesis and mitosis in cultured gastric mucosal cells when extracts of gastric mucosa are added *in vitro* and has attributed this tissue-specific inhibition to a gastric chalone. However, more than 15 years after this study no confirmation of these results has been published. Because nobody has ever succeeded in isolating or purifying a gastric chalone, this mechanism is becoming increasingly doubtful. Cellular exfoliation at the surface could perhaps be compared with any kind of wounding in other tissue. The subsequent mitotic reaction could represent normal nonspecific tissue repair, the mechanisms of which remain unknown.

Such nonspecific stimulation of epithelial cell division in the stomach may become an abnormally intense reaction when massive gastric exfoliation is induced by several chemical irritants. Desquamating agents like bile acids, ethanol, spicy pickles, or aspirin, for instance, may induce considerable mitotic activity in the stomach through this indirect mechanism. Chronically increased cell loss of unknown origin might be one of the factors causing accelerated epithelial cell renewal and subsequent maturation disorders in human atrophic gastritis. Regulation of the mucosal resistance to cell loss is an important research field to be explored further.[23]

IV. HORMONAL FACTORS

A. GASTRIN

An alternative approach to explain postprandial epithelial cell division in the oxyntic mucosa has been to investigate the effect of hormones released by feeding. There is a considerable body of evidence for a pharmacological trophic effect of gastrin or pentagastrin on the gastric mucosa. The first observation that gastrin is actually a mitogenic hormone for the oxyntic mucosa was done in the canine stomach.[24] In this study we observed that a latent period of approximately 8 h elapsed between infusion of gastrin and the synchronization of progenitor cells into the DNA synthesis phase. This lag period was similar to that found after feeding (Figure 2). Histamine, despite inducing the same acid hypersecretion, has no influence on mucosal cell turnover.[25] As

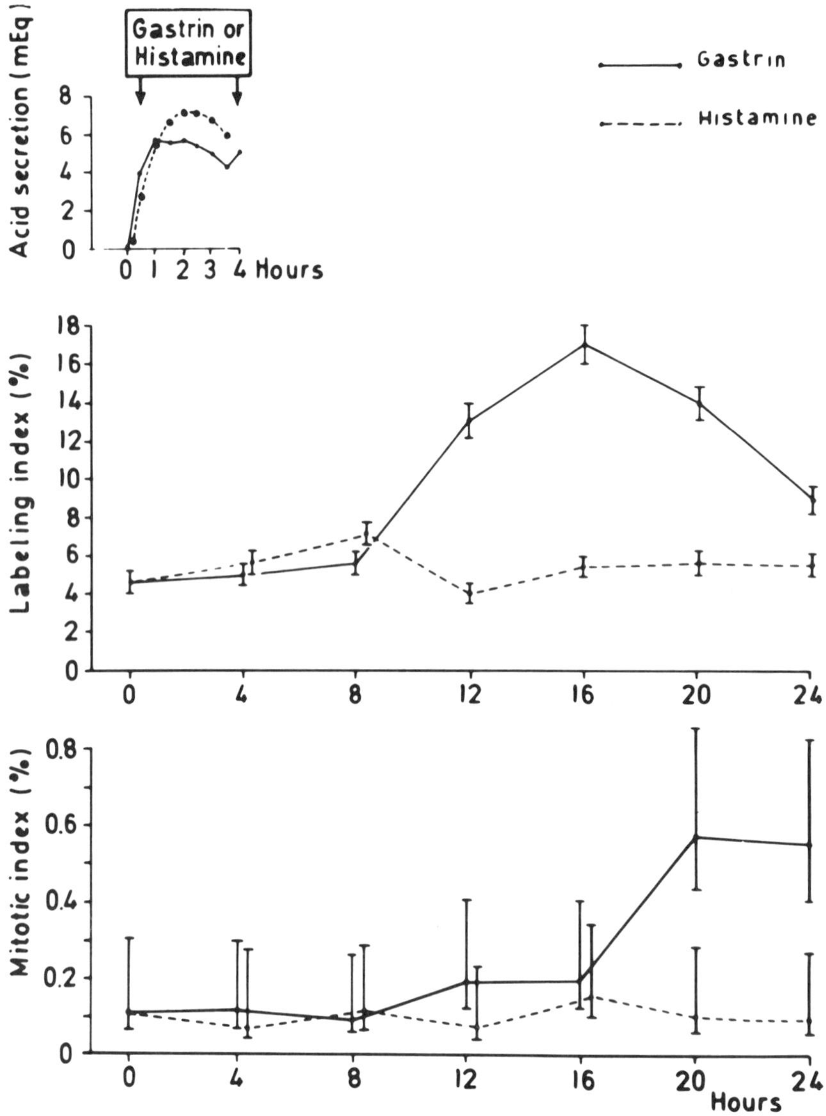

FIGURE 2. Effect of gastrin and histamine on gastric acid secretion and on proliferative parameters. Mean values in the fundic mucosa of four dogs. Histamine had no effect on the parameters of cell turnover. Note the burst of proliferative activity starting between 8 and 12 h after gastrin. (From Willems G., Vansteenkiste, Y., and Limbosch, J. M., Stimulating effect of gastrin on cell proliferation kinetics in canine fundic mucosa, *Gastroenterology*, 62, 637, 1972. © by Williams & Wilkins, 1972. With permission.)

after feeding, the increase in DNA synthesis induced by pentagastrin is preceded by an increase in thymidine kinase activity.[26] Previously, Crean had observed that chronic administration of pentagastrin to rats increased the total parietal cell mass.[7] On the basis of stimulation of RNA synthesis in gastric and duodenal mucosa by pentagastrin in rodents, Johnson hypothesized that gastrin was a trophic hormone not only for the stomach but also for the entire gastrointestinal tract.[27] Much controversy has been raised by this interesting and attractive theory because it was based mainly on studies using pharmacological doses of pentagastrin and sometimes controversial methods for estimating cell proliferation.[28-30] It soon became clear that gastrin had no trophic properties on the antral mucosa[31] and had only very weak influences on the small intestine.[28,30]

The physiological importance of the mitogenic properties of gastrin in the gut was disputed.[28,29]

Concerning the oxyntic progenitor cells there is little doubt that physiological levels of endogenous gastrin are able to stimulate their mitotic activity.[31-34] This trophic property has been proposed for use in promoting the healing of acute gastric ulcerations.[35] If the latter effect holds true in the human stomach, important clinical applications could be expected. Marked efficacy of drugs inducing anacidity as well as hypergastrinemia on the healing or prevention of fundic lesions in patients with hemorrhagic gastritis would be predicted. Besides direct stimulation of mitotic activity in the oxyntic gland progenitor cells, pentagastrin and endogenous gastrin were found to increase the parietal cell mass in rats as well as in patients with Zollinger Ellison syndrome.[7,36,37] Autoradiography indicated that gastrin increased not only the rate of replication in the immature progenitor cells, but also the rate of the maturation process of these cells into parietal cells in the oxyntic glands.[38] Thus the mechanism of parietal cell hyperplasia induced by gastrin was shown to be an active growth promoting effect instead of a passive accumulation of old parietal cells because of their prolonged survival. Morphogenesis of young parietal cells and mitotic stimulation of progenitors were actually two quite different aspects of the so-called trophic property of this hormone.

Recently, the existence of a third trophic mechanism by which gastrin promotes growth in the mucosa has been described, namely its ability to induce endocrine cell proliferation in the fundus.[12,13] This hypothetical mechanism could be elicited by the use of potent gastric antisecretory drugs. This point will be discussed in detail below.

B. CHOLECYSTOKININ

The trophic effect of cholecystokinin (CCK) and of its synthetic analogues on the pancreas is generally recognized.[39] Stimulation of DNA synthesis by CCK octapeptide or by caerulein has been observed also in the duodenum[40] and in the gallbladder,[41] but not in the gastric oxyntic mucosa,[40-42] while its direct effect in the intestine is questionable.[30] Recently we have observed significant stimulation not only of ^{3}H-thymidine uptake and mitotic activity, but also of the upward epithelial cell migration in the antral glands of rats given chronic caerulein.[42] This effect could not be attributed to increased bile and pancreatic reflux into the stomach, since it was not abolished by distal diversion of these secretions. Further studies are required for confirming this important effect of CCK; it is still far from certain that this effect is a physiological one. At this point, however, selective hormonal manipulation of the cell cycle in gastric cancers[43,44] should be considered, since most adenocarcinomas arise from the antral progenitor cells. That CCK may promote antral carcinogenesis through mitotic stimulation is another yet unproved, though potentially interesting, working hypothesis. Such fascinating clinical prospects should encourage more basic research in this field.

C. OTHER PEPTIDES

Bombesin can affect the release of several gut hormones. It is a known growth factor of oat cell carcinoma.[45] Its chronic administration was found to stimulate mitotic replication of G cells in the adult antral mucosa.[46] The physiological importance of this peptide is, however, far from being established. Injections of growth hormone releasing factor (GHRF) have been found to increase DNA synthesis and mitotic indices of oxyntic glands in one study.[47] Because the plasma gastrin levels were also increased in these animals, an indirect effect of gastrin could not be excluded. Epidermal growth factor (EGF) seems to stimulate proliferative activity in several epithelia. For almost every site in the gut, gastric mucosa included,[48] a stimulatory effect of EGF, as well as convincing data showing the absence of any effect,[49] have been published. As for the other peptides, the physiological significance of this possible mitogen for the stomach is not known. Its lack of specificity may represent a handicap for practical use. Enteroglucagon is considered to be the best candidate for explaining the potent mitogenic effect in the remnant gut segments after extended small intestinal resection,[50] but its trophic effect appears to be restricted to the small intestine and does not affect the stomach. Secretin counteracts the promoting effect of gastrin on progenitor cell DNA synthesis[51] as well as on parietal cell morphogenesis,[52] but it seems not to have a direct "antitrophic" activity.[53] Somatostatin, when administered systemically, inhibits cell proliferation in the stomach.[54] Somatostatin inhibits the release of so many hormonal peptides, including the trophic peptides described above, that direct influence on the progenitor cells is questionable. Because somatostatin could be a potent inhibitor of growth stimulation by gastrin or CCK, its use at pharmacological doses might be proposed to inhibit proliferation of certain cell types and abnormal growths induced by these hormones.

D. VAGAL FACTORS

Truncal vagotomy increases proliferative parameters in the oxyntic glands of rats,[55] whereas superselective vagotomy in humans does not.[56] It should be mentioned that serum gastrin levels in the patients from the latter study were not elevated, whereas vagotomy in rats induces hypergastrinemia and gastrin cell hyperplasia.[57] Although vagotomy does not modify the total parietal cell mass,[8] it decreases significantly the maturation rate of progenitor cells in young parietal cells.[55] Therefore, it has been postulated that the life cycle of parietal cells may be prolonged. After vagotomy, the parietal cell turnover would slow down and the parietal cell population would get older. Curiously, pentagastrin as well as gastrin seems to be unable to stimulate the morphogenesis of the parietal cells when the latter are vagally denervated.[8] Bilateral vagotomy in rats does not modify the fundic argyrophil cell mass, although these endocrine cells should also proliferate under the effect of the endogenous hypergastrinemia.[12,58] Unilateral vagotomy, however, has clear antitrophic effects on the oxyntic endocrine cells in the denervated area.[58] In the latter experimental model the animals have no hypergastrinemia and so the effect of vagotomy alone can be observed selectively.

In the antral mucosa, stimulation of DNA synthesis in G cells[59] has been shown to precede the G cell hyperplasia induced by truncal vagotomy. It is likely that G cell proliferation in these animals may be attributed to antral distention and gastric stasis rather than to a direct effect of denervation. Food in the stomach is indeed a potent stimulant of G cell replication.[60] In conclusion, vagal stimuli may exert a physiological growth promoting effect, mainly in specialized cells of the oxyntic mucosa. Most of the time this vagal trophic factor is not evident because it is counteracted by the hypergastrinemia usually induced by vagal denervation.

E. ANTISECRETAGOGUES

Increasingly potent inhibitors of acid secretion are being produced and there is evidence that a constant high intragastric pH may cause increased gastrin release. Because of the potential danger of increased mitotic activity, the possibility that these drugs may influence cell proliferation in the stomach has been explored.

Prolonged administration of metiamide[61] and also of antacids[62] has been found to induce hyperplasia of parietal cells in rats. No mitogenic effect could be observed when cimetidine was given to rats.[63] Unfortunately, the interpretation of data on progenitor cell activity in human ulcer patients under cimetidine is unreliable because kinetic parameters measured after inadequate fasting periods are not reliable.[64]

Very recently, we have observed that prolonged treatment with omeprazole, a new powerful and long acting antisecretagogue, stimulates progenitor cell activity in the oxyntic mucosa in rats.[65] Treatment with this drug also increases the number of ECL cells, eventually leading to ECL cell hyperplasia in the oxyntic mucosa.[65-67] That prolonged treatment with omeprazole or with potent H_2 blockers in rats was sometimes able to induce gastric ECL cell carcinoids was an alarming observation.[67,68] Because carcinogenic influences may require more than 20 years before causing cancer in humans, these observations have raised obvious questions about the future of the numerous patients in which chronic suppression of acid secretion by such drugs would be produced. Knowing the mitogenic factor triggered by these treatments appears to be important. Gastrin may be involved since administration of pentagastrin to rats induces ECL cell hyperplasia,[66] whereas endogenous hypergastrinemia is usually accompanied by an increase in ECL cell density.[12,13] In a study on the effect of high doses of omeprazole on cell proliferation in rats, we have observed a significant increase in mitotic activity in the ECL cell population.[65] With this treatment, not only mitotic figures were appearing in ECL cells, but the DNA synthetic index reached levels as high as 20 times their control values. In the same animals, the DNA synthetic index was only doubled in the crypt progenitor cells and serum gastrin was increased. There was actually a significant correlation between the proliferative parameter values in the ECL cells and the serum gastrin concentrations of the animals.

The extreme sensitivity of ECL cells, which was notably higher than that of the common progenitor cell, to the mitogenic influence of this treatment may explain why ECL carcinoids and no adenocarcinoma are observed after prolonged administration of such potent antisecretagogues. An alternative explanation is that ECL cells are not as quickly desquamated in the gastric lumen as the epithelial mucous cells.[15] ECL cells survive in the mucosa for a long time and might therefore represent more "stable" targets for potential carcinogens.

Although correlation with the serum gastrin level did not prove that gastrin is the mitogenic factor involved, the hypothesis that gastrin caused the ECL cell proliferation appears to be the most reasonable one at the present time. The observation of ECL cell hyperplasia and carcinoids in human patients with hypergastrinemic states,[12,13,69,70] and very recently, of their regression after antrectomy,[71] further supports this hypothesis.

F. MISCELLANEOUS FACTORS

Prostaglandins have been reported to be actively trophic for the gastric mucosa.[72] This observation was based mainly on measures showing mucosal and cell hypertrophy after exogenous administration of PGE_2. This effect has not been confirmed by recent studies,[79] but indomethacin was found to stimulate cell turnover in the fundus despite inhibition of endogenous prostaglandin production.[72] Actually, the mechanism by which prostaglandins act on the gastric mucosa might include passive accumulation of mucous cells due to depressed turnover, as well as an increase in the cell size[74] in the mucosa, rather than an active growth promotion through mitotic stimulation.[75] The latter mechanisms would resemble those occurring in the gastric mucosa after administration of carbenoxolone.[23]

Circadian variations in mitotic activity of gastric progenitor cells have been described in rats.[48] Similarly, diurnal variations in the DNA synthesis activity of zymogen cells,[2] gastrin cells,[9] and ECL cells[65] have been reported in small rodents. The interference of feeding[19] on proliferative activity was, however, not controlled in these studies. Yet, no periodic changes in proliferative activity have been reported to occur in the human stomach. As shown in a series of earlier studies it is probable that several other hormonal factors may influence the population kinetics of gastric cells. For instance cortisone decreases gastric epithelial exfolation and turnover and hypophysectomy results in atrophy of the oxyntic mucosa.[76] Pregnancy and lactation are accompanied by hyperplasia of the parietal and peptic cells of rats.[77] Intestinal resection has been claimed to stimulate parietal, zymogen gastrin, and somatostatin cell proliferation and these effects could involve still unknown trophic gastrointestinal peptides.[78] Additional studies are evidently required to improve our knowledge of the different inhibitory, stimulatory, or potentiating factors involved in the complex mechanism of cell renewal in the stomach. The clinical importance of this area of research is further emphasized by the recent implication of gastric cell kinetics in the medical treatment of peptic ulcer and, indirectly, in the promotion of gastric tumors.

REFERENCES

1. **Rubin, W., Sleisenger, M. H., and Jefferies, G. H.,** The normal gastric epithelia. A fine structural study, *Lab. Invest.*, 19, 598, 1968.
2. **Willems, G., Galand, P., Vansteenkiste, Y., and Zeitoun, P.,** Cell population kinetics of zymogen and parietal cells in the stomach of mice, *Z. Zellforsch.*, 134, 505, 1972.
3. **Ragins, H., Liu, S. M., and Ditbrenner, M.,** The origin and survival of gastric parietal cells in the mouse, *Anat. Rec.*, 162, 99, 1968.
4. **Helander, H. F.,** Ultrastructure and function of gastric parietal cells in the rat during development, *Gastroenterology*, 56, 35, 1969.
5. **Lehy, T., Zeitoun, P., Dubrasquet, P., and Bonfils, S.,** Etude histoclinique et ultra-structurale de la cicatrisation de l'ulcère de contrainte chez le rat, *Biol. Gastroenterol.*, 2, 107, 1970.
6. **Mastsuyama, M. and Suzuki, H.,** Differentiation of immature mucus cells into parietal, argyzophil and chief cells in stomach grafts, *Science*, 96, 385, 1970.
7. **Crean, G. P., Marshall, M. W., and Rumsey, R. D. E.,** Parietal cell hyperplasia induced by the administration of pentagastrin to rats, *Gastroenterology*, 57, 147, 1969.
8. **Crean, G. P., Gunn, A. A., and Rumsey, R. D. E.,** The effects of vagotomy on the gastric mucosa of the rat, *Scand. J. Gastroenterol.*, 4, 675, 1969.
9. **Lehy, T. and Willems, G.,** Population kinetics of antral gastrin cells in the mouse, *Gastroenterology*, 71, 614, 1976.
10. **Fujimoto, S., Hattori, J., Kimoto, K., Yamashita, S., Fujita, S., and Kawai, K.,** Tritiated thymidine autoradiography on origin and renewal of gastrin cells in pyloric area of hamsters, *Gastroenterology*, 78, 1136, 1980.
11. **Lehy, T.,** Self replication of somatostatin cells in the antral mucosa of rodents, *Cell Tissue Kinet.*, 15, 495, 1982.
12. **Borch, K., Renvall, H., Liedberg, G., and Andersen, B. N.,** Relations between circulating gastrin and endocrine cell proliferation in the atrophic fundic mucosa, *Scand. J. Gastroenterol.*, 21, 357, 1986.
13. **Bordi, C., Cocconi, G., Togni, R., Vezzadini, P., and Missale, G.,** Gastric endocrine cell proliferation, *Arch. Pathol.*, 98, 274, 1974.
14. **Hakanson, R., Bottcher, G., Ekblad, E., Panula, P., Simonsson, M., Dohlsten, M., Hallberg, T., and Sundler, F.,** Histamine in endocrine cells in the stomach. A survey of several species using a panel of histamine antibodies, *Histochemistry*, 86, 5, 1986.
15. **Tielemans, Y., Willems, G., Sundler, F., and Hakanson, R.,** Self-replication of enterochromaffin like cells in the mouse stomach. An autoradiographic study, *Cell Tissue Res.*, submitted.
16. **Inokuchi, H., Fujimoto, S., Hattori, T., and Kawai, K.,** Tritiated thymidine autoradiographic study on the origin and renewal of secretin cells in the rat duodenum, *Gastroenterology*, 89, 1014, 1985.
17. **Cheng, H. and Leblond, C. P.,** Origin, differentiation and renewal of the four main epithelial cell types in the mouse small intestine. III. Entero-endocrine cells, *Am. J. Anat.*, 141, 503, 1974.
18. **Grant, R., Grossman, M. F., and Ivy, A. C.,** Histological changes in the gastric mucosa during digestion and their relationship to mucosal growth, *Gastroenterology*, 25, 218, 1953.
19. **Willems, G., Vansteenkiste, Y., and Smets, P.,** Effects of food ingestion on the cell proliferation kinetics in the canine fundic mucosa, *Gastroenterology*, 61, 323, 1971.
20. **Majumdar, A. P. N.,** Regulation of gastric mucosal DNA synthesis during fasting and refeeding in rats, *Digestion*, 27, 36, 1983.
21. **Ito, S. and Lacy, E. R.,** Morphology of rat gastric mucosal damage, defense and restitution in the presence of luminal ethanol, *Gastroenterology*, 88, 250, 1985.
22. **Philpott, G. W.,** Tissue-specific inhibition of cell proliferation in embryonic stomach epithelium *in vitro*, *Gastroenterology*, 61, 25, 1971.
23. **Domschke, W., Domschke, S., Hagel, J., Demling, L., and Croft, D. N.,** Gastric epithelial cell turnover, mucus production, and healing of gastric ulcers with carbenoxolone, *Gut*, 18, 817, 1977.
24. **Willems, G.,** Effet stimulant de la gastrine sur la prolifération cellulaire dans les glandes fundiques chez le chien, *C. R. Soc. Biol.*, 169, 1473, 1971.
25. **Willems, G., Vansteenkiste, Y., and Limbosch, J. M.,** Stimulating effect of gastrin on cell proliferation kinetics in canine fundic mucosa, *Gastroenterology*, 62, 637, 1972.
26. **Majumdar, A. P. N.,** Pentagastrin stimulation of DNA synthesis and thymidine kinase activity in gastric mucosa of rats, *Hormone Res.*, 15, 37, 1981.
27. **Johnson, L. R. and Chandler, A. M.,** Stimulation of RNA synthesis by pentagastrin in gastric and duodenal mucosa, *Gastroenterology*, 60 (Abstr.), 680, 1971.
28. **Oscarson, J. E., Veen, H. F., Williamson, R. C. N., Ross, J. S., and Malt, R. A.,** Compensatory postresectional hyperplasia and starvation atrophy in small bowel: dissociation from endogenous gastrin levels, *Gastroenterology*, 72, 890, 1977.

29. **Solomon, T.,** Trophic effects of pentagastrin on gastrointestinal tract in fed and fasted rats, *Gastroenterology*, 91, 108, 1986.
30. **Wright, N.,** The control of cell proliferation in the gastrointestinal tract, in *The Biology of Epithelial Cell Population*, Wright, N. and Allison, M., Eds., Clarenden Press, Oxford, 1984, 843.
31. **Willems, G. and Vansteenkiste, Y.,** Resistance of the antral mucosa to the trophic effect of gastrin in the dog, *Biol. Gastroenterol. (Paris)*, 7, 237, 1974.
32. **Willems, G., Vansteenkiste, Y., and Smets, P.,** Cell proliferation in the mucosa of Heidenhain pouches after feeding in dogs, *Am. J. Dig. Dis.*, 17, 671, 1972.
33. **Willems, G., Gepts, W., and Bremer, A.,** Endogenous hypergastrinemia and cell proliferation in the fundic mucosa in dogs, *Am. J. Dig. Dis.*, 22, 419, 1977.
34. **Johnson, L. R. and Guthrie, P. D.,** Proglumide inhibition of trophic action of pentagastrin, *Am. J. Physiol.*, 246, G62, 1984.
35. **Takeuchi, K. and Johnson, L. R.,** Effect of cell proliferation and cell loss on aspirin induced gastric damages in the rat, *Am. J. Physiol.*, 243, G463, 1982.
36. **Lehy, T., Bonnefond, A., Dubrasquet, M., Nasca, S., Lewin, M., and Bonfils, S.,** Comparative effects of antrocolic transposition and antrectomy on fundic mucosa and acid secretion of the rat, *Gastroenterology*, 64, 421, 1973.
37. **Neuburger, P., Lewin, M., and Bonfils, S.,** Parietal and chief cell populations in four cases of the Zollinger-Ellison syndrome, *Gastroenterology*, 63, 937, 1972.
38. **Willems, G. and Lehy, T.,** Radioautographic and quantitative studies on parietal and peptic cell kinetics in the mouse. A selective effect of gastrin on parietal cell proliferation, *Gastroenterology*, 69, 416, 1975.
39. **Brants, F. and Morisset, J.,** Trophic effect of cholecystokinin-pancreozymin on pancreatic acinar cells from rats of different ages, *Proc. Soc. Exp. Biol. Med.*, 153, 523, 1976.
40. **Morisset, J. and Genik, P.,** Effects of acute and chronic administration of secretin and caerulein on rat duodenal and gastric growth, *Regul. Pept.*, 5, 11, 1983.
41. **Lamote, J., Putz, P., and Willems, G.,** Effect of cholecystokinin octapeptide, caerulein and pentagastrin on epithelial cell proliferation in the murine gallbladder, *Gastroenterology*, 83, 371, 1982.
42. **Caes, F. and Willems, G.,** Administration of caerulein to rats promotes antral epithelial cell renewal, *Cell Tissue Res.*, 236, 711, 1984.
43. **Tatsuta, M., Iishi, H., Baba, M., and Taniguchi, H.,** Effect of 6-OH tryptamine on gastric carcinogenesis and tetragastrin inhibition of carcinogenesis induced by *N*-methyl-*N*¢-nitro-*N*-nitrosoguanidine in Wistar rats, *Cancer Res.*, 49, 4199, 1989.
44. **Yasui, W., Sumiyoski, H., Ochiai, A., and Tuhara, E.,** Cholecystokinin inhibition of tumor growth and gastrin stimulated cyclic adenosine 3¢ 5¢-monophospate metabolism in human gastric carcinoma in nude mice, *Cancer Res.*, 46, 740, 1986.
45. **Cuttita, F., Carney, D., and Mulshire, J.,** Bombesin-like peptides can function as autocrine growth factor in human small cell cancer, *Nature*, 316, 823, 1985.
46. **Lehy, T., Accary, J. P., Labeille, D., and Dubrasquet, M.,** Chronic administration of bombesin stimulates antral gastrin cell proliferation in the rat, *Gastroenterology*, 84, 914, 1983.
47. **Lehy, T., Accary, J. P., Dubrasquet, M., and Lewin, J. M.,** Growth hormone-releasing factor (somatocrinin) stimulates epithelial cell proliferation in the rat digestive tract, *Gastroenterology*, 90, 646, 1986.
48. **Goodlad, R. A., Wilson, T. G., Lenton, W., Wright, N. A., Gregory, H., and McCullagh, K. G.,** Intravenous but not intragastric urogastrone-EGF is trophic to the intestinal epithelium of parenterally fed rats, *Gut*, 28, 573, 1987.
49. **Al Nafussi, A. I. and Wright, N. A.,** The effect of epidermal growth factor (EGF) on cell proliferation of the gastrointestinal mucosa in rodents, *Virchows Arch. (Cell Pathol.)*, 40, 63, 1982.
50. **Goodlad, R. A. and Wright, N. A.,** Cell proliferation, plasma enteroglucagon and plasma gastrin levels in starved and refed rats, *Virchows Arch. (Cell. Pathol.)*, 43, 55, 1983.
51. **Johnson, L. R. and Guthrie, P. D.,** Secretin inhibition of gastrin stimulated deoxyribonucleic acid synthesis, *Gastroenterology*, 67, 601, 1974.
52. **Stanley, M. D., Coalson, R. E., Grossmann, M. I., and Johnson, L. R.,** Influence of secretin and pentagastrin on acid secretion and parietal cell number in rats, *Gastroenterology*, 63, 264, 1972.
53. **Wiseman, D. A. and Johnson, L. R.,** Evidence that secretin does not have direct antitrophic effects on the rat stomach, *Proc. Soc. Exp. Biol. Med.*, 153, 277, 1976.
54. **Lehy, T., Dubrasquet, M., and Bonfils, S.,** Effect of somatostatin on normal and gastrin stimulated cell proliferation in the gastric and intestinal mucosa of the rat, *Digestion*, 9, 99, 1979.
55. **Ley, R., Willems, G., and Vansteenkiste, Y.,** Influence of vagotomy on parietal cell kinetics in the rat gastric mucosa, *Gastroenterology*, 65, 764, 1973.
56. **Guttierez, O., Lehy, T., and René, E.,** Epithelial cell proliferation in human fundic and antral mucosae. Influence of superselective vagotomy and relationship with gastritis, *Dig. Dis. Sci.*, 30, 1034, 1985.

57. **Delince, P., Willems, G., and De Graef, J.,** Antral gastrin cell proliferation after vagotomy in rats, *Digestion,* 18, 27, 1978.
58. **Hakanson, R., Vallgren, S., Ekelund, M., Rehfeld, J. F., and Sundler, F.,** The vagus exerts trophic control of the stomach in the rat, *Gastroenterology,* 86, 28, 1984.
59. **Assad, S. T., Grenett, H. E., Bowden, T. A., Garver, F. A., and Webster, P. D.,** Effects of vagotomy and pyloroplasty on the renewal of antral gastrin cells in the rat, *Gastroenterology,* 78 (Abstr.), 1133, 1980.
60. **Bertrand, P. and Willems, G.,** Induction of antral gastrin cell proliferation by refeeding of rats after fasting, *Gastroenterology,* 78, 918, 1980.
61. **Witzel, L., Halter, F., Olah, A. J., and Hacki, W.,** Effects of prolonged mitiamide medication on the fundic mucosa, *Gastroenterology,* 73, 797, 1977.
62. **Mazzacca, G., Cascione, F., Budillon, G., D'Agostino, L., Cimino, L., and Femiano, C.,** Parietal cell hyperplasia induced by long administration of antacids to rats, *Gut,* 19, 798, 1978.
63. **Eastwood, G. L. and Quimby, G. F.,** Effect of chronic cimetidine ingestion on fundic and antral epithelial proliferation in the rat, *Dig. Dis. Sci.,* 28, 61, 1983.
64. **Fich, A., Arber, N., Sestieri, M., Zajicek, G., and Rachmilewitz, D.,** Effect of misoprostol and cimetidine on gastric cell labeling index, *Gastroenterology,* 89, 57, 1985.
65. **Tielemans, Y., Sundler, F., Hakanson, R., and Willems, G.,** Omeprazole stimulated ECL cell replication in the rat, *Gastroenterology,* 92 (Abstr.), 1671, 1987.
66. **Blom, H.,** Effects of omeprazole on normal and regenerating gastric mucosa in the rat. A light and electron microscopic study, *Scand. J. Gastroenterol.,* 21 (Suppl. 118), 70, 1986.
67. **Ekman, L., Hansson, E., and Havu, N.,** Toxicological studies on omeprazole, *Scand. J. Gastroenterol.,* 20 (Suppl. 108), 53, 1985.
68. **Brittain, R., Jack, D., and Reeves, J.,** Pharmacological basis for the induction of gastric carcinoid tumors in the rat by loxtidine, an unsurmountable H_2 receptor blocking drug, *Br. J. Pharmacol.,* 85, 843, 1985.
69. **Borch, K., Renvall, H., and Liedberg, G.,** Gastric endocrine cell hyperplasia and carcinoid tumors in pernicious anemia, *Gastroenterology,* 88, 638, 1985.
70. **Larsson, H., Carlsson, E., and Mattsson, H.,** Plasma gastrin and gastric enterochromaffin like cell activation and proliferation, *Gastroenterology,* 90, 391, 1986.
71. **Richards, A. T., Hinder, R. A., and Harrison, A. C.,** Gastric carcinoid tumors associated with hyperplasia and pernicious anemia — regression of tumors by antrectomy, *S. Afr. Med. J.,* 72, 51, 1987.
72. **Reinhart, W., Muller, O., and Halter, F.,** Influence of long term 16,16 dimethylprostaglandin E_2 treatment on the rat gastrointestinal mucosa, *Gastroenterology,* 85, 1103, 1983.
73. **Baumgartner, A., Koeltz, H. R., and Halter, F.,** Indomethacin and turnover of gastric mucosal cells in the rat, *Am. J. Physiol.,* 250, G830, 1986.
74. **Helander, H. F., Johansson, C., Blom, H., and Uribe, A.,** Trophic actions of E2 prostaglandins in the rat gastrointestinal mucosa. A quantitative morphologic study, *Gastroenterology,* 89, 1393, 1985.
75. **Johansson, C., Uribe, A., and Rubio, C.,** Effect of oral prostaglandin E2 and DNA turnover in gastric and intestinal epithelia in the rats, *Eur. J. Clin. Invest.,* 16, 509, 1986.
76. **Crean, G. P., Rumsey, R. D. E., and Weeler, S. M.,** Further observations concerning the effects of hypophysectomy on gastric mucosa of the rat, *Gut,* 12, 721, 1971.
77. **Crean, G. P. and Rumsey, R. D. E.,** Hyperplasia of the gastric mucosa during pregnancy and lactation in the rat, *J. Physiol.* (*London*), 215, 181, 1971.
78. **Keuppens, F., Woussen-Colle, M. C., De Graef, J., and Willems, G.,** Increase in gastrin and somatostatin cell number in the antrum of dogs after small bowel resection, *Gastroenterol. Clin. Biol.,* 7, 287, 1983.
79. **Goodlad, A. J. A., Madgwick, A. J. A., Moffatt, M. R., Levin, S., Allan, J. L., and Wright, N. A.,** Prostaglandins and the gastric epithelium: effects of misoprostol on gastric epithelial cell proliferation in the dog, *Gut,* 30, 316, 1989.

Chapter 9

REGENERATION FOLLOWING GASTRIC INJURY

F. Halter

TABLE OF CONTENTS

I. ABSTRACT

The gastric mucosa has a great potential to heal both superficial and deep mucosal lesions. The healing process of superficial lesions (restitution) is rapidly achieved within minutes or hours predominantly by cell migration. In contrast, healing of deep lesions occurs over days to weeks since it requires formation of new mesenchymal and epithelial structures. The time relationship of this healing process conforms to an organic decay curve whereby a maximum healing phase is interspaced between an initial and late slow healing phase. While it is well established that luminal acid can delay the healing of both superficial and deep lesions, conflicting views exist as to which factors accelerate healing. Trophic effects of both gastrin and epidermal growth factor have an accepted potential to enhance healing of deep lesions. In contrast, evidence is accumulating that prostaglandins cannot accelerate restitution or healing of deep lesions by means other than inhibition of gastric acid.

The gastrointestinal mucosa, particularly that of the stomach, has a great potential to repair both superficial and deep epithelial lesions. While it has long been recognized that regeneration by mitotic reproduction from proliferating stem cells is an important factor in this healing process, it has been shown more recently that superficial lesions can rapidly be repaired by migration of underlying cells from the gastric pits. The latter process must be differentiated from the much slower evolution of wound healing (Table 1).

II. REPAIR OF SUPERFICIAL MUCOSAL LESIONS

A. REPAIR BY CELL MIGRATION

Grant[1] observed in 1945 that damage to the superficial third of the mucosa of cat stomachs exposed to 60% ethanol was repaired within 4 h. Although this was originally interpreted as cell regeneration, regeneration by mitotic proliferation would require at least 1 to 2 d.[2] More recently, Wallace and Morris[3] proposed that the reestablishment of rat gastric mucosal epithelial continuity after ethanol injury occurred by migration of underlying epithelial cells from the gastric pits. This phenomenon has been well established in *in vitro* and *in vivo* animal models[4-6] and has been termed gastric restitution by W. Silen and his group. Restitution is defined as "the process by which epithelial integrity and continuity are rapidly reestablished following injury, before cell proliferation or extensive inflammatory response occurs".[7] If concentrated ethanol is instilled into the rat stomach for 30 to 40 s and then removed, evidence of repair occurs as early as 3 min following the insult.[8] The denuded surface is repaired by upward migration of viable deep pit and glandular isthmus cells from areas adjacent to or just beneath the injured surface. Early restitution of gastric mucosa is characterized by coverage of the denuded area by flattened mucous cells that gradually become more cuboidal. Initially, the gastric pits are greatly shortened, and mucous neck cells and even oxynticopeptic cells are found very close to the gastric luminal surface. A prerequisite for restitution is that the mucosal damage is limited to the surface and part of the gastric pit and does not destroy the entire gastric gland. It has been postulated, but not proven, that the process requires an intact basal lumina.[9] Since gastric restitution also occurs, albeit at a slower rate *in vitro* in the absence of blood flow, and since cell proliferation is not required, it must be a process dependent upon inherent properties of the epithelial cell itself. Early epithelial coverage of denuded areas is accompanied by the near complete recovery of transmucosal potential difference to pre-injury values. Repair by restitution may take from 15 min to 3 h, depending on the severity of damage and the species studied. The process is relatively slow in humans compared to rats, probably because the distance between gland openings in the human stomach is greater.[10,11]

In contrast, extensive hemorrhagic lesions have a distinctly different healing sequence, and the repair process, which involves inflammation and true mucosal regeneration, requires many hours or days.[12]

TABLE 1
Definition of Repair

Restitution	Healing
Cell migration	Cell proliferation
Superficial injury	Deep injury
Minutes to hours	Days to weeks

Restitution of gastric mucosa is not restricted to injury by hypertonic damaging agents, but occurs also following injury induced by aspirin, both *in vitro* and *in vivo*. Indeed, Yeomans and co-workers[13,14] were the first to suggest that the initial restoration of gastric mucosa in rats luminally exposed to aspirin was due to rearrangement or migration of existing cells.

The process of restitution can be repeated several times in the same tissue after the initial restitution has occurred.[15-17] Restitution is probably a generalized property of gastrointestinal epithelial cells, as has been demonstrated in gallbladder, duodenum, and colon.[18-20]

1. Factors Influencing Gastric Restitution

a. Luminal Acid

Morris and Wallace[3] showed in experiments performed in an *in vivo* rat stomach preparation that hemorrhagic lesions only occur following 40% ethanol instillation only when 50 m*M* luminal acid was present, but not when saline bathed the surface. They demonstrated that luminal acid destroyed the basal lamina and thus "probably removed the substratum necessary for orderly epithelial reestablishment".

b. Mucoid Cap

Damaging agents applied topically to the gastric mucosa cause the rapid release of mucus which, along with fibrin, plasma, and cellular debris, forms a continuous cap over sites of damage and provides a favorable microenvironment for restitution to occur.[3,6,17] Dissolution or mechanical removal of this mucoid cap[15-17] leads to more extensive damage with a subsequent delay in epithelial restitution following gastric mucosal exposure to 50 or 70% ethanol or 1 *M* NaCl in 0.05 *M* HCl.

c. Calcium

Calcium depletion completely prevents the migration of epithelial cells of the bare basal lumina following injury with hypertonic NaCl.[21]

d. Prostaglandins

Since it has been claimed that prostaglandins exert a "cytoprotective" effect on the gastric mucosa,[22] the term "cytoprotection" has been uncritically translated by many into enhancement of the healing of mucosal damage. It is now well established that prostaglandins, when applied prior to exposure of the gastric mucosa to noxious agents, cannot prevent the necrosis of the superficial epithelial cells and therefore the term cytoprotection is inadequate and misleading.[23] Prostaglandin pretreatment has, however, consistently been shown to reduce or eliminate the areas of macroscopically visible necrotic lesions by minimizing deep mucosal congestion.[10] As elegantly demonstrated by Schmidt et al.,[24] the prevention of such deep lesions, respectively the preservation of larger proportions of intact epithelial cells in the gastric pits, enhances restitution. When rats were exposed to 1 ml of absolute ethanol the length of the injured surface epithelium was the same whether or not rats were pretreated with 10 μg/kg of 16,16-dimethyl PGE_2 or saline, but the deeper layers of the epithelium were protected by the prostanoid. In prostanoid pretreated animals the repair was similar to that in saline pretreated rats, but it was initiated earlier, was more widespread, and a considerably larger proportion of the mucosa was reepithelialized after 24 h. Prostaglandins thus influence restitution only indirectly. It has been shown that neither luminal nor nutrient administration of prostaglandins affects either favorably or adversely the process of restitution in frog mucosa following injury with 1 *M* NaCl.[25]

Those that persist in the use of the term "cytoprotection" must at least clearly distinguish the development of an acute injury from the recovery phase, over which prostaglandins exert essentially no direct effect.

B. REPAIR BY REEPITHELIZATION

Despite prolonged exposure of the gastric mucosa to noxious agents, during which enhanced cell shedding can be measured, the overall integrity of the gastric mucosa is maintained in most individuals. Damage that may occur involving deeper parts of the mucosa is almost always of focal distribution. A continuous repair process is in effect initiated by prolonged injury for which constant cell division, as well as restitution, are of prime importance in limiting further damage. Both in experimental animals and in man, prolonged exposure of the gastric mucosa to aspirin induces increased cell division.[26-28]

In a detailed morphometric study Baumgartner et al.[28] demonstrated that in the rat prolonged treatment with indomethacin, 2 mg/kg, b.d. for 2 weeks, is followed by an increase in cell shedding, acceleration of cell migration from the proliferative zone to the surface, and an increase in cell reproduction (Figure 1). In the gastric corpus, but not antrum, this enhanced cell reproduction overcompensates for the increased cell loss, resulting in an overall increase of the mucosal cell mass. It is likely that increased cell reproduction contributes to the increased tolerance of the gastric mucosa (gastric adaptation) observed both in experimental animals and in man during prolonged treatment with aspirin or indomethacin.[29,30]

III. HEALING OF GASTRIC WOUNDS

A. TIME COURSE OF HEALING

Experimental mucosal wounds generally extend through the entire gastric mucosa down to the tunica muscularis. It must be realized that repair is not a process restricted to epithelial cells. Necrotic tissue must be removed and recovery from chronic ulceration involves tissue rebuilding. This repair process requires a fully developed inflammatory response which leads to cellular infiltration, formation of an extracellular matrix, proliferation, and establishment of a new microvasculary supply. Macrophages may play an important role in the early healing phase[31] consistent with observations of the healing of skin lesions.[32]

The time course of healing, however, depends on the animal model and the experimental conditions.

Gastric wounds produced by surgical removal of mucosal flaps show a similar healing pattern to simple skin defects.[33,34] Studies performed by Halter et al.[33] in pigs suggest that the healing time relationship conforms to an organic decay curve (Figure 2). A maximum healing phase around the fifth day is interspaced between two slow healing phases, the initial "lag phase" and the so-called "remodeling phase".

The transition from one phase to the other is only partly understood, but expert histomorphometric studies have been performed to characterize the initial and late slow phase of the healing of experimental ulcers.[12,36-40]

B. HISTOLOGICAL FINDINGS DURING HEALING OF EXPERIMENTAL ULCERS

1. The Early Lag Phase

The early phase of healing includes formation of granulation tissue which replaces necrotic tissue elements. The wound margins are elevated, resulting from several factors including muscle contraction, infiltration with inflammatory elements, and bulging of distorted glands. In

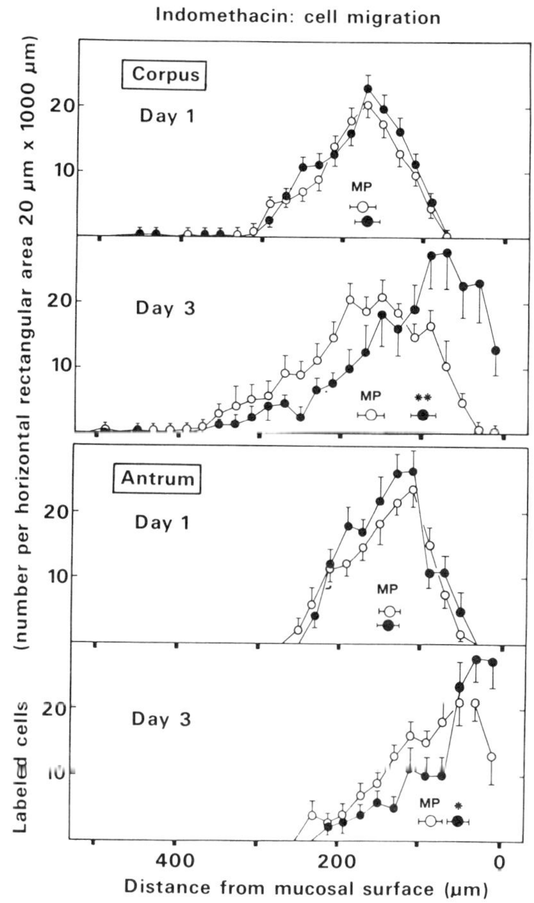

FIGURE 1 Histograms of labeled cells after 1 and 3 d of treatment with indomethacin (closed circles) or solvent (open circles). Animals were injected with (^{3}H)-thymidine intraperitoneally before treatment. MP, median position of labeled cells as distance from surface. Means ± SE of 5 rats. *$p < 0.05$; **$p < 0.01$. (From Baumgartner, A., Koelz, H. R., and Halter, F., *Am. J. Physiol.*, 250, G830, 1986. With permission.)

the initial phase, both submucosal edema and cyst-like formation distort the marginal gastric glands. The gastric pits in the margin are deeper than normal and the gland lumina are often dilated. Electromicroscopic histomorphometric studies[35] have revealed signs of immaturity in the epithelial cells of such glands, whereby it was not possible to establish whether the immaturity and differentiation of specialized cells were the consequence of the trauma or an indication of early repair.

2. The Rapid Healing Phase

It may be assumed that the transition from the initial lag phase to the phase of rapid healing is characterized by the movement of primitive epithelial cells from the ulcer margin towards the ulcer region, as well as by shrinkage of the ulcer bed, as granulation tissue is converted into fibrous tissue. Cell migration towards the center of the ulcer is initiated when an appropriate substructure is established, i.e., when the necrotic tissue is replaced by healthy granulation tissue. Migration is rapid and many more mitotic cells have been found in the ulcer margin during this phase.[37,38]

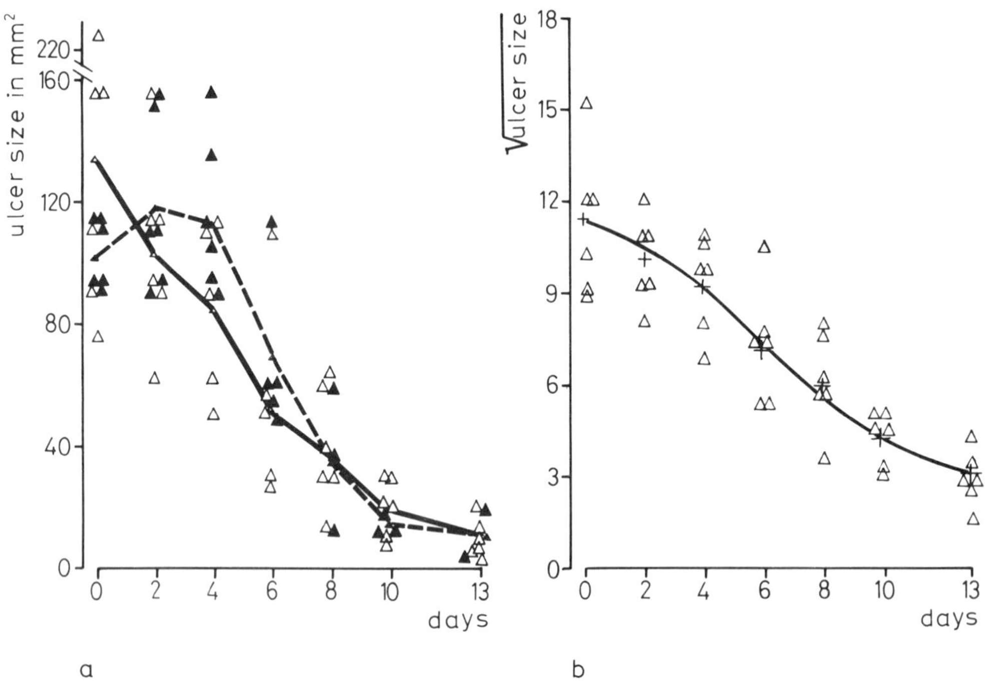

FIGURE 2. The healing of gastric wounds of pigs was endoscopically assessed at 2- to 3-d intervals. Treatment was either histamine 25 mg in 0.25 ml beeswax i.m. (▲) or 0.25 ml beeswax (Δ = controls). (a) Raw data. (b) Square root transformation of ulcer area of control animals. Eye-fitted regression. (From Halter, F., Barbezat, G. O., van Hoorn-Hickman, R., and van Hoorn, W. A., *Dig. Dis. Sci.*, 25, 916, 1980. With permission.)

3. The Remodeling Phase

The initial covering layer of the ulcer consists of cuboidal to low columnar epithelial cells containing a few PAS positive granules. The cells thus resemble superficial mucous cells.[35] In the rat model applied by Blom and Helander,[12] 1 month after lesions had been produced by cauterization, the entire mucosa was covered by tissue that histologically resembled the pyloric gland region in normal stomachs, with deep gastric pits and short glands. Before differentiated cells reappeared the mucosa became thinner, due to reduction of the depth of the gastric pits. At 60 d typical mucous neck cells as well as immature zymogen cells and small numbers of various types of endocrine cells were seen, but parietal cells appeared only after 90 d. At this stage parietal cells were immature but able to secrete hydrochloric acid.[39] Remodeling was still incomplete after 260 d since parietal cells occupied only 12% of the mucosal volume, which was less than in sham operated animals.

The healing kinetics described so far may not apply to all experimental ulcer models. Chronic antral ulcers such as those induced by constant duodeno-gastric reflux[41] may not heal until the pathogenetic perturbation is rectified. Wounds situated at the funde-antral junction seem to heal more slowly than those located in other parts of the stomach.[42] In addition, cysteamine-induced ulcers heal at a considerably slower pace,[43] and this seems to be related to the fact that these ulcers are deep and usually penetrate the submucosa. The question remains whether the depth of penetration and specific anatomic location within the stomach are the main reasons for slow healing and resistance to drug treatment in some human peptic ulcers.

a. Factors Influencing Healing of Gastric Wounds

i. Gastric Acid

Limited data are available from the literature concerning the extent to which gastric acid delays healing of experimental ulcers. As with peptic ulcers, experimental ulcers can heal in the

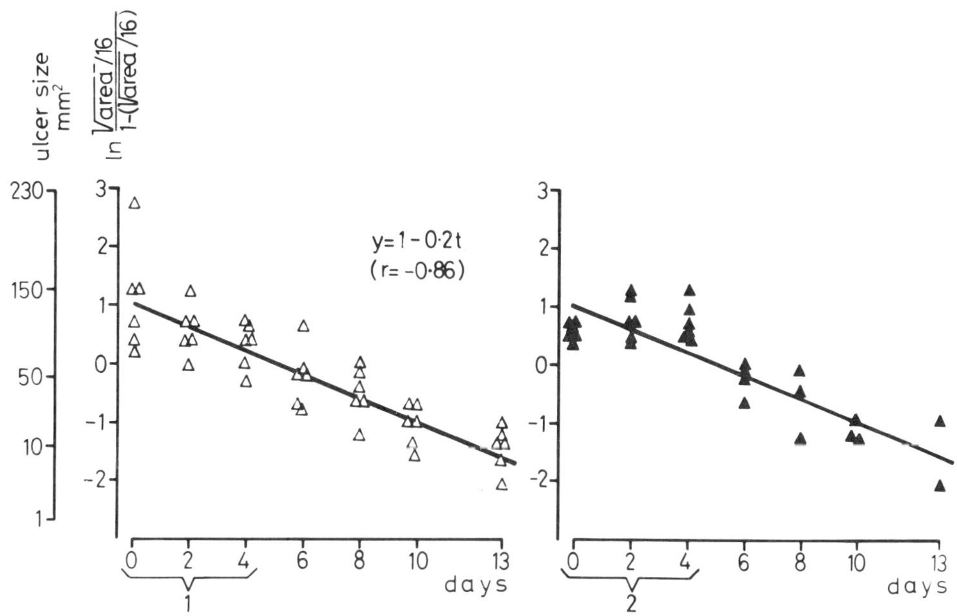

FIGURE 3. (a) Logit transformation of data of Figure 2. Calculated regression line. (b) Data of histamine-treated animals. The regression line is that of the control animals and thus identical with that of (a). (From Halter, F., Barbezat, G. O., van Hoorn-Hickman, R., and van Hoorn, W. A., *Dig. Dis. Sci.*, 25, 916, 1980. With permission.)

majority of cases without a reduction of gastric acid. Hyperacidity induced by chronic pentagastrin administration did not delay and occasionally improved the healing of experimental gastric or duodenal ulcers,[36,40] an observation likely attributable to the trophic effects of this peptide. However, pronounced and sustained hyperacidity induced by chronic administration of histamine in beeswax to pigs resulted in a delay of healing during the early lag phase.[33] The ulcer edges were undermined and some animals died of hemorrhage or perforation. It is possible that histamine-induced vasoconstriction contributed to some of these findings. It is of interest, however, that even with this aggressive treatment, healing showed a tendency towards compensation once the dangerous initial period had been overcome (Figure 3). This observation, as well as the findings of Blom and Erkoinen,[36] that chronic pentagastrin administration accelerated the reappearance of parietal cells during the remodeling phase, indicate that hyperacidity does not impair healing once the early lag phase is passed.

The role of acid in the healing of experimental ulcers is further underlined by the fact that both histamine-H_2-antagonists and omeprazole[44,45] can accelerate healing. Unfortunately, in none of these studies was ulcer healing systematically assessed over the whole life span of the ulcer, but the data are suggestive that suppression of acid favorably influences healing towards the end of the early lag phase,[45] as well as during the remodeling of gastric glands.[44]

ii. Gastrin

Since it is known that gastrin exerts trophic effects on the growth of the gastric corpus mucosa,[46] it has been speculated that this hormone may possibly accelerate the healing of peptic ulcers. Suggestive evidence in favor of this has been supplied by Blom and Erkoinen[36] and by Takeuchi and Johnson.[40] In these studies chronic pentagastrin administration accelerated the reappearance of parietal cells during the remodeling phase.[36] In addition, it counteracted the negative effects exerted by a liquid diet on healing of experimental gastric but not duodenal ulcers.[40] It was speculated that this effect might be attributed to induction of DNA synthesis in progenitor cells. This hypothesis is underlined by the observation that sustained acid inhibition

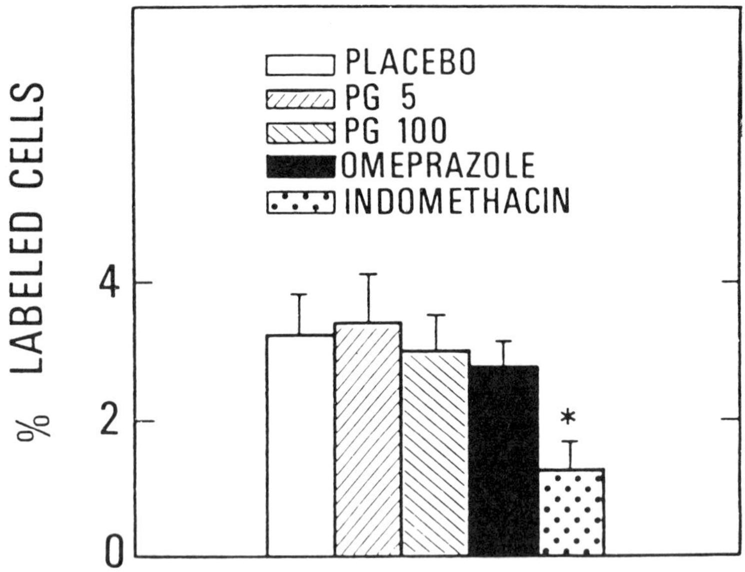

FIGURE 4. Percentage of radiolabeled cells measured in a 600-μm-wide region of corpus mucosa adjacent to the mucosal defect. Rats were treated for 5 d with intragastric 16,16-dimethyl prostaglandin E_2 [5μg/kg body weight, b.i.d. (PG 5) and 100 μg/kg body weight, b.i.d. (PG 100)], s.c. omeprazole (40 μmol/kg body weight), s.c. indomethacin (2 mg/kg body weight, b.i.d.), or placebo. Values are given as mean ± SEM (n = 8 animals for placebo and n = 6 animals for the other treatments). $*p < 0.05$ vs. all other treatments (one-way analysis of variance). (From Inauen, W., et al., *Gastroenterology*, 95, 636, 1988. With permission.)

by histamine-H_2 inhibitors and especially by omeprazole enhances healing of experimental ulcers in conjunction with pronounced hypergastrinemia. However, omeprazole accelerates healing significantly within the first 10 d,[45] but it is well established that H^+K^+ATPase inhibitors do not induce trophic effects of any significance before 4 weeks after onset of high dose treatment.[47,48] A further indication that trophic effects of gastrin are not directly involved in the early healing process comes from the observation that 10 d after the onset of omeprazole treatment, mitotic activity was not increased within the ulcer edge (Figure 4).[45] It is thus likely that acid suppression accelerates healing in the early phase through pathways other than increased cell production, e.g., by effects on the inflammatory process and the formation of a new microvasculature. In contrast, during potent acid inhibition, gastrin may well play an important factor in accelerating healing during the remodeling phase, as suggested by the studies of Blom.[36,44] It will, however, not be possible to fully clarify the role of gastrin on ulcer healing until potent and specific gastrin inhibitors become available.

iii. Prostaglandins

Some prostaglandins possess a unique combination of theoretically favorable properties to heal ulcers: they inhibit acid secretion, they are trophic to gastric mucosa, and they limit deep necrosis against various experimental injuries.[21,49-57] Although trophic effects have been observed in the rat after very low (so-called "cytoprotective") doses, a more marked increase of mucosal thickness results after high doses (Figure 5).[49] The trophic effects of prostaglandins are reflected in an increase of mucosal height and total mucosal volume, due to an increase of all epithelial cell types.[49] This can be observed as early as 1 week after treatment, even during low dose treatment.[50] Prostaglandins, despite their antisecretory capacity, do not increase plasma gastrin.[56,57]

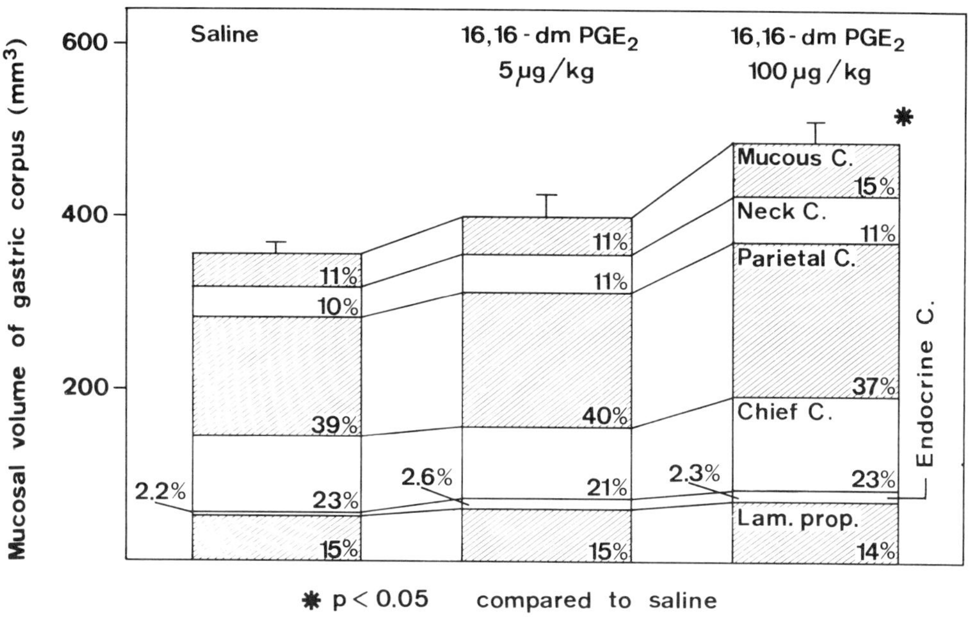

FIGURE 5. Mucosal volume of the gastric corpus with the percentage representation of surface and foveolar mucous cells, mucous neck cells, parietal cells, chief cells, and lamina propria after 21 d treatment with either saline, 5 µg/kg body weight of 16,16-dm PGE_2, 100 µg/kg body weight of 16,16-dm PGE, administered intragastrically every 8 h (n = 9, mean ± SEM). (From Reinhart, W. H., Müller, O., and Halter, F., *Gastroenterology*, 85, 1003, 1983. With permission.)

It must be emphasized that it is not justified to translate protective or trophic effects afforded by prostaglandins into the assumption that these compounds will accelerate healing of peptic or experimental ulcers. As shown by Inauen et al.[45] 10 d treatment with 5 or 100 µg/kg, b.i.d. of a synthetic PGE_2-analogue did not influence the healing of rat cryo-ulcers despite obvious trophic effects on the adjacent mucosa, while in the same study healing was significantly enhanced by high dose omeprazole treatment (Figure 6). The failure of prostaglandins to heal experimental ulcers in nonantisecretory doses, despite a remarkable trophic effect on the adjacent gastric tissue, is not unexpected since the trophic effects of prostaglandins predominantly result from a prolongation of the life cycle of the epithelial cells and not from increased cell production.[52-55]

Clinical studies demonstrate that some prostaglandins accelerate ulcer healing if given in antisecretory doses, but there is no evidence that they are effective in "cytoprotective" doses.[58,59] Any extrapolation from drug-induced phenomena, observed in conjunction with acute injury of gastric mucosa, on possible effects on healing must therefore be regarded as a misconception.

iv. Epidermal Growth Factor (EGF)

Similar to prostaglandins, the single chain polypeptide EGF has been claimed to exert protective, acid inhibitory, and trophic effects on the gastric mucosa.[60-63] As recently shown by Konturek et al.,[63] s.c. administration of 10 µg/kg EGF three times daily as well as oral EGF application (30 µg/kg) accelerates considerably the healing of both gastric and duodenal ulcers induced in the rat by acetic acid. Since oral EGF does not inhibit acid secretion and since enhanced healing is accompanied by an increase of both RNA and DNA concentrations of the ulcer-adjacent mucosa, this effect may be due primarily to the mitogenic properties of EGF (Table 2).

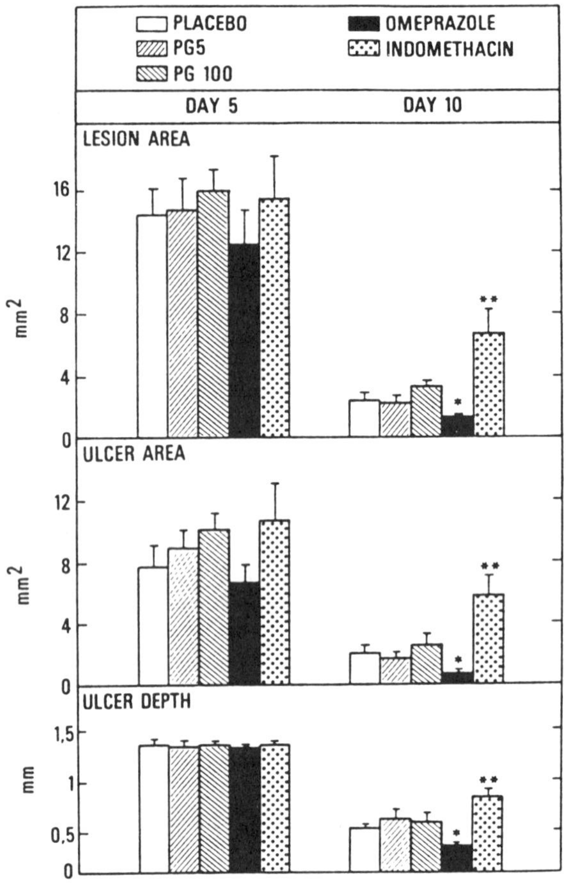

FIGURE 6. Microscopic analysis of mucosal injury. Lesion area, ulcer area, and ulcer depth were calculated from sectional slides analyzed by light microscopy. Rats were treated for 5 and 10 d with intragastric 16,16-dimethyl prostaglandin E_2 (5 μg/kg body weight), b.i.d. (PG 5) and 100 μg/kg body weight, b.i.d. (PG 100), subcutaneous omeprazole (40 μmol/kg body weight), subcutaneous indomethacin (2 mg/kg body weight, b.i.d.) or placebo. Values are given as mean ± SEM (n = 8 animals for placebo and n = 6 animals for the other treatments). $^*p < 0.05$, $^{**}p < 0.01$ vs. all other treatments (one-way analysis of variance). (From Inaven, W., et al., *Gastroenterology*, 95, 636, 1988. With permission.)

IV. POSSIBLE RELATION OF EXPERIMENTAL ULCER HEALING TO HUMAN PEPTIC ULCERS

Although, as described above, chronic experimental ulcers show a similar histological appearance to human peptic ulcers, one must recognize that both the location and etiology are different. The healing of a gastric wound in a rat or pig is obviously faster than a human peptic ulcer, but otherwise follows a similar pattern. Since neither the development of ulcers nor the early lag phase of healing are generally available for observation in humans, the healing time relationship in peptic ulcers appears to conform to an exponential function (Figure 7),[64] but it must be assumed that in reality, it also follows an organic decay curve. It is thus probably justified to assume that cautious analogies may be drawn from the healing dynamics of experimental ulcers to that of true peptic ulcers.

TABLE 2
Effects of Epidermal Growth Factor Given Orally (30 μg/kg ·d) on the Area of Acetic Acid-Induced Gastric and Duodenal Ulcerations in Rats with Intact and Resected Salivary Glands

	Ulcer area (mm^2)			
	Gastric ulcer		Duodenal ulcer	
	Day 7	Day 11	Day 7	Day 11
Intact salivary glands				
Control	7.8 ± 1.5	0.8 ± 0.4	5.9 ± 1.4	1.0 ± 0.4
ECF p.o.	0.4 ± 0.2[a]	0.2 ± 0.2	2.7 ± 0.4[a]	0.5 ± 0.4
Sialoadenectomy				
Control	10.2 ± 0.9	2.7 ± 0.3	10.5 ± 1.4	1.8 ± 0.6
ECF p.o.	3.5 ± 0.4[a]	0.2 ± 0.1[a]	4.1 ± 0.9[a]	0.8 ± 0.4

Note: ECF, epidermal growth factor. Data are the mean ± SEM of 10 tests on 10 rats.

[a] Significant (p <0.05) decrease below the control values.

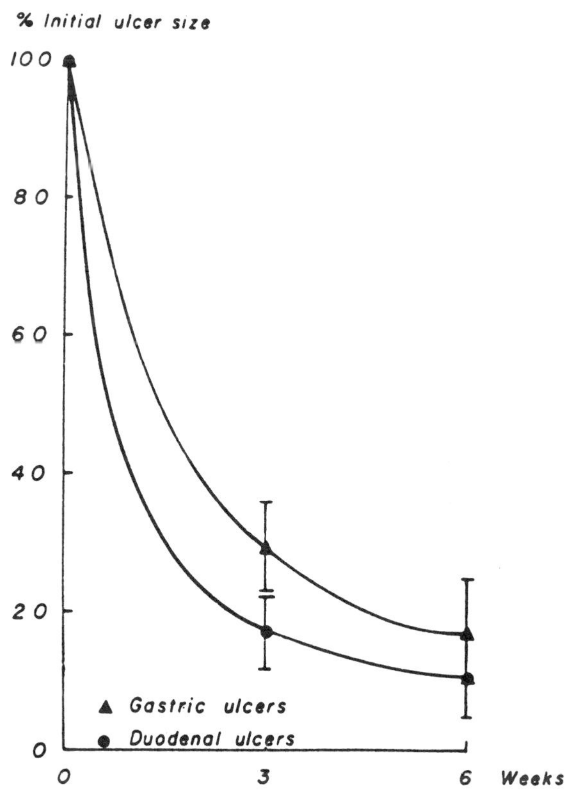

FIGURE 7. Healing time course of incompletely healed gastric and duodenal ulcers, given in average percentage of ulcer size (± SEM) as a function of the time. (From Scheurer, U., Witzel, L., Halter, F., Keller, M., Huber, R., and Galeazzi, R., *Gastroenterology*, 72, 838, 1977. With permission.)

REFERENCES

1. **Grant, R.**, Rate of replacement of the surface epithelial cells of the gastric mucosa, *Anat. Rec.*, 91, 175, 1945.
2. **Lipkin, M., Sherlock, P., and Bell, B.**, Cell proliferation kinetics in the gastrointestinal tract in man. II. Cell renewal in stomach, ileum, colon, and rectum, *Gastroenterology*, 83, 45, 721, 1963.
3. **Morris, G. P. and Wallace, J. L.**, The roles of ethanol and of acid in the production of gastric erosions in rats, *Virchow's Arch. B*, 38, 23, 1981.
4. **Svanes, K., Ito, S., Tacheuki, K., and Silen, W.**, Restitution of the surface epithelium of the in vitro frog gastric mucosa after damage with hyperosmolar NaCl, *Gastroenterology*, 82, 1409, 1982.
5. **Rutten, M. J. and Ito, S.**, Morphology and electrophysiology of guinea pig gastric mucosal repair in vitro, *Am. J. Physiol.*, 244(7), G171, 983.
6. **Lacy, E. R. and Ito, S.**, Ethanol-induced insult to the rat superficial gastric epithelium. A study to damage and rapid repair, in *Mechanisms of Mucosal Protection in the Upper Gastrointestinal Tract*, Allen, A., Eds., Raven Press, New York, 1984, 49.
7. **Silen, S. and Ito, S.**, Mechanisms for rapid re-epithelialization of the gastric mucosal surface, *Ann. Rev. Physiol.*, 47, 217, 1985.
8. **Lacy, E. R. and Ito, S.**, Rapid epithelial restitution of the rat gastric mucosa after ethanol injury, *Lab. Invest.*, 51, 573, 1984.
9. **Vracko, R.**, Basal lamina scaffold: anatomy and significance of orderly tissue structure a review, *Am. J. Pathol.*, 77, 314, 1974.
10. **Wallace, J. L., Morris, G. P., Krousse, E. J., and Graeves, S. E.**, Reduction by cytoprotective agents of ethanol-induced damage to the rat gastric mucosa: a correlated morphological and physiological study, *Can. J. Physiol. Pharmacol.*, 60, 1686, 1982.
11. **Morris, G.**, Are prostaglandins truly "cytoprotective"?, *J. Clin. Gastroenterol.*, 8, 326, 1986.
12. **Blom, H. and Helander, H.**, Quantitative ultrastructural studies on parietal cell regeneration in experimental ulcers in rat gastric mucosa, *Gastroenterology*, 80, 334, 1981.
13. **Yeomans, N. D.**, Electronmicroscopic study of the repair of aspirin-induced gastric erosions, *Am. J. Dig. Dis.*, 21, 533, 1977.
14. **Yeomans, N. D., St. John, D. I., and de Boer, W. G.**, Regeneration of gastric mucosa after aspirin-induced injury in the rat, *Am. J. Dig. Dis.*, 18, 773, 1973.
15. **Wallace, J. L. and Whittle, B. J. R.**, Role of mucus in the repair of gastric epithelial damage in the rat. Inhibition of epithelial recovery by mucolytic agent, *Gastroenterology*, 91, 603, 1986.
16. **Lacy, R.**, Gastric mucosal resistance to a repeated ethanol insult, *Scand. J. Gastroenterol.*, 20 (Suppl. 110), 63, 1985.
17. **Wallace, J. L. and Whittle, B. J. R.**, The role of extracellular mucus as a protective cap over gastric mucosal damage, *Scand. J. Gastroenterol.*, 21 (Suppl. 125), 79, 1986.
18. **Hudspeth, A. J.**, Establishment of tight junctions between epithelial cells, *Proc. Natl. Acad. Sci. U. S. A.*, 72, 2711, 1975.
19. **Feil, W., Wenzl, E., Vattay, P., Starlinger, M., Sokugoglu, T., and Schiessel, R.**, Repair of rabbit duodenal mucosa after acid injury in vivo and in vitro, *Gastroenterology*, 92, 1973, 1987.
20. **Feil, W., Lacy, E. R., Wong, M., Burger, D., Wenzl, E., Starlinger, M., and Schiessel, R.**, Rapid restitution of the rabbit and human colonic mucosa after superficial damage, *Gastroenterology*, 94, A124, 1988.
21. **Critchlow, J., Takeuchi, K., Ito, S., Magee, D., and Silen, W.**, Effect of calcium depletion on restitution of gastric mucosal injury, *Gastroenterology*, 82, 1038, 1982.
22. **Robert, A., Nezamis, J. E., Lancaster, C., and Hanchar, A. J.**, Cytoprotection by prostaglandins in rats. Prevention of gastric necrosis produced by alcohol, HCl, NaOH, hypertonic NaCl, and thermal injury, *Gastroenterology*, 77, 433, 1979.
23. **Lacy, E. R. and Ito, S.**, Microscopic analysis of ethanol damage to rat gastric mucosa after treatment with a prostaglandin, *Gastroenterology*, 83, 619, 1982.
24. **Schmidt, K. L., Bellard, R. L., Smith, G. S., Henagan, J. M., and Miller, T. A.**, Influence of prostaglandin on repair of rat stomach damaged by absolute ethanol, *J. Surg. Res.*, 41, 367, 1986.
25. **Svanes, K., Cirtchlow, J., Takeuchi, K., Magee, D., Ito, S., and Silen, W.**, Factors influencing reconstitution of frog gastric mucosa: role of prostaglandins, in *Mechanisms of Mucosal Protection in the Upper Gastrointestinal Tract*, Allen, A., Eds., Raven Press, New York, 1984, 49.
26. **Eastwood, G. L. and Quimby, G. F.**, Effect of chronic aspirin ingestion on epithelial proliferation in rat fundus, antrum and duodenum, *Gastroenterology*, 82, 852, 1982.
27. **Kuwayama, H. and Eastwood, G. L.**, Effects of water immersion, restraint, stress and chronic indomethacin ingestion on gastric antral and fundic epithelial proliferation, *Gastroenterology*, 88, 362, 1985.
28. **Baumgartner, A., Koelz, H. R., and Halter, F.**, Indomethacin and turnover of gastric mucosal cells in the rat, *Am. J. Physiol.*, 250, G830, 1986.

29. **St. John, D. B. F., Yeoman, D. D., McDermott, F. T., and de Baer, W. G. R. M.,** Adaptation of the gastric mucosa to repeated administration of aspirin in the rat, *Am. J. Dig. Dis.*, 18, 881, 1973.
30. **Graham, D. Y., Smith, J. L., and Dobbs, S. M.,** Gastric adaptation occur with aspirin administration in man, *Dig. Dis. Sci.*, 28, 1, 1983.
31. **Koo, J., Ho, J., Lam, S. K., Wong, J., and Ong, G. B.,** Selective coating of gastric ulcer by tripotassium dicitrato bismuthate in the rat, *Gastroenterology*, 82, 864, 1982.
32. **Peacock, E. E. and Van Winkle, W.,** Surgery and biology of wound repair, in *Inflammation and the Cellular Response to Injury*, W. B. Saunders, Philadelphia, PA, 1970, chap 1.
33. **Halter, F., Barbezat, G. O., van Hoorn-Hickman, R., and van Hoorn, W. A.,** Healing dynamics of traumatic gastric mucosal defects in the normal and hyperacid stomach, *Dig. Dis. Sci.*, 25, 916, 1980.
34. **Peacock, E. E.,** Wound healing and wound care, in *Principles of Surgery*, Schwartz, S. J., Ed., McGraw-Hill, New York, 1969, 226.
35. **Helander, H. F.,** Morphological studies on the margin of gastric corpus wounds in the rat, *J. Submicrosc. Cytol.*, 15, 627, 1983.
36. **Blom, H. and Erkoinen, T.,** Trophic effect of pentagastrin on normal and regenerating parietal cells, *Gastroenterology*, 87, 537, 1984.
37. **Helpap, B., Hattori, T., and Gedigk, P.,** Repair of gastric ulcer. A cell kinetic study, *Virchows Arch. Pathol. Anat.*, 392, 159, 1981.
38. **McMinn, R. M. H. and Johnson, L. R.,** The cytology of mucosal regeneration in experimental gastric ulcer, in *Pathophysiology of Peptic Ulcer*, Skornya, S. C., Ed., Lippincott, Philadelphia, PA, 1963, 369.
39. **Blom, H.,** Immature parietal cells in healing gastric wounds respond to stimulation. An experimental study in rats, *Digestion*, 24, 29, 1986.
40. **Takeuchi, K. and Johnson, L. R.,** Effect of cell proliferation on healing of gastric and duodenal ulcers in rats, *Digestion*, 33, 92, 1986.
41. **Kaminishi, M., Sadatsuki, H., Johjima, Y., Oohara, T., and Kondo, Y.,** A new model of production of chronic gastric ulcer by duodenogastric reflux in rats, *Gastroenterology*, 92, 1913, 1987.
42. **Wong, J. and Loewenthal, J.,** Chronic gastric ulcer in the rat produced by wounding at the fundo-antral junction, *Gastroenterology*, 71, 416, 1976.
43. **Poulsen, S. S., Olsen, P. S., and Kirkegaard, P.,** Healing of cysteamine-induced duodenal ulcers in the rat, *Dig. Dis. Sci.*, 30, 101, 1985.
44. **Blom, H.,** Cimetidine and parietal cell regeneration in experimental wounds in rat gastric mucosa. A light and electron-microscopical study, *Scand. J. Gastroenterol.*, 18, 853, 1983.
45. **Inauen, W., Wyss, P. A., Kayser, A., Baumgartner, A., Schûrer-Maly, C. C., Koelz, H. R., and Halter, F.,** Influence of prostaglandins, omeprazole and indomethacin on healing of experimental ulcers in the rat, *Gastroenterology*, 95, 636, 1988.
46. **Johnson, L. R.,** The trophic action of gastrointestinal hormones, *Gastroenterology*, 70, 278, 1976.
47. **Larsson, H., Carlsson, E., Mattsson, H., Lundell, L., Sundler, F., Sundell, G., Wallmark, B., Watanabe, T., and Hakanson, R.,** Plasma gastrin and gastric enterochromaffinlike cell activation and proliferation. Studies with omeprazole and ranitidine in intact and antrectomized rats, *Gastroenterology*, 90, 391, 1986.
48. **Inauen, W., Eigenmann, F., Koelz, H. R., and Halter, F.,** Omeprazole reduces trophic effects on parietal cells by endogenous gastrin and treatment with atropine, *Gastroenterology*, 90, 1471, 1986.
49. **Reinhart, W. H., Müller, O., and Halter, F.,** Influence of long-term 16,16-dimethyl prostaglandin E_2 treatment on the rat gastrointestinal mucosa, *Gastroenterology*, 85, 1003, 1983.
50. **Meyrat, P., Kappeler, M., and Halter, F.,** A simple gastrocopy technique in the rat, *Dig. Dis. Sci.*, 29, 327, 1984.
51. **Halter, F., Meyrat, P., Fritsche, R., Müller, O., and Koelz, H. R.,** Both topical and systemic treatments with 16,16-dimethyl prostaglandin are trophic to rat gastric mucosa, *Scand. J. Gastroenterol.*, 19, 101, 47, 1984.
52. **Fich, A., Arber, N., Sestieri, M., Zajicek, G., and Rachmilewitz, D.,** Effect of misoprostol and cimetidine on gastric cell labelling index, *Gastroenterology*, 89, 57, 1985.
53. **Tytgat, G. N. J., Offerhaus, G. J. A., van Minnen, A. J., Everts, V., Hensen-Logmans, S. C., and Samson, G.,** Influence of oral 15(R)-15-methyl prostaglandin E_2 on human gastric mucosa, *Gastroenterology*, 90, 1111, 1986.
54. **Johansson, C., Uribe, A., Rubio, C., and Isenberg, J. I.,** Effect of oral prostaglandin E_2 on DNA turnover in gastric and intestinal epithelia of the rat, *Eur. J. Clin. Invest.*, 16, 509, 1986.
55. **Uribe, A., Rubio, C., and Johansson, C.,** Cell kinetics of rat gastrointestinal mucosa. Autoradiographic study after treatment with 15(R)15-methyl-prostaglandin E_2, *Scand. J. Gastroenterol.*, 21, 246, 1986.
56. **Ippoliti, A. F., Isenberg, J. I., Maxwell, V., and Walsh, J. H.,** The effect of 16,16-dimethyl prostaglandin E_2 on meal-stimulated gastric acid secretion and serum gastrin in duodenal ulcer patients, *Gastroenterology*, 70, 488, 1976.
57. **Konturek, S. J., Robert, A., Hanchar, A. J., and Nezamis, J. E.,** Comparison of prostacyclin and prostaglandin E_2 on gastric secretion and gastrin release and mucosal blood flow in dogs, *Dig. Dis. Sci.*, 25, 673, 1980.

58. **Hawkey, C. J. and Walt, R. P.,** Prostaglandins for peptic ulcer: a promise unfulfilled, *Lancet*, 2, 1084, 1986.
59. **Lauritsen, K. and Rask-Madsen, J.,** Prostaglandins and clinical experience in peptic ulcer disease, *Scand. J. Gastroenterol.*, 21 (Suppl. 125), 174, 1986.
60. **Dembinski, A., Gregory, H., Konturek, S. J., and Polanski, M.,** Trophic action of epidermal growth factor on the pancreas and gastroduodenal mucosa in rats, *J. Physiol. (London)*, 325, 35, 1982.
61. **Konturek, S. J., Cieszkowsik, M., Jaworek, J., Brzozowski, T., and Greogry, H.,** Effects of epidermal growth factor on gastrointestinal secretions, *Am. J. Physiol.*, 246, G580, 1984.
62. **Dembinski, A., Drozdowicz, D., Gregory, H., Konturek, S. J., and Warzecha, Z.**, Inhibition of acid formation by epidermal growth factor in the isolated rabbit gastric glands, *J. Physiol. (London)*, 378, 347, 1986.
63. **Konturek, S. J., Dembinski, A., Warzecha, Z., Brzozowsik, T., and Gregory, H.,** Role of epidermal growth factor in healing of chronic gastroduodenal ulcers in rats, *Gastroenterology*, 94, 1300, 1988.
64. **Scheurer, U., Witzel, L., Halter, F., Keller, M., Huber, R., and Galeazzi, R.,** Gastric and duodenal ulcer healing under placebo treatment, *Gastroenterology*, 72, 838, 1977.

Intestine

Chapter 10

FETAL AND POSTNATAL DEVELOPMENT OF THE SMALL AND LARGE INTESTINE: PATTERNS AND REGULATION

Daniel Ménard and Raymond Calvert

TABLE OF CONTENTS

I. INTRODUCTION

The goal of this chapter is twofold: first, to give an overview of the developmental changes related to digestive functions and cellular proliferation associated with morphogenesis in the small and large intestine, and second, to summarize our understanding of the regulatory mechanism(s) involved. Because of space limitations, this review, while not attempting to be comprehensive, focuses on the major recent advances relating to patterns and regulation of normal gastrointestinal tract development, including man. The interested reader will find detailed descriptions of the morphological and functional changes occurring during fetal and postnatal development in available reviews.[1-10]

II. LABORATORY ANIMALS

A. FETAL DEVELOPMENT

The morphogenetic processes which accompany organogenesis and functional differentiation of the small intestine are highly ordered. In the mouse fetus, 4 d before birth the intestinal mucosa is still composed of a stratified epithelium. Within the next few days, villi differentiate and morphological and functional maturation of most cell types occurs.

Although our understanding of the control mechanisms which regulate these processes is still very limited, considerable advances have been made during the last few years. The use of cell and organ culture methods has permitted characterization of some of the morphogenetic events in controlled environments. The ontogenesis of the small intestine is very complex, but it can be divided into two main events which are intimately bound: (1) morphogenesis and (2) functional maturation of the mucosa.

1. Morphogenetic Events

a. Formation of Villi

The most striking events observed during the development of the small intestine are the formation of villi and crypts. When 15-d fetal mouse duodenum is cultured in Trowell T8 medium for 48 h, the stratified epithelium gradually becomes simple columnar or cuboidal. At the end of the culture period the epithelium remains flat and quite often mesenchymal cells form small epithelioid aggregates[11] close to the basal lumina (Figure 1). These islets are regularly distributed under the epithelium.

If glutamine is added to Trowell T8 medium, villi will differentiate.[12] This observation indicates that duodenal explants are able to express their phenotype without any extrinsic influence, provided that basic nutrients are present. If Leibovitz L-15 medium (which contains glutamine) is used, villi will also differentiate. Hence the formation of villi is part of the intrinsic program of the duodenal mucosa. The formation of islets by mesenchymal cells was also observed after 24 h of culture using Leibovitz L-15 medium.[12] It appears as if the outgrowths on the surface epithelium, which will produce the villi, are initiated by these mesenchymal epithelioid aggregates. With Trowell T8 medium, villi do not differentiate but mesenchymal cells are still able to form the aggregates; since villi do not form, it is possible that the mesenchymal aggregates are unable to initiate the process or that the epithelium is unresponsive.

What is the importance of epithelial-mesenchymal interactions during the formation of villi? In the absence of surrounding mesenchyme, chick endoderm isolated from the small intestine and wrapped in the vitelline membrane can be cultured and will self-differentiate into digestive tract epithelium but villi will never form.[13] On the other hand, in the absence of endoderm, 15-d fetal mouse duodenal explants cultured in a medium which allows the differentiation of villi and crypts in intact explants[14] did not express the formation of villi.[101] Furthermore, when the mesenchyme is grafted alone, it involutes.[15] Hence, conversely the mesenchyme by itself cannot form villi.

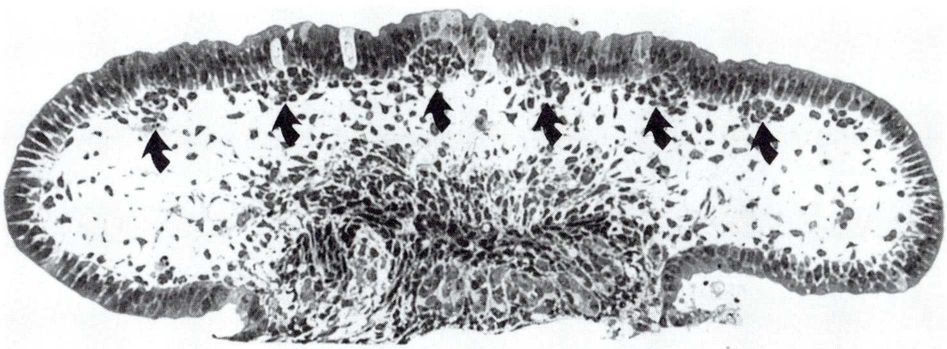

FIGURE 1. Duodenal explant cultured for 48 h with Trowell T8 medium. Notice the presence of regularly distributed mesenchymal cell aggregates (arrows) under the flat epithelium. (Magnification × 250.)

In order to observe the formation of villi, endoderm and mesenchyme have to be present and most probably, they have to interact. Since the pioneer work of Le Douarin[16] we know that the foregut endoderm has inductive properties. The somatopleural mesenchyme which normally gives rise to striated muscle can be induced to form connective tissue and smooth muscle cells when chick foregut endoderm is grafted into the somatopleural mesenchyme.

The Strasbourg group also stressed the importance of endodermal inductive activity on various heterologous mesenchymal cell types.[17] Indeed, chick intestinal endoderm associated with fibroblastic cell cultures derived from rat skin or lung and grafted into the coelomic cavity induces the formation of smooth muscle cells in heterospecific mesenchymes and villi differentiate. Furthermore, the mesenchyme isolated from the small intestine influences extraembryonic endoderm from allantois[18,19] or yolk sac[20] to form villi. Although the intestinal endoderm seems very resistant to inductive influences of heterotopic mesenchymes, Lacroix et al.[21] have shown that human fetal gastric mesenchymal cells are able to induce avian intestinal endoderm to form gastric structures. The role of mesenchyme during differentiation of the gut has been reviewed recently.[22]

In conclusion, in the presence of endoderm or mesenchyme alone, villi do not differentiate. It appears that the intrinsic program for the formation of villi is stored both in the endoderm and mesenchyme, and in each case the inductive effect can be instructive or permissive.

b. Formation of Crypts

The formation of crypts is far less spectacular than that of villi and this is probably the reason why it has drawn less attention. In the mature intestinal epithelium, crypts are described as tubular downgrowths between the bases of villi and they extend close to the muscularis mucosae. Functionally the proliferative activity is confined to the crypts and the stem cell zone is found at the bottom of these crypts.[23]

In the fetal rat, a significant number of epithelial cells located in the upper third of the villi are able to take up tritiated thymidine 2 d before birth.[24] Just before birth, proliferation is confined to incompletely formed crypts. At the end of gestation, a factor seems to be responsible for the cessation of cell proliferation on the villi. Kedinger et al.[25] have designed a model in order to analyze the effect of fetal gut mesenchyme on the cytodifferentiation of crypt cells and of embryonic progenitor cells. Using a rat intestinal crypt cell line IEC-17 (established by Quaroni and Isselbacker)[26] or primary cell cultures prepared from isolated 14-d-old fetal rat intestinal endoderm, they have shown that when these cells are associated with fetal rat gut mesenchyme and grafted under the kidney capsule of adult rats, villi and crypt differentiate. This experiment clearly demonstrates the permissive inductive effect of the mesenchyme on the morphogenesis of both villi and crypts.

We have shown that rat amniotic fluid, at 19 d of gestation, contains a factor able to induce the formation of crypts in undifferentiated mouse duodenal explants.[14] It now appears that the formation of crypts is intrinsically programmed and results from the mutual interaction between the endoderm and mesenchyme. In organ culture, the hypothetical crypt differentiation factor accelerates the process. Furthermore, all the hormones tested *in vitro* had no effect on the formation of crypts. Nevertheless, in 17-d fetal mouse duodenal explants cultured in the presence of dexamethasone for 48 h, labeling index was significantly reduced for cells located over cell position five but was unchanged for cells found at the bottom of the short crypts.[27] Hence, functionally, although glucocorticoids have no effect on the formation of crypts, they might contribute to the acquisition of adult proliferative patterns.

2. Functional Maturation of the Mucosa

Recent investigations have disclosed that the functional maturation of the intestinal mucosa in the fetus occurs in response (1) to a biological clock, (2) to extrinsic factors such as hormones, and (3) to epithelial-mesenchymal interactions.

a. Biological Clock

When mouse[28] or rat[29] fetal small intestine is implanted under the kidney capsule or under the skin in adult hosts, the brush border enzymes develop according to an intrinsic program. In the rat this program is already established as early as 17 d of gestation.[29] Administration of cortisone at age 10 d induces a precocious development of sucrase activity in normal jejunum and in grafts of jejunum.[28] These data indicate that the maturation of the intestinal mucosa occurs according to an intrinsic program which triggers the initiation processes while the full expression of the phenotype is modulated by hormones.

b. Extrinsic Factors

The involvement of hormones in the regulation of enzymatic maturation of the fetal small intestine has been established by means of *in vivo* and *in vitro studies*. Injection of cortisone and/or triiodothyronine to pregnant rats induces the precocious appearance of sucrase, of maltase and of enterokinase activities in the fetal intestinal mucosa. Epidermal growth factor injected in pregnant mice induces increases in alkaline phosphatase and trehalase activities of the brush border and also in endoplasmic reticulum-bound glucose-6-phosphatase activity.[30]

Simon-Assmann et al.[31] were the first to show a direct and specific effect of glucocorticoids on the maturation of fetal rat small intestine in organ culture. Using a completely synthetic medium, we observed that glucocorticoids increase specifically alkaline phosphatase, maltase, trehalase, and sucrase activities in fetal mouse small intestine. Furthermore, in presence of dexamethasone (an analog of glucocorticoids) the percentage of goblet cells in duodenal explants increases significantly, at a level exceeding that observed *in vivo*.[32,33] On the other hand, the direct implication of thyroxine and insulin in prenatal development of the gut functions of the mouse appears to be of minor importance.[34]

Other extrinsic factors can also influence the functional maturation of the fetal mouse small intestine. A single injection of lactose into the amniotic sac of rats on day 17 of gestation induced an increase in lactase activity in fetal jejunum.[35] However, the role of substrates in the normal physiology of the fetal intestine is probably not determinant.

c. Epithelial-Mesenchymal Interactions

The inductive effect of the mesenchyme on the functional maturation of the endoderm is permissive and/or instructive. First, Kedinger et al.[36] have shown that the endodermal tissue carries the specific characteristics of its future biochemical differentiation. Using xenoplastic recombinations, they observed that fetal rat intestinal endoderm associated with chick intestinal mesenchyme develops enzyme patterns characteristic of rat intestine: high lactase and low sucrase activities. The inverse recombination (chick endoderm with rat mesenchyme) exhibits the chick pattern: high sucrase and low lactase activities.

Second, an instructive induction is observed when chick intestinal mesenchyme is associated with stomach endoderm grafted into the coelomic cavity[37] or kept in organ culture.[38] Intestinalization of chick stomach endoderm was monitored by biochemical determination of sucrase and maltase activities[37] and by immunolabeling of sucrase.[38] Mesenchymal cells are not only promoters for intestinal endodermal cell differentiation,[39] but, moreover, epithelial-mesenchymal interactions mediate the response of endodermal cells to the action of hormones.[40] When heterospecific recombinants (rat intestinal fibroblast with chick endoderm, rat fibroblasts from intestinal muscular layers, skin, or lung with chick endoderm, rat intestinal mesenchyme with chick endoderm) are cultured for 48 h in the presence of dexamethasone after a grafting period of 10 d, chick endoderm associated with heterospecific mesenchymes shows a tremendous increase of alkaline phosphatase activity, even higher than that mediated by homologous chick mesenchyme, while no stimulation is noted with fetal rat intestinal mesenchyme. These data indicate that the hormonal effects on the functional maturation of the small intestine could be mediated by epithelial-mesenchymal interactions.

B. POSTNATAL DEVELOPMENT

1. Brush Border Enzymes

The neonatal rodent small intestine is mainly characterized by the presence of high lactase activity and the absence of sucrase.[8] Indeed, lactase, which hydrolyzes milk lactose and is already detectable in late fetal life, attains maximal activity during the first 2 weeks after birth and then declines, reaching adult values by the end of the fourth week. On the other hand, sucrase activity is undetectable during the first 2 weeks, but appears on approximately day 15 and rises rapidly, reaching adult levels by the end of the fourth week. Maltase, glucoamylase, and trehalase have low activities during the first 2 weeks, then undergo a substantial increase during the next 2 weeks. These enzymatic modifications are correlated with quantitative changes in the protein bands of brush border microvillus membranes as studied by gel electrophoresis.

During the first 2 weeks of life, the neonatal rodent small intestine has an immature mode of protein digestion. Many brush border peptidase activities such as leucylnaphthylamidase, γ-glutamyltranspeptidase and tripeptidase are present at low levels.[7,41,42] In general, these hydrolytic activities increase during the third postnatal week, each with specific patterns depending on the enzymic activity and/or the intestinal segment studied. Finally, enteropeptidase activity, the key enzyme in the activation of the pancreatic proteolytic enzymes, present at birth, remains low during the first 15 days of life and rises rapidly reaching adult levels.[43] In addition to these enzymes, many investigations have focused on the marked increase in duodenal alkaline phosphatase activity that occurs during the third postnatal week.[7] Incidentally, the pioneer work by Florence Moog on the development of intestinal function was established with this enzyme, though its function is yet to be understood.

2. Cellular Proliferation

By birth, the establishment of the villus-crypt axis in rodent small intestine is under progress and it is characterized by villi lined with a single layer of columnar epithelial cells and by the presence of proliferative cells localized exclusively in the lower regions of the crypts. Cell proliferation patterns in the small intestine appear to differ from species to species and have been recently reviewed.[5] The following are basic characteristics obtained with rats and mice. During the development of rat intestine, crypt cell division occurs at a low rate. Indeed, mitotic and labeling indices and RNA/DNA ratios are significantly higher in rats older than 21 d as compared to suckling rats. The cell length of the cycle duration markedly decreases at the weaning period, the decline being mainly attributable to a shorter G_1 phase. Finally, cellular migration, as measured by the transit of labeled cells from the crypt to the villus tip, is increased during weaning in both rat and mouse small intestine.

The neonatal large intestine has not been studied as thoroughly as the small intestine. One study on rat colon reports a relatively constant epithelial labeling index measured on days 1, 4, 10, 15, 20, and 21 after birth.[44] However, another suggests a different pattern whereby the ^{3}H-thymidine incorporation overshoots the adult levels with a peak occurring at day 14, followed by a sharp decline between days 14 and 20 towards adult values.[45] At the same time, epithelial cell labeling and mitotic indices decrease. Obviously, more studies are needed to fully elucidate the developmental cell proliferation pattern in the colonic mucosa.

3. Regulation of Developmental Changes

a. Diet

Because of the temporal correlation between brush border enzymic changes in the intestine and the onset of weaning, the possible regulatory influence of the diet seemed obvious. It became clear, however, that this dietary change could not be considered the primary cue initiating the various enzymic changes occurring during the third postnatal week.[3,6] Since the precocious appearance of sucrase activity induced by sucrose administration to suckling rats can be prevented by adrenalectomy, it can be suggested that the enzymic modification is a stress response rather than a dietary response. This conclusion is supported by the lack of stimulatory effect of various sugars on sucrase and maltase activities when they are added to cultured intestinal explants from 6-d-old rats.[46] However, even though the dietary change of weaning does not initiate the intestinal changes, diet does modulate the levels of enzymic activities to be reached. This is clearly illustrated with sucrase activity for which a high or low plateau activity will be elicited by a high or low carbohydrate diet, respectively.[47]

b. Glucocorticoid Hormones

Since the pioneer work of Moog in 1953, glucocorticoid hormones have been the most studied modulator of intestinal development.[7] The circulating concentrations of these hormones (corticosterone being the principal glucocorticoid in rodents) increase at the beginning of the third week of life and occur 48 h before the enzymic changes are normally initiated in the small intestine.[3] This lag time is consistent with the fact that glucocorticoid action in the intestinal mucosa is mediated via crypt epithelial cells. Indeed, as in the case for the normal developmental rise of sucrase activity, the premature appearance of sucrase observed after glucocorticoid treatment results from the replacement of sucrase-free villus cells by a new population exhibiting sucrase activity.[4,8,48]

Glucocorticoid administration to suckling rodents has a broad effect on all digestive functions of the intestinal cells including disaccharidase, glucoamylase, alkaline phosphatase, and peptidase activities. It is also generally agreed that these treatments do not affect intestinal cell proliferation although a recent report states otherwise.[49] The use of organ culture convincingly demonstrates the direct action of glucocorticoid on cultured intestinal explants from neonatal rats and mice by eliciting a premature appearance of sucrase and a precocious increase of the other brush border hydrolytic activities.[50,51] The small intestine loses its responsiveness to glucocorticoids at about 16 to 17 d, at which time it no longer responds to elevation of glucocorticoid concentrations.[8] Therefore, glucocorticoids are able to modulate the developing intestine during a rather short period of time of approximately 4 d.

The primary site of control (transcription, translation) of the various small intestinal enzymes has recently been addressed.[52] Using the cDNA of rabbit prosucrase-isomaltase (proSI), it was shown that the levels of sucrase-isomaltase mRNA in the small intestine of baby rabbits parallel those of sucrase and isomaltase activities, thereby providing strong indication that the primary control site is at the level of transcription. Interestingly, glucocorticosteroid treatment of baby rabbits leads to a parallel increase of both sucrase and isomaltase activities and of sucrase-isomaltase mRNA, further suggesting that glucocorticoid hormones modulate the sucrase-isomaltase activity at the transcriptional level.[53] The control of the SI gene contrasts with the control of aminopeptidase N, another brush border enzyme which has been found to be at translational level.[54] Whether or not these different gene controls can be applied to all intestinal disaccharidases, including lactase, shall remain an open question until appropriate molecular tools become available.

Finally, the importance of epithelial-mesenchymal interactions has been stressed in the preceding section, though its exact nature is not fully understood. Recently, dexamethasone-induced precocious maturation of suckling rat intestinal disaccharidase activity was shown to be associated with a differential regulation of the gene expression of procollagen types I, III, and IV, fibronectin, and laminin.[55] Whether these two phenomena are affected simultaneously by glucocorticoids or are related in a "cause-effect" fashion remains to be established. Future studies will be definitely oriented in this direction.

c. Thyroxine

Since the first evidence presented by Yeh and Moog in 1974 that the thyroid gland played an unsuspected role in rodent intestinal maturation,[7] thyroxine has received considerable attention considering that its circulating level rises significantly during the second week of life.[3] Administration of thyroxine or triiodothyronine to intact animals causes premature decline of intestinal lactase and precocious increase of sucrase, maltase, and glucoamylase. On the other hand, hypothyroidism delays the usual enzymatic modifications observed at weaning. The conclusion that thyroxine played a direct role in the development of intestinal sucrase was hampered by the demonstration that (1) its administration causes an increase in plasma corticosterone and (2) that hypothyroidism almost abolishes the normal developmental rise in corticosterone.[56] Subsequent investigation clearly showed that there is no stimulation of sucrase and maltase when thyroxine is given during the first two postnatal weeks, if serum corticosterone levels are not allowed to increase.[57] This indirect effect of thyroxine is further supported by the fact that in contrast to glucocorticoids, it is unable to induce sucrase and maltase activities in cultured intestinal explants.[50,51]

Thyroxine (T4) may, however, play a role in the modulation of jejunal lactase activity since its administration to hypophysectomized suckling rats restores the normal developmental pattern of lactase.[8] This hormonal effect cannot be secondary to glucocorticoid secretion since the circulating concentrations of corticosterone are unaffected by T4 in these hypophysectomized animals. The possibility, however, that the thyroxine effect on lactase activity may result from an altered rate of cellular proliferation and migration must be kept in mind. Indeed, thyroxine treatment is able to increase intestinal cell proliferation and there is evidence that the regulation of lactase expression takes place in two stages during neonatal development, involving reduced lactase synthesis and increased enterocyte replacement.[58] The fact that trehalase activity is unaffected by thyroxine treatment in suckling mice[59] suggests that thyroxine may have a more limited action, compared to glucocorticoids, on disaccharidase development.

The recent reports of Yeh et al.[60] reopen the debate as to the direct action of thyroxine in sucrase development. Indeed, the administration of a single dose of thyroxine to intact or adrenalectomized rats induced a precocious expression of sucrase-isomaltase (SI) protein without any increase in serum corticosterone. Furthermore, the jejunal SI protein, induced by T4, showed significantly lower sucrase hydrolytic activity and a higher isoelectric point than the one induced by cortisone.

d. Insulin

Several lines of evidence suggest a role for endogenous insulin in the regulatory mechanism of the intestinal development. Its circulating concentrations increase steadily from 10 to 25 d after birth.[61] Moreover, ^{125}I-insulin binding to isolated epithelial cells from suckling mouse intestine also increases steadily from 12 to 40 d,[62] therefore paralleling the increase of circulating insulin levels.

Administration of insulin to 8-d-old mice causes a premature appearance of sucrase activity with a precocious increase of all brush border enzymic activities. Epithelial cell proliferation is also stimulated.[63] These effects cannot be secondary to insulin-induced elevation of endogenous glucocorticoid hormones since the same phenomena are observed when insulin is added alone to serum-free cultured intestinal explants.[64] Moreover, insulin has a shorter lag time than that of cortisone and when both hormones are administered simultaneously the resulting effects are additive.[65] A direct influence of insulin has also been shown using chick embryo intestine in culture in which brush border sucrase, maltase, and alkaline phosphatase activities were increased.[66] Moreover, insulin and thyroxine acted synergistically on alkaline phosphatase as was shown for the differentiation and proliferation of epithelial cells of suckling mouse intestine.[65] The extent to which endogenous insulin modulates the spontaneous development of the small intestine, and perhaps the colon, after birth is still open for investigation.

e. Epidermal Growth Factor

Since Malo and Ménard[67] first reported that exogenous EGF administered to 8-d-old mice was able to induce the precocious appearance of intestinal sucrase and to stimulate DNA synthesis in both the small intestine and colon, this growth factor raised marked interest and numerous investigations on its implication in the development of the gastrointestinal tract.[68]

EGF content in mouse milk increases steadily from birth to a concentration of 427 μg/ml at day eight of lactation.[69] These high levels are maintained until the approach of weaning when values decrease to 130 μg/ml at day 22. EGF content (38 ng/ml) in rat milk is much lower than that of the mouse and could explain some discrepancies regarding the biological effects of EGF on proliferation and differentiation of the small intestine between suckling rats and mice.[67,70,71] Furthermore, since the effect on the developing intestine varies with the route of administration (orogastric vs. intraperitoneal) it is suggested that both systemic and luminal EGF play a role in regulating growth and maturation of the GI tract.[72] Studies are needed in order to determine the endogenous source(s) of this growth factor during postnatal periods.

Any possible effect of EGF on the developing gut is limited to the presence of functional EGF receptors since the first step of its action is its binding to specific membrane receptors. Such receptors are present in neonatal gut, including regional differences between both proximal and distal colonic binding patterns and that of the small intestine.[62,73] Indeed, epithelial cells isolated from the colon bind two to four times more [^{125}I]-EGF than do corresponding small intestinal cells. A further two-fold increase in EGF binding in the distal colon compared to that in the proximal colon is also observed.[73] Interestingly, EGF has a greater stimulatory effect on colonic DNA synthesis in suckling mice and rats compared to that in the small intestine.[74,75]

In order to investigate whether or not EGF has a direct action on cell proliferation and differentiation, EGF was added to serum-free organ culture.[76] In all experimental conditions tested, the addition of EGF to cultured 8-d-old mouse small intestine does not influence brush border enzymic activities, DNA content or [^{3}Hb]-thymidine incorporation into DNA. A similar observation was made using organ culture of adult mouse and suckling rat small intestine.[69] This nonresponse to EGF observed *in vitro* certainly does not take any significance away from the biological effects demonstrated *in vivo*. Rather, it points to the possibility that EGF induces a systemic *in vivo* reaction, which then influences the neonatal gastrointestinal tract. Incidentally, EGF can be absorbed in a biological active form from the intestinal lumen to the general circulation,[77] therefore stressing the need for further studies on the systemic influences of EGF during development. The use of organ culture has also led to the observation that fetal mouse small intestine was directly influenced by EGF.[69] These observations clearly suggest that the influence of EGF depends on the developmental stage at which it is applied. Whether these conclusions can be extrapolated to mammals with longer gestational periods, including man, remains to be confirmed.

III. HUMAN

For many years now, the development of the human gut has drawn considerable attention and detailed descriptions of the morphological and functional changes occurring during gestation are readily available.[1,2,10,78] Unlike that of small rodents, the human fetal small intestine organizes villi during the end of the first trimester *in utero;* they are covered with well-differentiated enterocytes exhibiting a full complement of brush border hydrolytic activities. During this period, the fetal colon also exhibits morphological and biochemical characteristics similar to that of the small intestine. These features are transient however since they will disappear between 30 weeks of gestation and birth. Therefore, the neonatal human gut is relatively mature compared to that of laboratory animals where final maturation is completed at a time when important dietary modifications (weaning) are taking place. These distinct developmental rates and sequences as well as important physiological differences stress the caution that one has to take in extrapolating animal data to the human.

A. DIGESTIVE FUNCTIONS

Sucrase, maltase, isomaltase, and trehalase activities are all present at 10 weeks gestation.[2] Although variations in their developmental profiles have been reported, mainly resulting from differences in starting material (homogenate vs. mucosal scraping of parts of or the entire small intestine, isolated brush borders, isolated epithelial cells) and/or expression of enzymic activities, it is generally agreed that by 14 weeks of gestation, intestinal digestive activities represent between 70 and 100% of the levels found in adult intestine.[79,80] Interestingly, the proximo-distal gradient of α-glucosidases activity established by 14 weeks gestation parallels the development of a fully functional Na^+-D-glucose cotransport system in both jejunal and ileal brush border membrane vesicles.[81]

In recent years, the SI complex, one of the primary intestinal brush border enzymes, has been extensively studied.[82] Whenever SI is synthesized and inserted into the membrane in the absence of pancreatic proteases, it is found exclusively in the form of a single polypeptide chain (220 kDa) and not as subunits. This aspect is clearly illustrated in human fetal gut where between 12 and 30 weeks of gestation SI is present only as the high molecular weight precursor.[83,84] From 30 weeks until birth, the proform and the two subunits are present, coinciding with a functional maturation of the exocrine pancreas. This has also been shown in human small intestinal explants maintained in organ culture.[82] An important aspect of the development of SI activity is related to its main regulatory site of biosynthesis. Whereas the biosynthesis of most proteins is regulated at the transcription level, a recent report by Danielson et al.,[54] suggesting that the control of the biosynthesis of brush border aminopeptidase N was located at the translation level, prompted Sebastio et al. to study this aspect in humans.[53] In this study, the appearance of SI mRNA closely parallels that of sucrase and isomaltase activities and follows a proximal-distal gradient (proximal small intestine > distal small intestine > colon). These data indicate that in spite of the different timing in development between humans and rodents, the biosynthesis of sucrase-isomaltase in man is also most likely to be controlled at the level of transcription and therefore contrasts with the control of aminopeptidase N.

Lactase activity, a key enzyme in the hydrolysis of milk lactose, is also present at 10 weeks gestation.[2,80] However, contrary to the other disaccharidases, lactase remains very low during gestation, its activity representing 30%, between 26 and 34 weeks, of that found in full term newborns. Toward the end of gestation, a marked enhancement of lactase activity occurs, bringing levels, at term, of two or four times the values found in normal infants 2 to 11 months of age. Because of this late gestational rise in lactase activity, the intestinal mucosa of premature infants born between 26 to 34 weeks of gestation is better equipped to deal with hydrolysis of α-glucosides than of lactose. Although biosynthesis, glycosylation, and intracellular transport of intestinal lactase-phlorizin hydrolase are well characterized in laboratory animals,[85] these intracellular processes as well as the regulatory mechanism(s) involved in ontogenesis of human lactase remain to be established.

Alkaline phosphatase is detectable in 11-week-old fetal jejunum and rises fourfold by 23 weeks of gestation.[2] The exact developmental profile from 23 weeks until birth is still undefined but this activity will rise since it is considerably lower at 23 weeks than that observed in the mature jejunum. A recent study illustrated the difficulty in studying alkaline phosphatase development and therefore the care one must take in interpreting such patterns.[86] We found that the distribution of alkaline phosphatase differs markedly from sucrase and trehalase activities as only 50% of this activity is of epithelial origin in the proximal intestine compared to 90 to 99% of total disaccharidases. This percentage further decreases to half this value in the distal intestine. This is in contrast to the situation in adults in whom the major component of ALP is epithelial.[87] The physiological significance of alkaline phosphatase activity as well as of its differential distribution among the epithelium and stroma during development remains to be elucidated.

The human fetal colon also exhibits alkaline phosphatase and all disaccharidase activities including low but definite lactase activity as early as 13 weeks of gestation.[10] Enzymic activities in isolated epithelial cells are lower than in the corresponding intestinal epithelium with ratios of 1:3–4 at 15 weeks of gestation and up to 1:10 at 20 weeks.[86] The distribution of the colonic alkaline phosphatase activity again differs markedly from all disaccharidases, as only 7 to 12% of total tissue activity is found in the epithelium. Of interest, during colonic development, is the presence of SI complex which is the usual marker of enterocytic differentiation. Before 24 weeks of gestation, the colonic SI complex is also expressed exclusively in the form of high molecular weight prosucrase-isomaltase.[84] Available data would also suggest that the control of the biosynthesis of the colonic SI is likely to be found at the level of transcription.[53] Subsequent development leads to the disappearance of the complex in the fetal colon while it remains expressed in the small intestine. These different patterns raise numerous questions. The use of cell lines from human colonic carcinomas (which express SI activities)[53] and of organ cultures of human fetal colon[88] should help to better understand the regulatory mechanism(s) involved, as well as the physiological significance of the presence of this enzymic complex during colonic maturation.

The overall developmental patterns for colonic digestive enzymes are not well defined and unfortunately the ontogenic profiles available were described with a limited number of fetuses.[89] Obviously, there is a need for more investigations regarding the development of brush border enzymes in the different regions of the colon.

B. CELLULAR PROLIFERATION

Based on the presence of numerous mitotic figures at all levels of the stratified epithelium during the first trimester, a high proliferative capacity of intestinal epithelial cells was suspected.[78] Until recently no quantitative data were available during morphogenesis of the human gut. The use of serum-free organ culture[90] enabled us to identify proliferating cells by autoradiographic techniques and to establish the labeling indices in all tissue layers of the fetal jejunum and colon between 8 and 19 weeks of gestation.[91,92]

Cells that incorporate ^{3}H-thymidine are found in the epithelial cells of the mucosa, in the mesenchymal cells of the developing submucosa, and in the developing smooth muscle cells of the external muscular coat of all ages studied. As long as the epithelium is stratified in both jejunum and colon, the proliferative pool is very large based on the presence of numerous labeled nuclei throughout the thickness of the epithelium. The highest epithelial labeling indices (up to 28%) are found during this period. With the formation of villi, the proliferative pools focus more and more in the intervillus regions, at which time the labeled epithelial nuclei are exclusively located in the developing crypts. Unlike the small intestine, the colonic labeling index is similar in both the epithelium and the mesenchyme from 8 up to 11 weeks gestation. By 12 to 13 weeks, the intestinal cell proliferation pattern (labeling index: epithelium >mesenchyme > muscular coat) already noted at 8 to 9 weeks is now established in the colon. Up to 14 weeks, the mesenchymal cell proliferation is high although steadily decreasing and reaches at 15 weeks the intestinal mesenchymal level attained at 11 to 12 weeks. A similar profile is also noted for the muscular layers.

Therefore, the developmental pattern of cell proliferation in the fetal colon is very similar to that of the small intestine, but reaches the same tissue distribution and cellular proliferation levels 1 month later. This delay correlates very well with the morphogenetic modifications in the two segments of the gut.[10] The parallelism of the evolution of the cell proliferation patterns of the epithelium and of the mesenchyme, in both parts of the gut, is intriguing and certainly deserves further study taking into account the close relationship between the epithelium and the mesenchyme during development. Nevertheless, more studies are needed to clearly establish the whole gestational profile of cell proliferation in the developing human gut.

C. REGULATORY MECHANISM

The role of hormones (glucocorticoids, thyroxine, insulin) and of growth factors (EGF) in the modulation of the postnatal development of the rodent gut has been discussed in the preceding section. However, the extent to which these data could be extrapolated to human development remained a matter of speculation until a reliable serum-free organ culture technique was established for the human fetal small intestine and colon.[88,90]

1. Glucocorticoid Hormones

The fact that human fetal serum hydrocortisone levels double during the last 4 weeks of gestation[93] has often been interpreted as an indication of a possible specific modulatory influence of glucocorticoid hormones on human fetal intestine, especially in relationship to the perinatal increase of lactase activity.[80]

Using cultured human fetal intestines aged 12 to 14 weeks in serum-free medium, the influence of hydrocortisone on the digestive functions and on cellular proliferation was investigated.[94] In this study the hormonal levels selected were representative of different gestational periods: (1) 12.5 ng/ml, 15 to 17 weeks; (2) 25 ng/ml, 35 weeks; (3) 50 ng/ml, 37 to 40 weeks. The lowest dosage used did not influence the various digestive activities or cell proliferation. However, the addition of 25 and 50 ng/ml induces a significant rise of lactase and alkaline phosphatase activities without influencing sucrase, trehalase, and glucoamylase. It is tempting to correlate these data with the *in utero* situation. Indeed, the low lactase activity, which is relatively stable during the gestational period, increases during the last 4 to 6 weeks of gestation in which period the circulating level of hydrocortisone doubles. The influence of glucocorticoid hormones was further supported by the reported twofold increase of brush border lactase activity observed after the addition of dexamethasone to cultured human fetal intestine (aged 8 to 13 weeks).[95]

The possibility of culturing human fetal colon in identical conditions[88] enabled us to evaluate the involvement of glucocorticoid hormones on this segment of the GI tract. So far, the addition of 50 ng/ml to 14- to 17-week colon does not affect any of the brush border digestive enzymes.[96] The basic mechanism underlying this nonresponse to hydrocortisone remains to be determined. However, the potential action of glucocorticoids was recently studied in a human colonic carcinoma cell line (Caco-2) which expresses sucrase-isomaltase.[53] These preliminary data indicate a twofold increase of SI mRNA which is similar to the rise of sucrase activity following the addition of dexamethasone. The discrepancy between these two models could be explained many ways, one of them being perhaps the difficulty to define the fetal age corresponding to confluent Caco-2 cells and to correlate them with a given fetal colonic tissue. Nevertheless, these data urge more investigations regarding the involvement of glucocorticoid hormones in the development of digestive functions in human gut.

The potential involvement of hydrocortisone in the regulatory mechanism of cell proliferation in the small intestine was simultaneously studied using radioautography and quantitation of DNA synthesis.[94] An important increase of ^{3}H-thymidine incorporation into total DNA was recorded following the addition of the hormone. A specific response of the epithelial cells was demonstrated by the establishment of the labeling index. Since no data on cellular kinetics during late human gut development are available, it is not known whether there is a normal increase of epithelial cellular proliferation during the last weeks of gestation. Hydrocortisone remains without effect on colonic DNA synthesis,[96] indicating a differential response of the two segments to glucocorticoids.

2. Epidermal Growth Factor

a. EGF Receptors

Measurable concentrations of specific EGF binding sites are present in epithelial cells of both human fetal small intestine and colon as early as 12 weeks gestation.[97] [^{125}I]-EGF binding is highest between 12 and 14 weeks of age and decreases steadily afterwards in both GI segments. However, compared to the small intestine, colonic EGF binding is substantially higher (1.5- to 2.5-fold) in all age groups studied and may be related to an increased concentration of high- and low-affinity binding sites in colonic epithelial cells.

The time-frames in which these EGF-binding patterns occur during development are of particular interest since they coincide with the period in which both tissues exhibit their highest epithelial labeling indices and undergo major morphological modifications. Since the onset of villus formation and epithelial differentiation in the colon occurs 3 to 4 weeks later than in the small intestine, there may exist a relationship between the fact that levels of EGF binding in the colon at 17 weeks coincide with peak levels noted in younger age intestine (12 to 14 weeks). Unfortunately, no data are available concerning the characteristics of EGF receptors and cell proliferation in older fetuses as well as in newborns or adults.

b. Biological Effect

The addition of EGF to fetal small intestine (11 to 14 weeks of gestation) in organ culture at concentrations ranging from 25 to 100 ng/ml decreases ^{3}H-thymidine incorporation into total DNA, the decrease being already noted within 24 hours.[98] This decrease of DNA synthesis contrasts with the stimulation of ^{3}H-thymidine incorporation into DNA induced by hydrocortisone. Obviously, these two hormones are both able to directly influence the intestinal tissue but in opposite ways. On the other hand, EGF inhibits in a dose-dependent manner the tissue accumulation of sucrase, trehalase, and glucoamylase activities without affecting alkaline phosphatase. However, the significant increase of lactase activity induced by EGF clearly demonstrates the differential effects of this polypeptide hormone on brush border enzymes. The observed increase in lactase activity is quite interesting, and it appears that both hydrocortisone[94] and EGF are able to induce a rise of lactase activity in human small intestine, but without synergistic action.

The influence of EGF (1, 10, and 100 ng/ml) on human fetal colon was recently investigated.[96] After 5 d of culture, a significant decrease of ^{3}H-thymidine incorporation into total DNA was observed. This overall reduction in DNA synthesis was in part attributed to the epithelium itself as reflected by a fall in epithelial labeling index. At the brush border membrane level, EGF induced a significant decrease of all digestive enzymes including lactase, all of these enzymic modifications occurring between the third and fifth day of culture, as observed for the small intestine.

Therefore, EGF is able to influence proliferation and differentiation of both developing human small intestine and colon. The exact physiological meaning of these data remains to be established. The amounts of EGF used are representative of those found in human colostrum (20 to 438 ng/ml) and mature milk (20 to 110 ng/ml).[69] The reported data would indicate that milk

EGF could influence the immature human small intestine, such as the one found in premature infants. However, one has to postulate that the intestinal tissue will remain sensitive to EGF throughout the gestational period and even beyond, which is for the moment totally uncertain. A mean concentration of EGF in human amniotic fluid (12 to 16 weeks) of 0.53 ng/ml with a range of 0.2 to 1.02 ng/ml was recently reported.[99] Such concentrations would be able to influence cell proliferation at least in the developing colon.[96] The recent immunohistochemical localization of EGF in the kidney and Paneth cells of the small intestine of a 20-week fetus deserves special attention.[100] The Paneth cells which develop early in the second trimester are closely associated with the undifferentiated cells of the proliferative pool in the developing crypts. If these Paneth cells are indeed able to synthesize an active form of EGF, then the appearance of endogenous sources of this factor fittingly coincides with the high number of EGF receptors [97] and a decrease of the labeling index observed between 12 and 18 weeks of gestation.[91] Whether or not there is a physiological relationship between these phenomena remains to be verified.

IV. FUTURE STUDIES

From the preceding sections, it is apparent that future studies will be directed towards an understanding of the exact relationship between mesenchymal cells and the epithelium (endoderm) during development. Do mesenchymal cells produce factors providing the signals for differentiation? What are the cellular processes that lead to the onset of differentiation in response to an extracellular signal? Investigation should also be focused on the characterization of hormones and factors involved in the modulation of differentiation. The development of molecular probes such as specific cDNA should provide essential data regarding the site of control (transcription or translation) of these modulators. Although numerous basic developmental patterns throughout gestation remain to be established for human gastrointestinal tissues, future studies will definitely be oriented towards the use of organ and cell cultures as vital tools for the study of various aspects related to functional development of the human gut.

REFERENCES

1. **Deren, J. J.,** Development of intestinal structure and function, in *Handbook of Physiology*, Vol. 3, Code, C. F., Ed., American Physiological Society, Washington, DC, 1968, 1099.
2. **Grand, R. J., Watkins, J. B., and Torti, F. M.,** Progress in gastroenterology. Development of the human gastrointestinal tract, a review, *Gastroenterology*, 70, 790, 1976.
3. **Henning, S. J.,** Postnatal development: coordination of feeding, digestion, and metabolism, *Am. J. Physiol.*, 241, G199, 1981.
4. **Koldovsky, O.,** Role of nutritional, hormonal, and intrinsic factors in expression of normal perinatal developmental pattern of intestinal disaccharidases in mammals, in *Textbook of Gastroenterology and Nutrition in Infancy*, Lebenthal, E., Ed., Raven Press, New York, 1981, 77.
5. **Klein, R. M. and McKenzie, J. C.,** The role of cell renewal in the ontogeny of the intestine. I. Cell proliferation patterns in adult, fetal, and neonatal intestine, *J. Pediatr. Gastroenterol. Nutr.*, 2, 10, 1983.
6. **Lebenthal, E. and Lee, P. C.,** Interactions of determinants in the ontogeny of the gastrointestinal tract: a unified concept, *Pediatr. Res.*, 17, 19, 1983.
7. **Moog, F.,** Perinatal development of the enzymes of the brush border membrane, in *Textbook of Gastroenterology and Nutrition in Infancy*, Lebenthal, E., Ed., Raven Press, New York, 1981, 139.
8. **Henning, S. J.,** Ontogeny of enzymes in the small intestine, *Annu. Rev. Physiol.*, 47, 231, 1985.
9. **Kedinger, M., Haffen, K., and Simon-Assmann, P.,** Control mechanisms in the ontogenesis of villus cells, in *Molecular and Cellular Basis of Digestion*, Desnuelle, P., Sjostrom, H., and Noren, O., Eds., Elsevier, Amsterdam, 1986, 323.
10. **Ménard, D.,** Growth promoting factors and the development of the human gut, in *Human Gastrointestinal Development*, Vol. 1, 2nd ed., Lebenthal, E., Ed., Raven Press, New York, 1989, 123.
11. **Calvert, R. and Micheletti, P. A.,** Selection of a chemically defined medium for culturing fetal mouse small intestine, *In Vitro*, 17, 331, 1981.

12. **Beaulieu, J.-F. and Calvert, R.,** Permissive effect of glutamine on the differentiation of fetal mouse small intestine in organ culture, *Differentiation,* 29, 50, 1985.
13. **Mizuno, T. and Ishizuya, A.,** Etude ultrastructurale de l'endoderme du tube digestif de l'embryon de poulet après autodifférenciation in vitro, *C.R. Séances Soc. Biol.,* 176, 889, 1982.
14. **Calvert, R., Lehoux, J.-G., Arsenault, P., and Ménard, D.,** Extracts of rat amniotic fluid contain a potent inducer of intestinal crypt formation, *Anat. Rec.,* 205, 27, 1983.
15. **Haffen, K., Kedinger, M., Simon, P. M., and Raul, F.,** Organogenetic potentialities of rat intestinal epithelioid cell cultures, *Differentiation,* 18, 97, 1981.
16. **La Douarin, N. and Bussonnet, C.,** Détermination précoce et rôle inducteur de l'endoderme pharyngien chez l'embryon de poulet, *C.R. Acad. Sci. (D) Paris,* 263, 1241, 1966.
17. **Lacroix, B., Kedinger, M., Simon-Assmann, P., and Haffen, K.,** Enzymatic response to glucocorticoids of the chick intestinal endoderm associated with various mesenchymal cell types, *Biol. Cell,* 54, 235, 1985.
18. **Yasugi, S. and Mizuno, T.,** Heterotypic differentiation of chick allantoic endoderm under the influence of various mesenchymes of the digestive tract, *Wilhem Roux Dev. Biol. Arch.,* 174, 107, 1974.
19. **Yasugi, S.,** Chronological changes in the inductive ability of the mesenchyme of the digestive organs in avian embryos, *Dev. Growth Differ.,* 21, 343, 1979.
20. **Masui, T.,** Differentiation of the yolk-sac endoderm under the influence of the digestive-tract mesenchyme, *J. Embryol. Exp. Morphol.,* 62, 277, 1981.
21. **Lacroix, B., Kedinger, M., Simon-Assmann, P. M., and Haffen, K.,** Effect of human fetal gastroenteric mesenchymal cells on some development aspects of animal gut endoderm, *Differentiation,* 28, 129, 1984.
22. **Haffen, K., Kedinger, M., and Simon-Assmann, P.,** Mesenchyme-dependent differentiation of epithelial progenitor cells in the gut, *J. Pediatr. Gastroenterol. Nutr.,* 6, 14, 1987.
23. **Cheng, H. and Leblond, C. P.,** Origin, differentiation and renewal of the four main epithelial cell types in the mouse small intestine. I. Columnar cell, *Am. J. Anat.,* 141, 461, 1974.
24. **Hermos, J. A., Mathan, M., and Trier, J. S.,** DNA synthesis and proliferation by villous epithelial cells in fetal rats, *J. Cell Biol.,* 50, 255, 1971.
25. **Kedinger, M., Simon-Assmann, P. M., Lacroix, B., Marxer, A., Hauri, H. P., and Haffen, K.,** Fetal gut mesenchyme induces differentiation of cultured intestinal endodermal and crypt cells, *Dev. Biol.,* 113, 474, 1986.
26. **Quaroni, A. and Isselbacher, K. J.,** Cytotoxic effects and metabolism of benzo[a]pyrene and 7,12-dimethylbenz[a]anthracene in duodenal and ileal epithelial cell cultures, *J. Natl. Cancer Inst.,* 67, 1353, 1981.
27. **Beaulieu, J.-F. and Calvert, R.,** Hormonal regulation of epithelial cell proliferation in the fetal mouse duodenum in vitro, *Anat. Rec.,* 217, 250, 1987.
28. **Ferguson, A., Gerskovich, V. P., and Russell, R. I.,** Pre- and postweaning disaccharidase patterns in isografts of fetal mouse intestine, *Gastroenterology,* 64, 292, 1973.
29. **Montgomery, R. K., Sybicki, M. A., and Grand, R. J.,** Autonomous biochemical and morphological differentiation in fetal rat intestine transplanted at 17 and 20 days of gestation, *Dev. Biol.,* 87, 76, 1981.
30. **Calvert, R., Beaulieu, J.-F., and Ménard, D.,** Epidermal growth factor (EGF) accelerates the maturation of the fetal mouse intestinal mucosa in utero, *Experientia,* 38, 1096, 1982.
31. **Simon-Assmann, P. M., Kedinger, M., Grenier, J.-F., and Haffen, K.,** Control of brush border enzymes by dexamethasone in the fetal rat intestine cultured in vitro, *J. Pediatr. Gastroenterol. Nutr.,* 1, 257, 1982.
32. **Beaulieu, J.-F. and Calvert, R.,** Effect of dexamethasone on the fetal mouse small intestine in organ culture, *Anat. Rec.,* 210, 61 1984.
33. **Beaulieu, J.-F. and Calvert, R.,** Influence of dexamethasone on the maturation of the fetal mouse intestinal mucosa in organ culture, *Comp. Biochem. Physiol.* A, 82A, 91, 1985.
34. **Beaulieu, J.-F. and Calvert, R.,** Role of thyroxine and insulin on the development of the fetal mouse duodenum in organ culture, *Can. J. Physiol. Pharmacol.,* 64, 1137, 1986.
35. **Lev, R., Bender, M. S., and Appleton A. D.,** The biochemical and histochemical demonstration of lactase induction in fetal rat intestine by intra-amniotic injection of lactose, *Histochemistry,* 61,255, 1979.
36. **Kedinger, M., Simon, P. M., Grenier, J.-F., and Haffen , K.,** Role of epithelial-mesenchymal interactions in the ontogenesis of intestinal brush-border enzymes, *Dev. Biol.,* 86, 339, 1981.
37. **Haffen, K., Kedinger, M., Simon-Assmann, P. M., and Lacroix, B.,** Mesenchyme-dependent differentiation of intestinal brush-border enzymes in the gizzard endoderm of the chick embryo, in *Embryonic Development, Part b, Cellular Aspects,* Alan R. Liss, New York, 1982, 261.
38. **Ishizuya-Oka, A. and Mizuno, T.,** Chronological analysis of the intestinalization of chick stomach endoderm induced in vitro by duodenal mesenchyme, *Wilhelm Roux Arch. Dev. Biol.,* 194, 301, 1985.
39. **Kedinger, M., Simon-Assmann, P., Alexandre, E., and Haffen, K.,** Importance of a fibroblastic support for in vitro differentiation of intestinal endodermal cells and their response to glucocorticoids, *Cell Diff.,* 20, 171, 1987.
40. **Lacroix, B., Kedinger, M., Simon-Assmann, P., and Haffen, K.,** Enzymatic response to glucocorticoids of the chick intestinal endoderm associated with various mesenchymal cell types, *Biol. Cell,* 54, 235, 1985.

41. **Ménard, D., Malo, C., and Calvert, R.,** Development of γ-glutamyltranspeptidase activity in mouse small intestine: influence of cortisone and thyroxine, *Biol. Neonate*, 40, 70, 1981.
42. **Vaeth, G. F. and Henning, S. J.,** Postnatal development of peptidase enzymes in rat small intestine: influence of hormones, *J. Pediatr. Gastroenterol. Nutr.*, 1, 111, 1982.
43. **Arsenault, P. and Ménard, D.,** Development of enteropeptidase activity in mouse small intestine: influence of hormones, *Can. J. Physiol. Pharmacol.*, 63, 472, 1985.
44. **Maskens, A. P.,** Histogenesis of colon glands during postnatal growth, *Acta Anat.*, 100, 17, 1978.
45. **Buts, J. P., de Meyer, R., and Kolanowski, J.,** Ontogeny of cell proliferation and DNA synthesis in rat colon: role of glucocorticoids, *Am. J. Physiol.*, 244, G469, 1983.
46. **Raul, F., Kedinger, M., Simon, P. M., Grenier, J. F., and Haffen, K.,** Comparative in vivo and in vitro effect of mono- and disaccharides on intestinal brush border enzyme activities in suckling rats, *Biol. Neonate*, 39, 200, 1981.
47. **Henning, S. J. and Guerin, D. M.,** Role of diet in the determination of jejunal sucrase activity in the weanling rat, *Pediatr. Res.*, 15, 1068, 1981.
48. **Hauri. H. P., Quaroni, A., and Isselbacker, K. J.,** Monoclonal antibodies to sucrase/isomaltase: probes for the study of postnatal development and biogenesis of the intestinal microvillus membrane, *Proc. Natl. Acad. Sci. U.S.A.*, 77, 6629, 1980.
49. **Yeh, K. Y., Yeh, M., and Holt, P. R.,** Hormonal regulation of adaptative intestinal growth in artificially reared rat pup, *Am. J. Physiol.*, 253, G802, 1987.
50. **Simon, P. M., Kedinger, M., Raul, F., Grenier, J. F., and Haffen, K.,** Organ culture of suckling rat intestine: comparative study of various hormones on brush border enzymes, *In Vitro*, 18, 339, 1982.
51. **Arsenault, P. and Ménard, D.,** Comparative study of the effect of hydrocortisone and thyroxine on suckling mouse small intestine in organ culture, *Comp. Biochem. Physiol.*, 77A, 721, 1984.
52. **Sebastio, G., Hunziker, W., Ballabio, A., Aurrichio, S., and Semenza, G.,** On the primary site of control in the spontaneous development of small intestinal sucrase-isomaltase after birth, *FEBS Lett.*, 208, 460, 1986.
53. **Sebastio, G., Hunziker, W., O'Neill, B., Malo, C., Ménard, D., Auricchio, S., and Semenza, G.,** The biosynthesis of intestinal sucrase-isomaltase in human embryo is most likely controlled at the level of transcription, *Biochem. Biophys. Res. Comm.*, 149, 830, 1987.
54. **Danielsen, E. M., Cowell, G. M., Sjostrom, H., and Noren, O.,** Translational control of an intestinal microvillar enzyme, *Biochem. J.*, 235, 447, 1986.
55. **Walsh, M. J., Leleiko, N. S., and Sterling, K. M.,** Regulation of types I, III, and IV procollagen mRNA synthesis in glucocorticoid-mediated intestinal development, *J. Biol. Chem.*, 262, 10814, 1987.
56. **D'Agostino, J. B. and Henning, S. J.,** Role of thyroxine in coordinate control of corticosterone and CBG in postnatal development, *Am. J. Physiol.*, 242, E33, 1982.
57. **Martin, G. R. and Henning, S. J.,** Relative importance of corticosterone and thyroxine in the postnatal development of sucrase and maltase rat small intestine, *Endocrinology*, 111, 912, 1982.
58. **Smith, M. W. and James, P. S.,** Cellular origin of lactase decline in postweaned rats, *Biochim. Biophys. Acta*, 905, 503, 1987.
59. **Malo, C. and Ménard, D.,** Effects of cortisone and thyroxine on intestinal trehalase activity in infant mouse, *Experientia*, 35, 874, 1979.
60. **Yeh, K. Y., Yeh, M., and Holt, P. R.,** Developmental changes and effects of hormones on sucrase-isomaltase structure and function in infant rat, *Gastroenterology*, 94, A509, 1988.
61. **Blasquez, E., Montoya, E., and Quijada, C. L.,** Relationship between insulin concentrations in plasma and pancreas of fetal and weanling rats, *J. Endocrinol.*, 48, 553, 1970.
62. **Gallo-Payet, N., Pothier, P., and Hugon, J. S.,** Ontogeny of EGF receptors during postnatal development of mouse small intestine, *J. Pediatr. Gastroenterol. Nutr.*, 6, 114, 1987.
63. **Ménard, D., Malo, C., and Calvert, R.,** Insulin accelerates the development of intestinal brush border hydrolytic activities of suckling mice, *Dev. Biol.*, 85, 150, 1981.
64. **Arsenault, P. and Ménard, D.,** Insulin influences the maturation and proliferation of suckling mouse intestinal mucosa in serum-free organ culture, *Biol. Neonate*, 46, 229, 1984.
65. **Malo, C. and Ménard, D.,** Synergistic effects of insulin and thyroxine on the differentiation and proliferation of epithelial cells of suckling mouse small intestine, *Biol. Neonate*, 44, 177, 1983.
66. **Moog, F. and Goellner, J. J.,** Chick embryo intestine in culture: influence of insulin and other hormones on sucrase, maltase, and alkaline phosphatase, *J. Pediatr. Gastroenterol. Nutr.*, 1, 401, 1982.
67. **Malo, C. and Ménard, D.,** Influence of epidermal growth factor on the development of suckling mouse intestinal mucosa, *Gastroenterology*, 83, 28, 1982.
68. **Weaver, L. T. and Walker, W. A.,** Epidermal growth factor and the developing human gut, *Gastroenterology*, 94, 845, 1988.
69. **Ménard, D. and Arsenault, P.,** Epidermal and neural growth factors in milk: effects of epidermal growth factor on the development of the gastrointestinal tract, in *Biology of Human Milk, Vol. 15, Nestlé Nutrition Workshop Series*, Hanson, L. A., Ed., Nestec, Vevey/Raven Press, New York, 1988, 105.

70. **Oka, Y., Ghishan, F. K., Greene, H. L., and Orth, D. N.,** Effect of mouse epidermal growth factor/urogastrone on the functional maturation of rat intestine, *Endocrinology*, 112, 940, 1983.
71. **Majumdar, A. P. N.,** Postnatal undernutrition effect of epidermal growth factor on growth and function of the gastrointestinal tract in rats, *J. Pediatr. Gastroenterol. Nutr.*, 3, 618, 1984.
72. **O'Loughlin, E. V., Chung, M., Hollenberg, M., Hayden, J., Zahavi, I., and Gall, O. G.,** Effect of epidermal growth factor on ontogeny of the gastrointestinal tract, *Am. J. Physiol.*, 249, G674, 1985.
73. **Ménard, D., Pothier, P., and Gallo-Payet, N.,** Epidermal growth factor receptors during postnatal development of the mouse colon, *Endocrinology*, 121, 1548, 1987.
74. **Arsenault, P. and Ménard, D.,** Stimulatory effect of epidermal growth factor on deoxyribonucleic acid synthesis in the gastrointestinal tract of suckling mouse, *Comp. Biochem. Physiol.*, 86B, 123, 1987.
75. **Pollack, P. F., Gada, T., Colony, P. C., Edmond, J., Thornburg, W., Kosc, M., and Koldovsky, O.,** Effects of enterally fed epidermal growth factor on the small and large intestine of suckling rat, *Regul. Peptides*, 17, 121, 1987.
76. **Ménard, D., Arsenault, P., and Gallo-Payet, N.,** Epidermal growth factor does not act as primary cue for inducing developmental changes in suckling mouse jejunum, *J. Pediatr. Gastroenterol. Nutr.*, 5, 949, 1986.
77. **Thornburg, W., Matrisian, L., Magun, B., and Koldovsky, O.,** Gastrointestinal absorption of epidermal growth factor in suckling rats, *Am. J. Physiol.*, 246, G80, 1984.
78. **Colony, P. C.,** Successive phases of human fetal intestinal development, in *Nutritional Adaptation of the Gastrointestinal Tract of the Newborn*, Kretchmer, N. and Minkowski, A., Eds., Vevey-Raven Press, New York, 1983, 3.
79. **Antonowicz, I., Chang, S. K., and Grand, R. J.,** Development and distribution of lysosomal enzymes and disaccharidases in human fetal intestine, *Gastroenterology*, 67, 51, 1974.
80. **Antonowicz, I. and Lebenthal, E.,** Developmental pattern of small intestinal enterokinase and disaccharidase activities in the human fetus, *Gastreoenterology*, 72, 1299, 1977.
81. **Malo, C. and Berteloot, A.,** Proximo-distal gradient of Na-dependent D-glucose transport activity in the brush border membrane vesicles from the human fetal small intestine, *FEBS Lett.*, 220, 201, 1987.
82. **Semenza, G., Sebastio, G., Hunziker, W., Malo, C., Ménard, D., and Auricchio, S.,** Biosynthesis of human small intestine sucrase-isomaltase, in *Brush Border Membrane Symposium*, Lentze, M. J. and Sterchi, E. E., Eds., Thieme Verlag, Stuttgart, 1988, 43.
83. **Skovbjerg, H.,** High molecular weight pro-sucrase-isomaltase in human fetal small intestine, *Pediatr. Res.*, 16, 948, 1982.
84. **Triadou, N. and Zweibaum, A.,** Maturation of sucrase-isomaltase complex in human fetal small and large intestine during gestation, *Pediatr. Res.*, 19, 136, 1985.
85. **Buller, H. A., Montgomery, R. K., Sasak, W. V., and Grand, R. J.,** Biosynthesis, glycosylation and intracellular transport of intestinal lactase-pholorizin hydrolase in rat, *J. Biol. Chem.*, 262, 17206, 1987.
86. **Ménard, D. and Pothier, P.,** Differential distribution of digestive enzymes in isolated epithelial cells from developing human fetal small intestine and colon, *J. Pediatr. Gastroenterol. Nutr.*, 6, 509, 1987.
87. **Lev, R. and Griffiths, W. C.,** Colonic and small intestinal alkaline phosphatase. A histochemical and biochemical study, *Gastroenterology*, 82, 1427, 1982.
88. **Ménard, D. and Arsenault, P.,** Human fetal colon in organ culture, *Anat. Embryol.*, 176, 441, 1987.
89. **Raul, F., Lacroix, B., and Aprahamian, M.,** Longitudinal distribution of brush border hydrolasees and morphological maturation in the intestine of the preterm infant, *Early Hum. Dev.*, 13, 225, 1986.
90. **Ménard, D. and Arsenault, P.,** Explant culture of human fetal small intestine, *Gastroenterology*, 88, 691, 1985.
91. **Arsenault, P. and Ménard, D.,** Cell proliferation in developing human jejunum, *Biol. Neonate*, 51, 297, 1987.
92. **Arsenault, P. and Ménard, D.,** Cell proliferation during morphogenesis of human colon, *Biol. Neonate*, 55, 137, 1989.
93. **Murphy, B. E. P.,** Human fetal cortisol levels related to gestational age, *Am. J. Obstet. Gynecol.*, 114, 276, 1982.
94. **Arsenault, P. and Ménard, D.,** Influence of hydrocortisone on human fetal small intestine in organ culture, *J. Pediatr. Gastroenterol. Nutr.*, 4, 893, 1985.
95. **Simon-Assmann, P., Lacroix, B., Kedinger, M., and Haffen, K.,** Maturation of brush border hydrolases in human fetal small intestine maintained in organ culture, *Early Hum. Dev.*, 13, 65, 1986.
96. **Ménard, D. and Arsenault, P.,** Influence of epidermal growth factor (EGF) and hydrocortisone on developing human colon, *Gastroenterology*, 94, A298, 1988.
97. **Pothier, P. and Ménard, D.,** Presence and characteristics of epidermal growth factor receptors in human fetal small intestine and colon, *FEBS Lett.*, 228, 113, 1988.
98. **Ménard, D., Arsenault, P., and Pothier, P.,** Biologic effects of epidermal growth factor in human fetal jejunum, *Gastroenterology*, 94, 656, 1988.
99. **Weaver, L. T., Freiberg, E., Israel, E. J., and Walker, W. A.,** Epidermal growth factor in human amniotic fluid, *Gastroenterology*, 95, 1436, 1988.
100. **Poulsen, S. S., Nexo, E., Olsen, P. S., Hess, J., and Kirkegaard, P.,** Immunohistochemical localization of epidermal growth factor in rat and man, *Histochemistry*, 85, 389, 1986.
101. Personal, Unpublished data.

Chapter 11

REGULATION OF INTESTINAL MUCOSAL GROWTH

Gary M. Levine

TABLE OF CONTENTS

I. REGULATION OF INTESTINAL MUCOSAL GROWTH

Intestinal absorption of nutrients and conservation of fluid and electrolytes are crucial in order to maintain homeostasis. The small and large intestinal epithelium is required to adapt to a variable diet and feeding frequency so that the organism may be capable of rapidly responding to an ever changing environment. The purpose of this chapter will be to describe the mechanism regulating mucosal structure and function in the adult animal.

During the last two decades, we have developed a much clearer understanding of the process of intestinal adaptation. The regulation of mucosal growth involves numerous nutritional, endocrine, paracrine, neural, and vascular factors. The primary physiologic stimulus for this process is dietary intake. While nutrients are the principal stimulus, nonnutritive luminal contents also play a role, particularly in the colon. In addition to documenting the importance of luminal nutrition, we have begun to elucidate the "second and third messengers", which transmit luminal signals to the proliferating crypt cells, regulating intestinal turnover. Additional insight into the process of intestinal adaptation has been provided by recent studies which have focused on the regulation of transport functions rather than measurements of intestinal mass and epithelial cell turnover.

II. DIETARY MANIPULATION

Dietary manipulation (fasting, parenteral nutrition, hyperphagia, isolated nutrient infusion) as well as surgical procedures (intestinal bypass, transposition, and resection) have provided compelling evidence that feeding is the stimulus which underlies the regulation of mucosal growth. This thesis is most simply demonstrated in studies examining the effects of fasting, parenteral feeding, and hyperphagia.

A. FASTING

A quarter of a century ago, Brown and co-workers[1] demonstrated that fasting decreased small intestinal renewal, measured by radioactive thymidine incorporation into DNA and delayed migration of enterocytes up the villus column. Fasting-induced atrophy begins as soon as the organism misses one feeding cycle.[2] The loss of intestinal mass is rapid for several days and then plateaus.[3] After 6 d of starvation in the rat, small intestinal DNA content decreased by over one-third and protein content decreased by one-half.[4] The most recent study of fasting-induced atrophy demonstrated that the decline in gut mass is not a uniform phenomenon throughout the small intestine.[5] After 3 d of starvation, it was observed that duodenal villus height and crypt depth decreased by approximately 25%, blunting the proximal-distal gradient. In the ileum, villus height remained the same or increased depending on the strain of rat studied, but the absolute villus height and cell number in the ileum were still 50% less than that seen in the fasted duodenum. Concomitant with the decreased mucosal mass of fasting is a fall in glucose transport and disaccharidase activity.[2] The decline in sugar transport is caused by a decrease in the maximal transport capacity without a change in the affinity of the carrier.[6]

Intestinal crypt cell proliferation is triggered within hours of refeeding. The cell cycle time decreases due to a shortening of the rest phase.[7] Villus height increases as the cell production rate exceeds exfoliation. As a result, within 24 h of refeeding, the rodent intestine has returned towards normal.[3,7]

B. EFFECTS OF PARENTERAL NUTRITION

The parenterally nourished animal has provided an excellent model for investigating the mechanisms controlling intestinal adaptation. TPN (total parenteral nutrition) eliminates the multiple negative metabolic factors associated with fasting and effectively "disconnects" the intestinal tract from normal dietary signals while maintaining whole body nitrogen balance and growth. With this technique, it quickly became apparent that parenteral nutrition in the rat,[8,9] dog,[10] or rabbit[11] produced intestinal atrophy. TPN fed rats were compared to controls receiving the identical chemically defined diet intragastrically. Intestinal atrophy occurs at a rapid pace, similar to that seen in fasting animals, tending to plateau at a basal level equivalent to that seen after 48 to 72 h of fasting in the rat. TPN causes a loss of the normal proximal to distal gradient of intestinal mass and leads to a decrease in total sugar, amino acid, and fat absorption.[6,12-14]

III. HYPERPHAGIA

Since fasting produces intestinal atrophy, it should be no surprise that hyperphagia resulting from lactation, diabetes, hyperthyroidism, and hypothalamic lesions leads to a marked increase in intestinal growth.

A. LACTATION

The marked increase in food intake necessary to sustain milk production is a powerful stimulus for intestinal growth. During lactation there is a progressive increase in crypt cell labeling index and expansion of the proliferation zone.[15] As a result, by 15 d postpartum in the rat, there is a 25% increase in mucosal thickness and a doubling of small intestinal weight per unit body weight. Although the serum gastrin concentration is elevated, it is not responsible for the mucosal growth since intestinal proliferation correlates with food intake, but not serum gastrin concentration. In addition, antrectomized lactating rats have unimpaired adaptation.[16] Extraluminal factors may be involved in the hyperplastic response seen in lactating animals. Villus height and crypt depth increase in bypassed jejunal loops of lactating rats.[17] However, these changes may not represent hyperplasia since the values for tissue weight per unit length did not change. In the ileum of lactating rats with a bypassed jejunum, there was a significant increase in villus height, crypt depth, and tissue weight compared to nonlactating controls. In those rats, the ileum became the main absorptive surface and showed the most impressive increase in parameters of intestinal hyperplasia. Increased functional capacity accompanied mucosal growth as exemplified by a greater *in vivo* glucose absorption.[17] These data suggest that while luminal nutrition is not the only determinant of hyperplasia seen in lactation, it is the most important.

More recently, it has been shown that polyamine synthesis is involved in lactation induced hyperplasia. Ornithine decarboxylase (ODC) activity increases approximately 20-fold during lactation. When ODC is inhibited by difluoromethylornithine (DFMO), adaptation is inhibited.[18]

Other experimental approaches to hyperphagia include exposure to cold,[19] to the hypothalamic lesions satiety center,[20] and experimental hyperthyroidism.[21] In the case of hyperthyroidism, withdrawal of excessive thyroid hormone reverses hyperphagia and returns intestinal mass to normal.

B. DIABETES

Experimental diabetes produced by alloxan or streptozotocin has been one of the most extensively studied models of hyperplasia. Intestinal weight and sugar transport capacity approximately double in diabetic rats.[22] In a series of carefully controlled studies, Miller et al.[23] studied crypt cell proliferation and migration in diabetic rats. There was a twofold greater proliferative activity in diabetic rats which persisted during pair feeding experiments. These data suggest that factors other than increased luminal nutrition are responsible for the intestinal adaptation of diabetes. Furthermore, the bypassed intestines of diabetic rats have increased disaccharidase activity, demonstrating that hyperglycemia plays a role in regulating intestinal growth in diabetes.[24] Accompanying hyperplasia is a dramatic increase in sugar absorption. Mucosal hypertrophy accounts for approximately 30% of the increase, the remainder being due to a greater number of sugar binding sites in the intestine.[22] Treatment with insulin returns food intake, intestinal mass, and absorptive activity to normal.[25] As can be gleaned from these studies, it is difficult to unravel the individual contributions of luminal and extraluminal hormonal factors in causing mucosal growth in diabetes.

IV. INTESTINAL RESECTION

Loss of intestinal mass results in compensatory structural and functional adaptation in the residual bowel. The degree of hyperplasia in the remaining bowel varies directly with the length of intestinal resection.[26] The adaptation that occurs after resection is not a unique phenomenon and probably shares many of the same mechanisms governing intestinal mucosal growth in general.

As long ago as 1912, Flint published the results of seminal experiments in dogs and clinical observations in man that had undergone intestinal resection. Not only did he show that villus hyperplasia occurred, but he inferred that enhanced absorptive function occurred since postoperative malabsorption corrected over time.[27] Subsequent research has clearly showed that DNA synthesis increases within hours after resection, followed by a concomitant increase in crypt cell mitotic rats, epithelial migration up the villus, and villus lengthening.[28-30]

Initially, mucosal enzyme and transport specific activities are decreased as immature cells occupy the enlarging villus.[31] Over time, specific activity returns to normal and the increased mucosal function results from a multiplicative effect of increasing mucosal mass.[32] Recently, Freeman et al. demonstrated that an adaptive increase in brush border membrane sodium dependent glucose transport was not detectable in the residual bowel 2 weeks after surgery, but had occurred at 6 weeks.[33] Urban and Michel carried out a study in which they compared the hyperplastic response with changes in transport function in the rat. Whereas mucosal hyperplasia reaches a maximum within 2 weeks, increments in absorptive function are delayed.[34] For example, at 2 weeks the specific activity of sodium absorption decreased by 35% in the duodenum and 58% in the ileum but absorption was unchanged, per centimeter, offset by an increase in mucosal mass. At 4 weeks, the specific activities of sodium, chloride, and galactose absorption returned to normal, but the 60% greater total transport capacity mirrored the extent of mucosal hyperplasia. Compensatory increases in amino acid and peptide absorption in the residual bowel have been observed.[35] Adaptation of protein absorption occurs rapidly, returning to normal 1 week after resection in the rat.[36]

The primary event initiating hyperplasia is increased food intake. Menge et al. found that adaptation was more marked in rats given free access to food. Resected rats that were pair fed with control animals did not develop as striking hyperplasia.[37] Further evidence for the importance of feeding comes from studies showing that parenterally nourished animals did not undergo hyperplasia following extensive intestinal resection.[38,39]

The biochemical events mediating adaptation have begun to be elucidated. Immediately after resection, ornithine decarboxylase activity increases and results in synthesis of higher concentrations of mucosal polyamines. Enhanced DNA synthesis soon follows.[40] Feeding also leads to the appearance of at least two distinct growth promoting peptides within the mucosa of the small intestine.[41] Further research is needed to better characterize the role of these messengers.

Feeding induced secretion of several hormones may play a role in post resection hyperplasia. Enteroglucagon levels increase after resection and Thirty-Vella bypass[42,43] as a result of hypertrophy of enteroglucagon containing neuroendocrine cells in the remaining bowel mucosa.[44] Epidermal growth factor (EGF) also may play a role, since hyperphagia probably leads to increased salivary secretion of this hormone.[45] In addition, it has been suggested that growth hormone is involved in compensatory hyperplasia. Plerocercoid growth factor (an analogue of growth hormone lacking gluconeogenic effects) has been shown to stimulate adaptation after intestinal resection in the rat.[46] Gastrin is not involved in mediating adaptation because antrectomized rats have a normal hyperplastic response to resection.[47]

The effects of different dietary components on post resectional adaptation also have been

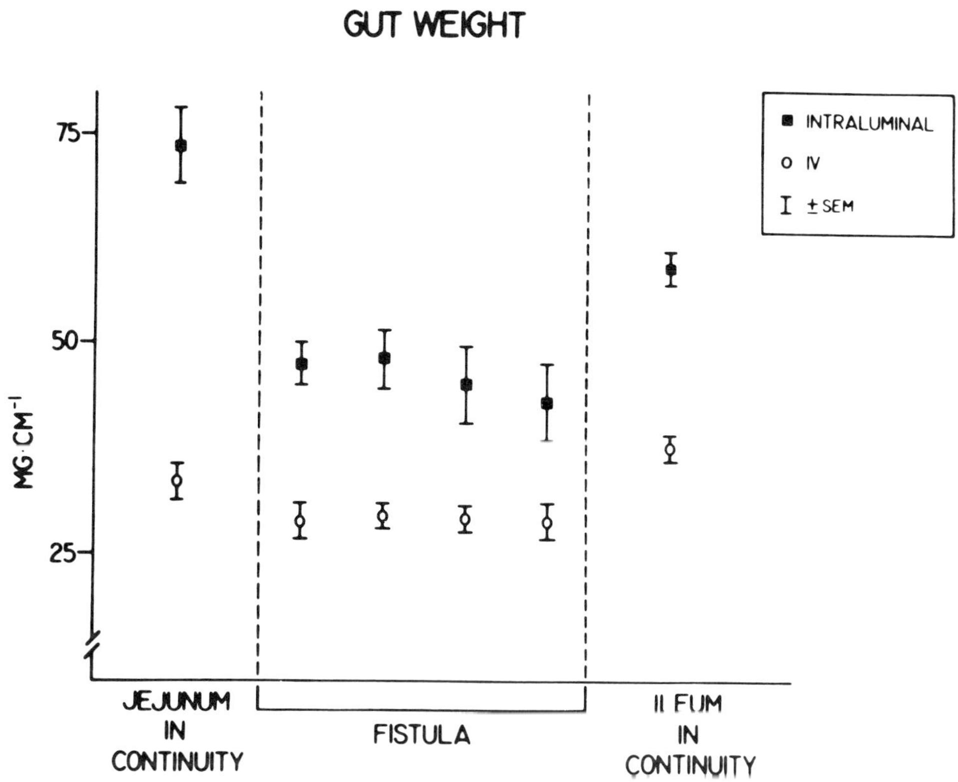

FIGURE 1. The relative contribution of luminal and indirect factors in regulating gut growth can be demonstrated in Thirty-Vella fistula rats undergoing either luminal or intravenous administration of a chemically defined diet. Intragastric infusion of the diet produces profound hypeplasia in the gut in continuity. Luminal nutrients also stimulate gut growth to a lesser degree in the fistula excluded from luminal contents. (From Dworkin, L. D., Levine, G. M., Farber, N. J., and Spector, M. H., *Gastroenterology*, 71, 626, 1976. With permission.)

studied. Long-chain triglycerides enhance intestinal adaptation more than the administration of equicaloric protein, carbohydrate, or medium chain triglycerides.[48] Dietary fiber also plays a role in producing adaptation. Pectin supplementation of a liquid diet enhanced mitotic activity as well as mucosal weight, protein, and DNA content after resection.[49] It was postulated that bacterial fermentation of pectin resulted in the production of short-chain fatty acids which stimulated mucosal growth. A subsequent study has shown that intravenous administration of short-chain fatty acids can produce a modest but significant increase in ileal mass after resection.[50]

V. EXTRALUMINAL FACTORS

Several lines of evidence suggest that extraluminal factors triggered by intraluminal events are partly responsible for intestinal growth. First, it has been demonstrated that infusion of either sugars or amino acids into the distal small intestine causes hyperplasia of the proximal intestine.[51,52] A second line of evidence supporting the role of indirect factors is derived from studies creating a Thirty-Vella fistula isolated from the luminal stream. Dworkin et al.[53] compared two groups of rats with excluded loops 1 week after receiving a chemically defined diet either intragastrically or parenterally. The excluded loop of the luminally fed animals demonstrated a greater mucosal mass compared to the excluded loop of parenterally fed animals. However, this degree of hyperplasia did not approach the threefold greater intestinal mass seen in the intestine exposed to luminal contents (Figure 1). The third line of evidence supporting extraluminal factors derives from a study of parabiotic animals in which one member of the parabiotic pair underwent intestinal transection or resection. As would be expected, intestinal hyperplasia occurred in these rats. In the parabiotic, nonoperated rat, there was a significant increase in mucosal DNA synthesis. However, because of the short term of the experiment, no increase in intestinal mass was found.[54] The mechanism by which extraluminal factors produce mucosal growth is unknown. Among the possibilities are hormonal stimulation, neural innervation, and changes in mucosal blood flow.

A. HORMONES

Several hormones have been proposed as extraluminal growth factors for the intestine. The most extensively studied has been gastrin. Although shown to be an important trophic factor for the stomach, pancreas, and colon, gastrin is not a significant trophic factor for the small intestine.[16,47,55] On the other hand, enteroglucagon may play a role in adaptation. Gleeson et al. described a patient in whom an enteroglucagon secreting tumor produced intestinal hyperplasia. Infusion of extracts from this tumor into rats caused hyperplasia.[56]

Feeding releases neurotensin, another hormone which is widely distributed throughout the gut within neuroendocrine cells. Injection of neurotensin into rats significantly increased intestinal mass and disaccharidase activity.[57]

More recently, EGF has been implicated in the control of intestinal growth. EGF is secreted by the salivary, biliary, and duodenal epithelium.[45,58] The secretion of high concentrations of EGF into the proximal intestine helps to create the normal proximal-distal gradient of intestinal mass. In addition, EGF may produce hyperplasia seen at the site of transposition of the duodenal papilla,[59] as well as part of the trophic response to pancreatobiliary secretions.[60] When it is infused into the lumen of adult rats, EGF produces an increase in ornithine decarboxylase activity followed by increased incorporation of thymidine into DNA and a greater crypt cell labeling index. In addition to the local effect seen at the site of infusion, a lesser degree of stimulation was seen in the proximal jejunum, suggesting that EGF was absorbed and transported in the blood.[61]

In addition to growth factors, it has been proposed that chalones or growth inhibitors control intestinal proliferation.[62,63] The mechanism postulated by these studies involves increased secretion of the chalone during fasting and inhibition of its release during feeding.

B. NEURAL AND VASCULAR FACTORS

The presence of luminal contents triggers a larger number of neural and vascular changes. Adrenergic nerves terminate around the intestinal crypts and may be responsible, in part, for regulating proliferation.[64] Alpha adrenergic stimulation increases proliferation and enterocyte migration out of the crypts, while beta adrenergic stimulation and sympathectomy decrease crypt cell turnover.[65] The inhibitory effects of sympathectomy are overcome by luminal nutrition.[66]

Another way in which luminal contents exert indirect control of intestinal growth is through changes in blood flow. Raw bile or bile salt-nutrient mixtures, but not pancreatic enzymes, are potent initiators of intestinal hyperemia.[67] Shepard and Riedel compared the ability of various nutrients to regulate mucosal blood. Glucose increases blood flow, whereas amino acids and triglycerides do not. However, when fatty acids are solubilized with bile salts, they become potent vasodilators.[68]

VI. LUMINAL FACTORS

A large body of experimental evidence has accumulated which strongly suggests that luminal events are the principal stimulus for small intestinal growth. Several hypotheses have been advanced to explain this phenomenon. Dowling initially proposed "luminal nutrition" as the underlying regulator of intestinal growth.[69] This theory posed that the presence of nutrients within the gut was essential for adaptation. Clarke popularized the concept of the "work-load hypothesis", attributing intestinal growth to the consequence of absorptive or secretory function, regardless of whether or not nutrients were involved.[70] The stimulation of endogenous secretions by feeding is yet another concept proposed to explain intestinal growth.

A. WORK-LOAD HYPOTHESIS

Clarke proposed this hypothesis based on experiments in which he examined the morphology and crypt cell metaphase accumulation rate in starved rats with a self-emptying loop.[70] A group of the animals had distilled water infused into the self-emptying loop in order to determine whether or not a noncaloric luminal stimulus would affect parameters of intestinal growth. If one examines these data critically, it is apparent that there was approximately 10% greater villus height and crypt depth in the infused loop compared to controls that did not receive an infusion. However, the only significant difference between the groups was a greater number of colchicine-arrested metaphases in the most proximal part of the isolated sac exposed to the greatest volume of infusate. When isolated loops were infused with different sugars and saline, it was observed that sodium chloride, galactose, and glucose all produce a modest but nonsignificant increase in villus height and inconsistent increase in the number of crypt metaphases. Alpha methylglucoside and mannose did not change villus height, but alpha methylglucoside led to a significant increase in crypt metaphases compared to the controls. Menge and colleagues were unable to reproduce Clarke's findings when they infused glucose, galactose, or electrolyte solutions into an isolated intestinal loop.[37] Only glucose increased villus height whereas the nonmetabolizable sugar, galactose, or the infusion of electrolytes had no effect. Other studies which provide evidence for work load as a regulator of intestinal growth include observations that the mid-jejunal mass of parenterally fed rats was significantly greater after luminal isotonic saline infusion.[13,71] The degree of mucosal growth is modest and equal to that seen after infusion of 5% dextrose. Passive movements of fluid induced by the presence of hypertonic solutions do not significantly stimulate hyperplasia.[52]

B. ENDOGENOUS SECRETIONS

Feeding stimulates the stomach, liver, pancreas, and small intestine to secrete. Altmann was the first to demonstrate that either pancreatic secretions or bile were able to produce intestinal hyperplasia.[72] He performed a series of elegant experiments in which the pancreatic ampulla or the bile duct was reimplanted into the distal intestine. Near the implantation site, the villus size index increased significantly. Diverting bile from the proximal small intestine reduced proliferative activity and ^{14}C thymidine incorporation into DNA. These changes can be reversed by the infusion of sodium taurocholate.[73]

Hughes et al.[10] studied beagle dogs maintained on chronic TPN. They gave half of the dogs a daily infusion of secretin and cholecystokinin, which was sufficient to stimulate pancreatic secretion. After 6 weeks, the dogs receiving the hormones had maintained their intestinal mass and sugar absorption compared to the control dogs, who developed the characteristic atrophy seen in parenterally fed animals. This phenomenon was confirmed in the rat where secretin plus CCK-octapeptide significantly increased parameters of mucosal mass.[60] The mucosal growth produced by secretin and CCK is due to stimulation of pancreatic or biliary secretions and not to direct effects of the hormones on intestinal receptors. Parenterally nourished rats prepared with a bypassed loop of jejunum received secretin, CCK-octapeptide, or both.[74] Bypassed segments of jejunum did not undergo mucosal hyperplasia compared to the segments that were in continuity with luminal contents.

Further research needs to be done exploring the mechanism by which endogenous secretions increase cell turnover. As stated previously, EGF is a promising candidate as a mediator of this phenomenon.[61]

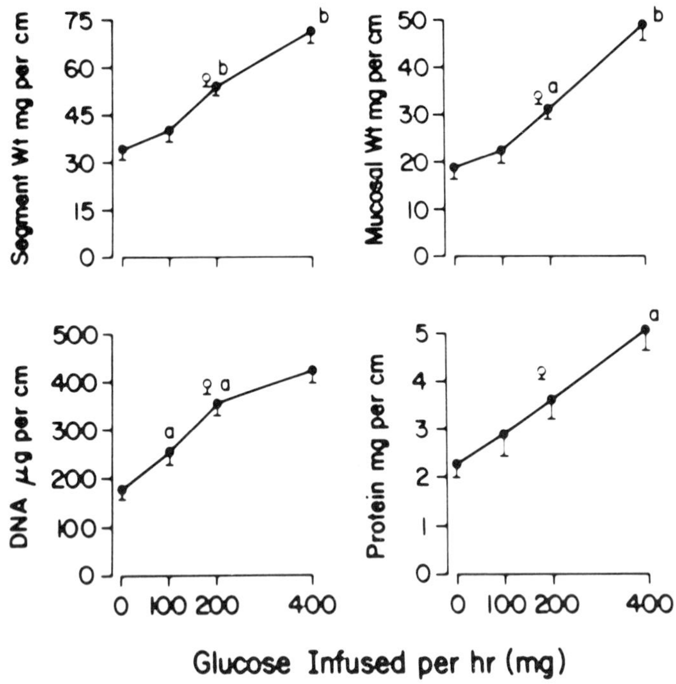

FIGURE 2. There is a direct correlation between the amount of glucose infused into the midintestine and gut growth. (a) $p < 0.05$; (b) $p < 0.01$ compared with previous value. (From Weser, E., Babbitt, J., and Vandeventer, A., *Dig. Dis. Sci.*, 30, 675, 1985. With permission.)

C. LUMINAL NUTRITION

Each of the three principal nutrients, carbohydrates, protein, and fat, are trophic to the gut.

1. Carbohydrates

The relative potency of various sugars in stimulating mucosal growth has been extensively investigated utilizing parenterally nourished rats. The infusion of dextrose intragastrically or into the mid-ileum causes profound gut hyperplasia.[51] There is a linear correlation between the quantity of dextrose infused into the gut and mucosal growth (Figure 2).[52] Disaccharides are more trophic to the intestine than monosaccharides. A 5% solution of glucose, galactose, or fructose did not increase intestinal mass compared to saline, whereas a 5% sucrose, maltose, or lactose solution did. It was apparent that absorption was needed for the trophic effect of sugar since the infusion of the glucosidase inhibitor acarbose blunted or shifted the trophic effects of disaccharides further down the gut.[75] Mannitol, a nonabsorbable sugar, is ineffective in producing growth while the infusion of a nonmetabolizable sugar, 3-O methyl glucose, does promote growth.[51,71]

Studies investigating the dynamics of the effects of luminal sugars on absorption have documented that relatively low luminal concentrations of dextrose stimulate maximal transport capacity (V_{max}) without affecting the affinity constant (K_m). When the gut is exposed to higher concentrations of sugar, there is further augmentation in absorption because of greater intestinal mass and enhanced permeability. The change in permeability is nonspecific since unrelated compounds are also absorbed at a greater rate.[71] Exposure of the gut to 3-O methyl glucose does not increase transport capacity either *in vitro*[71] or *in vivo*.[12]

Another approach to understanding how nutrients control mucosal growth has involved the use of supplemented diets. This type of experiment introduces multiple variables because when a high concentration of one dietary component is added, the concentration of other components invariably decreases. Mice raised on a high carbohydrate diet had approximately 15% greater intestinal mass, primarily of the proximal intestine, compared to those raised on a carbohydrate-free ration. High carbohydrate fed mice had enhanced sugar uptake *in vitro*, but switching these animals to a carbohydrate-free diet led to a rapid decrease in glucose uptake and a concomitant increase in amino acid uptake on the complementary carbohydrate-free high protein diet.[76] Thus, it appears that there is no "pre-programming" of intestinal function. These experiments provide additional evidence that luminal exposure is the most important factor for up- and down-regulating glucose and amino acid absorption. The time course of these changes is such that it takes several days for transport and enzyme activities to decrease after withdrawal of carbohydrate from the diet, whereas exposure to the substrate results in a rapid increment of transport capacity within 1 d.[77] The ability of various sugars to induce sugar transport has been compared. Fructose was found to be the best inducer of sugar transport, not only of its own facilitated absorption, but also of the active transport mediated absorption of aldohexoses. Glucose and maltose also stimulate absorption to a significant degree. Although galactose and 3-O methyl glucose increase absorption, they do so in a nonspecific manner related to their induction of a hyperphagic state.[78]

2. Amino Acids

Studies in parenterally nourished rats have demonstrated that the infusion of relatively low concentrations of amino acids (3 to 5%) stimulates mucosal growth to a greater degree than either isotonic saline or isocaloric dextrose.[51] This phenomenon was further studied when individual amino acids were infused. It was apparent that each amino acid had its own trophic potential. Histidine was a better stimulator of mucosal growth than valine or glycine in both jejunum and ileum. However, glycine did cause hyperplasia in the ileum.[79] Neither glutamine nor aspartic acid, both of which can serve as metabolic fuels for the small intestine, nor the essential amino acids lysine and valine produce any greater hyperplasia than a nonmetabolizable amino acid such as aminoisobutyric acid (Figure 3).[13]

Infusion of amino acids also appears to regulate uptake systems. The presence of at least three or four distinct amino acid transport systems adds complexity to understanding how luminal amino acids regulate their own transport. The several amino acid transport systems are not necessarily influenced by their own preferential substrate. For example, lysine and aspartic acid treatment increased the *in vitro* uptake of all three classes of amino acids tested, whereas valine, a branched chain essential amino acid, had no significant effect on stimulating either its own transport or the transport of other classes of amino acids.[13] Stein et al. supplemented the diet of mice with several different amino acids in order to further clarify the nonsubstrate specific nature of stimulation of amino acid transport. They found that an acidic amino acid, aspartate, was the best inducer of the basic amino acid transporter, whereas the basic amino acid, arginine, but not lysine (which shares the same transporter), was a good inducer of the acidic amino acid transporter.[80] Further studies pointed out that intact protein, peptides, and amino acids were equally effective in increasing amino acid and peptide transport.[81]

The mechanism by which amino acids cause mucosal growth is unknown. Jain et al. examined the effects of infusion of glycine, alanine, and glucose on ornithine decarboxylase activity.[82] Although glycine is not trophic to the jejunum,[79] it did greatly increase ODC activity compared to alanine and glucose, compounds which are known to stimulate growth. In addition to amino acids, dimethylamine, ethylamine, and putrescine, compounds normally present in the diet, stimulate mucosal growth.[83,84] Heat lyophilization, which removes dietary amines, prevented the normal stimulation of mucosal growth. The addition of dimethylamine to a lyophilized diet restores growth through the induction of ODC activity.[85]

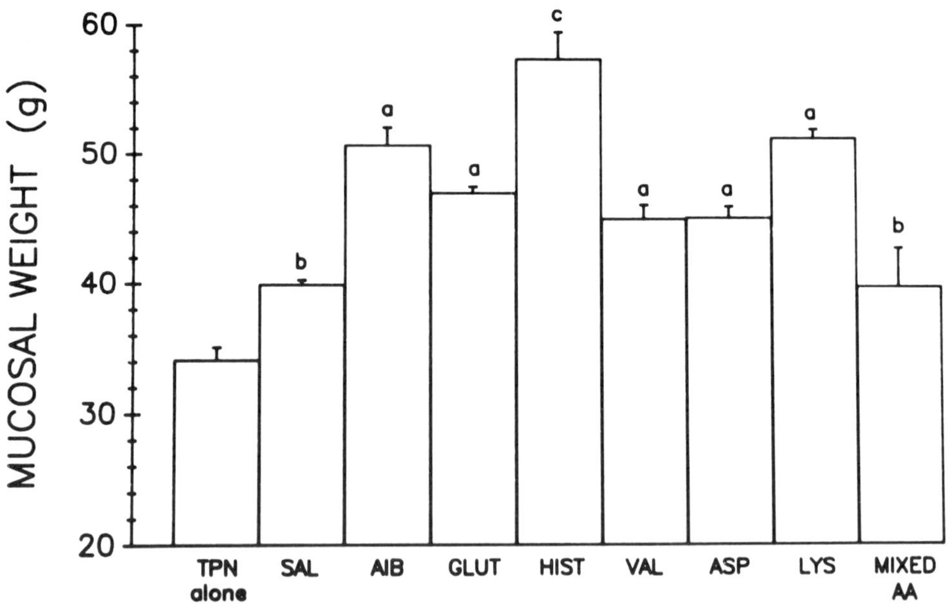

FIGURE 3. The infusion of individual amino acids in the midjejunum of TPN rats is trophic compared to isotonic saline (SAL) or a mixed amino acid solution (MAA, Travasol®). Histidine (HIST) is the most trophic. The metabolizable amino acids glutamine (GLUT) and aspartic acid (ASP) are no more trophic than nonmetabolizable aminoisobutyric acid (AIB) or the essential amino acids lysine (LYS) and valine (VAL). Note that saline produces modest hyperplasia compared to TPN alone. (a) $p < 0.01$ vs. saline and TPN; (b) $p < 0.01$ vs. TPN; (c) $p < 0.01$ vs. all other groups. (Adapted from References 13, 51, and 79.)

3. Dietary Fat

The presence of fat in a mixed diet has a relatively modest effect on mucosal growth. A chow diet containing 5% fat led to greater mucosal weight throughout the gut compared to a liquid diet containing 1% fat.[86] Balint et al. compared the effects of feeding conventional chow (4% fat by weight) to a high-fat diet containing 45% fat.[87] The villus height in the distal jejunum was greater in the high-fat group, without a significant difference in DNA, suggesting that a high-fat diet produced intestinal hypertrophy. A conflicting result was reported by Jacobs et al.[88] who found that feeding rats a high-fat lard diet or high-fat corn oil diet led to a decreased jejunal mucosal weight compared to control rats fed a 5% fat diet. The corn oil diet also produced mucosal atrophy in the distal ileum. Similar conclusions were reached in another study which compared conventional chow to two high-fat diets containing a predominance of either saturated or unsaturated fatty acids. In the jejunum, neither high-fat diet influenced measurements of villus architecture, although villus cell size was significantly greater with a concomitant fall in cell number. In the ileum, the high-saturated-fat diet led to a decrease in muscosal surface area although inconsistent increases in villus height were seen with the highly unsaturated fat diet.[89] Changes in the ileum may be difficult to interpret since the relative absorption of saturated vs. unsaturated fat may have exposed mucosa to different degrees of stimulation.

A further piece of evidence mitigating against fat as an important regulator of mucosal growth derives from a study in which a variety of liquid and low residue diets were compared for their ability to promote intestinal ^{3}H-thymidine incorporation and total DNA content. There was no relationship between the fat content of the diet and parameters of intestinal growth.[90] The mechanism by which fat increases gut mass is unknown, but does not appear to involve polyamines.[91] It should be recognized that feeding a high-fat diet improves fat absorption and mucosal fat esterification capacity even though total mucosal mass may not increase markedly.[14,87]

4. Fiber

Dietary fiber plays a minor role in regulating small intestinal growth and function. The addition of kaolin to a standard diet does not produce small intestinal hyperplasia although it does increase *in vitro* glucose and water absorption.[92] Of the water soluble, metabolizable fibers, guar gum markedly increased mucosal mass,[93] but pectin did not.[94,95] Another study found that rats fed a diet supplemented with either cellulose or pectin demonstrated a modest increase in villus height as well as a 3-O methyl glucose and aminoisobutyric acid absorption.[96] However, the fiber fed rats consumed more diet and the changes observed may have been due to increased luminal exposure to carbohydrate and protein rather than to the effects of the fiber itself. Other types of fiber which have not significantly affected the small intestine include oat bran, wheat bran, and alpha cellulose.[93,95]

VII. COLONIC ADAPTATION

In comparison to the small intestine, colonic mucosal growth requires the presence of nonnutritive bulk or fiber. Prolonged feeding of the nonnutritive agent kaolin increased colonic, weight, primarily due to muscle hypertrophy.[92] Feeding chemically defined diets lacking in bulk produces a 40 to 50% decrease in colonic weight, indices of mitotic activity and DNA synthesis.[86,90] Addition of alpha cellulose to a liquid diet restores colonic mass towards normal.[90,95] A diet containing 20% wheat bran has also been shown to increase cecal mucosal weight and DNA synthesis compared to a fiber free diet. Lesser but significant increases were seen in the distal colon of fiber fed rats.[96]

Since the colon, especially the cecum, is an area of significant bacterial fermentation of fiber, it can be inferred that production of short chain fatty acids may be one of the mechanisms producing colonic hyperplasia, since these compounds are a significant energy source for the colon.[97] Following small intestinal resection, colonic hyperplasia occurs probably as a result of increased amounts of non-absorbed carbohydrate and dietary fiber entering into the colon, resulting in production of increased amounts of short chain fatty acids.[49] Urban and colleagues examined the effect of small bowel resection on colonic adaptation. Within 2 to 4 weeks cecal and colonic weight significantly decreased along with colonic absorption.[98]

The trophic hormone gastrin has been shown to stimulate colonic mucosal growth. The infusion of pentagastrin into rats leads to increased colonic DNA and protein content.[55] The fall in colonic weight seen after feeding various liquid, chemically defined diets correlates with a decline in serum gastrin levels.[99] However, the importance of gastrin has been questioned since the addition of bulk to a liquid diet increased DNA synthesis and colonic growth without raising serum gastrin levels.[100]

Strong evidence for the presence of indirect stimulation of colonic proliferation was provided by Delvaux et al.[101] These investigators created a transverse colostomy in rats, thereby causing the distal colon to be excluded from luminal contents. After a 48 h fast, the animals were refed and indices of proliferation activity determined. The rates of ^{3}H-thymidine incorporation into DNA increased sevenfold in the proximal functioning colon and threefold in the excluded colon. The degree of these responses gives an indication of the comparative importance of luminal and indirect factors in stimulating colonic growth.

In interpreting studies in which fiber or bulk were added to the diet, it is important to recognize the differences that exist in these additives. Insoluble or inert bulking agents such as cellulose or kaolin are not as good stimulators of colonic growth as the water soluble, metabolizable fibers such as pectin, guar, and bran. Problems in interpreting data rise from the type of control group used. In studies in which a liquid, fiber-free diet is used as a control, it has been possible to show positive effect of almost any type of additive. However, when supplementation of standard diets is performed, the effects of the insoluble dietary additives are not as apparent.

Our knowledge of the regulation of gastrointestinal growth has increased dramatically in the past decades. However, many areas still remain to be investigated. Among the questions that are still unanswered are the relationships between intestinal metabolism and growth, the role of locally released hormones or growth factors in regulating intestinal turnover as well as the importance of second or third messengers such as prostaglandins or polyamines.

REFERENCES

1. **Brown, H. O., Levine, M. L., and Lipkin, M.,** Inhibition of intestinal epithelial cell renewal and migration induced by starvation, *Am. J. Physiol.*, 205, 868, 1963.
2. **McManus, J. P. A. and Isselbacher, J.,** Effect of fasting versus feeding on the rat small intestine: morphological, biochemical, and functional differences, *Gastroenterology*, 59, 214, 1970.
3. **Clarke, R. M.,** The time course of changes in mucosal architecture and epithelial cell production and cell shedding in the small intestine of the rat fed after fasting, *J. Anat.*, 120, 321, 1975.
4. **Steiner, M., Bourges, H. R., Freedman, L. S., et al.,** Effect of starvation on the tissue composition of the small intestine in the rat, *Am. J. Physiol.*, 215, 75, 1968.
5. **Holt, P. R., Wu, S. , and Yeh, K.-Y.,** Ileal hyperplastic response to starvation in the rat, *Am. J. Physiol.*, 251, G124, 1986.
6. **Kotler, D. P., Levine, G. M., and Shiau, Y.-F.,** Effects of nutrients, endogenous secretions, and fasting on in vitro glucose uptake, *Am. J. Physiol.*, 238, G227, 1980.
7. **Aldewachi, H. S., Wright, N. R., Appleton, D. R., and Watson, A. J.,** The effect of starvation and refeeding on cell kinetics population in the rat small intestine, *J. Anat.*, 119, 105, 1975.
8. **Levine, G. M., Deren, J. J., Steiger, E., and Zinno, R.,** Role of oral intake in maintenance of gut mass and disaccharidase activity, *Gastroenterology*, 67, 975, 1974.
9. **Johnson, L. R., Copeland, E. M., Dudrick, S. J., Lichtenberger, L. M., and Castro, G. A.,** Structural and hormonal alterations in the gastrointestinal tract of parenterally fed rats, *Gastroenterology*, 68, 1177, 1975.
10. **Hughes, C. A., Bates, T., and Dowling, R. H.,** Cholecystokinin and secretin prevent the intestinal mucosal hypoplasia of total parenteral nutrition in the dog, *Gastroenterology*, 75, 34, 1978.
11. **Eastwood, G. L.,** Small bowel morphology and epithelial proliferation in intravenously alimented rabbits, *Surgery*, 82, 613, 1977.
12. **Kotler, D. P., Levine, G. M., and Shiau, Y.-F.,** Effects of luminal nutrition and metabolic status on in vivo glucose absorption, *Am. J. Physiol.*, 240, G432, 1981.
13. **Levine, G. M.,** Nonspecific adaptation of jejunal amino acid uptake in the rat, *Gastroenterology*, 91, 49, 1986.
14. **Kotler, D. P., Shiau, Y.-F., and Levine, G. M.,** Effects of luminal contents on jejunal fatty acid esterification in the rat, *Am. J. Physiol.*, 28, G414, 1980.
15. **Cairnie, R. B. and Bentley, R. E.,** Cell proliferation studies in the intestinal epithelium of the rat, *Exp. Cell. Res.*, 46, 428, 1967.
16. **Lichtenberger, L. M. and Trier, J. S.,** Changes in gastrin levels, food intake and duodenal mucosal growth during lactation, *Am. J. Physiol.*, 237, 98, 1979.
17. **Elias, E. and Dowling, R. H.,** The mechanism for small bowel adaptation in lactating rats, *Clin. Sci.*, 51, 427, 1976.
18. **Yang, P., Baylin, S. B., and Luk, G. D.,** Polyamines and intestinal growth: absolute requirement for ODC activity in adaptation during lactation, *Am. J. Physiol.*, 247, G553, 1984.
19. **Heroux, D. L. and Gridgeman, N. T.,** The effect of acclimation on the size of organs and tissues of the rat, with special reference to modes of expression of results, *Can. J. Biochem. Physiol.*, 36, 209, 1958.
20. **Brobeck, J. R., Tepperman, J., and Long, C. N. H.,** Experimental hypothalamic hyperphagia in the albino rat, *Yale J. Biol. Med.*, 15, 831, 1943.
21. **Levin, R. J. and Smyth, D. H.,** The effect of the thyroid gland on intestinal absorption of hexoses, *J. Physiol. (London)*, 169, 755, 1963.
22. **Schedl, H. P. and Wilson, H. D.,** Effect of diabetes on intestinal growth and hexose transport in the rat, *Am. J. Physiol.*, 220, 1739, 1971.
23. **Miller, D. L., Hanson, W., Schedl, H. P., and Osborne, J. W.,** Proliferation rate and transit time of mucosal cells in small intestine of the diabetic rat, *Gastroenterology*, 73, 1326, 1977.
24. **Olsen, W. A. and Korsmo, H.,** Enhancement of intestinal sucrase activity in experimental diabetes: the role of intraluminal factors, *J. Lab. Clin. Med.*, 85, 832, 1975.
25. **Fedorak, R. N., Chang, E. B., Madara, J. L., and Field, M.,** Intestinal adaptation to diabetes, altered Na dependent nutrient absorption in streptozocin treated chronically diabetic rats, *J. Clin. Invest.*, 79, 1571, 1987.

26. **Nygaard, K.,** Resection of the small intestine in rats. I. Nutritional status and adaptation of fat and protein absorption, *Acta Clin. Scand.*, 132, 73, 1966.
27. **Flint, J. M.,** The effects of extensive resection of the small intestine, *Bull. Johns Hopkins Hosp.*, 23, 127, 1912.
28. **Dowling, R. H. and Gleeson, M. H.,** Cell turnover following small bowel resection and by-pass, *Digestion*, 8, 176, 1973.
29. **Obertop, H., Nundy, S., Malamud, D., and Malt, R. A.,** Onset of cell proliferation in the shortened gut. Rapid hyperplasia after jejunal resection, *Gastroenterology*, 72, 267, 1977.
30. **Hanson, W., R., Osborne, J. W., and Sharp, J. G.,** Compensation by the residual intestine after intestinal resection in the rat. Influence of post operative time interval, *Gastroenterology*, 72, 701, 1977.
31. **Weser, E. and Hernandez, M. H.,** Studies of small bowel adaptation after intestinal resection in the rat, *Gastroenterology*, 60, 69, 1971.
32. **McCarthy, D. M. and Kim, Y. S.,** Changes in sucrase, enterokinase, and peptide hydrolase after intestinal resection, *J. Clin. Invest.*, 52, 942, 1973.
33. **Freeman, J. H., Ellis, S. T., Johnston, G. A., Kwan, W. C., and Quamme, G. A.,** Sodium-dependent d-glucose transport after proximal small intestinal resection in rat, *Am. J. Physiol.*, 255, G292, 1988.
34. **Urban, E. and Michel, A. M.,** Separation of adaptive mucosal growth and transport after small bowel resection, *Am. J. Physiol*, 244, G295, 1983.
35. **Garrido, A. B., Jr., Freeman, H. J., Chung, Y. C., and Kim, Y. S.,** Amino acid and peptide absorption after proximal small intestinal resection in the rat, *Gut*, 20, 114, 1979.
36. **Curtis, K. J., Sleisenger, M. H., and Kim, Y. S.,** Protein digestion and absorption after massive small bowel resection, *Dig. Dis. Sci.*, 29, 834, 1984.
37. **Menge, H., Werner, H., Lorenz-Meyer, H., and Riecken, E. O.,** The nutritive effect of glucose on the structure and function of jejunal self-emptying blind loops in the rat, *Gut*, 16, 462, 1975.
38. **Levine, G. M., Deren, J. J., and Yezdimir, E.,** Small-bowel resection oral intake is the stimulus for hyperplasia, *Am. J. Dig. Dis.*, 21, 542, 1976.
39. **Feldman, E. J., Dowling, R. H., McNaughton, J., and Peters, T. J.,** Effects of oral versus intravenous nutrition on intestinal adaptation after small bowel resection in the dog, *Gastroenterology*, 70, 712, 1976.
40. **Luk, G. D. and Baylin, S. B.,** Polyamines and intestinal growth — increased polyamine biosynthesis after jejunectomy, *Am. J. Physiol.*, 245, G656, 1983.
41. **Grey, V. L. and Morin, C. L.,** Evidence for a growth-stimulating fraction in the rat proximal intestine after small bowel resection, *Gastroenterology*, 89, 1305, 1985.
42. **Besterman, H. S., Adrian, T. E., Malinson, C. N., Christofides, N. D., Sarson, D. L., Pera, A., Lombardo, L., Modigliani, R., and Bloom, S. R.,** Gut hormone release after intestinal resection, *Gut*, 23, 854, 1982.
43. **Sagor, G. R., Al-Mukirtar, M. Y. T., Ghatei, M. A., Wright, N. A., and Bloom, S. R.,** Evidence for a humoral mechanism after small intestinal resection. Exclusion of gastrin but not enteroglucagon, *Gastroenterology*, 84, 902, 1983.
44. **Buchan, A. M. J., Griffiths, C. J., Moreds, J. F., and Polak, J. M.,** Enteroglucagon cell hyperfunction in rat small intestine after gut resection, *Gastroenterology*, 88, 8, 1985.
45. **Li, A. K. C., Schattenkerk, M. E., Huffman, R. G., Ross, J. S., and Malt, R. A.,** Hypersection of submandibular saliva in male mice: trophic response in small intestine, *Gastroenterology*, 84, 949, 1983.
46. **Hart, M. H., Phares, C. K., Erdman, S. H., Grandjean, C. J., Park, J. H. Y., and Vanderhoof, J. A.,** Augmentation of postresection mucosal hyperplasia by plerocercoid growth factor (PGF), analog of human growth hormone, *Dig. Dis. Sci.*, 32, 1275, 1987.
47. **Dembinski, A. B. and Johnson, L. R.,** Role of gastrin in gastrointestinal adaptation after small bowel resection, *Am. J. Physiol.*, 243, G16, 1982.
48. **Grey, V. L., Garofalo, C., Greenberg, G. R., and Morin, C. L.,** The adaptation of the small intestine after resection in response to free fatty acids, *Am. J. Clin. Nutr.*, 40, 1235, 1984.
49. **Koruda, M. J., Rolandelli, R. H., Settle, R., Saul, S. H., and Rombeau, J. L,** The effect of a pectin supplemental elemental diet on intestinal adaptation to measure small bowel resection, *JPEN*, 10, 343, 1986.
50. **Koruda, M. J., Rolandelli, R. H., Settle, R. G., Zimmaro, D. M., and Rombeau, J. L.,** Effect of parenteral nutrition supplemented with short chain fatty acids on adaptation to massive small bowel resection, *Gastroenterology*, 95, 715, 1988.
51. **Spector, M. H., Levine, G. M., and Deren, J. J.,** Direct and indirect effects of dextrose and amino acids on gut mass, *Gastroenterology*, 72, 706, 1977.
52. **Weser, E., Babbitt, J., and Vandeventer, A.,** Relationship between enteral glucose load and adaptive mucosal growth in the small bowel, *Dig. Dis. Sci.*, 30, 675, 1985.
53. **Dworkin, L. D., Levine, G. M., Farber, N. J., and Spector, M. H.,** Small intestinal mass of the rat is partially determined by indirect effects of intraluminal nutrition, *Gastroenterology*, 71, 626, 1976.
54. **Williamson, R. C., Buchholtz, T. W., and Malt, R. A.,** Hormonal stimulation of cell proliferation in small bowel after transection and resection in rats, *Gastroenterology*, 75, 249, 1978.

55. **Johnson, L. R., Lichtenberger, L. M., Copeland, E. M., Dudrick, S. J., and Castro, G. A.,** Action of gastrin on gastrointestinal structure and function, *Gastroenterology*, 68, 1184, 1975.
56. **Gleeson, M. H., Bloom, S. R., Polak, J. M., Henry, K., and Dowling, R. H.,** Endocrine tumor in kidney affecting small bowel structure, motility, and absorptive function, *Gut*, 23, 773, 1971.
57. **Wood, J. G., Hoang, H. D., Bussjaeger, L. J., and Solomon, T.,** Neurotensin stimulates growth of small intestine of rats, *Am. J. Physiol.*, 255, G813, 1989.
58. **Cohen, S., Fava, R., and Sawyer, S. T.,** Purification and characterization of epidermal growth factor receptor/protein kinase from normal mouse liver, *Proc. Natl. Acad. Sci. U.S.A.*, 79, 6237, 1982.
59. **Altmann, G. G. and Leblond, C. P.,** Factors influencing villus size in the small intestine of adult rats as revealed by transposition of intestinal segments, *Am. J. Anat.*, 127, 15, 1970.
60. **Weser, E., Heller, R., and Tawil, T.,** Stimulation of mucosal growth in the rat ileum by bile and pancreatic secretions after jejunal resection, *Gastroenterology*, 73, 524, 1977.
61. **Ulshen, M. H., Lyn-Cook, L. E., and Raasch, R. H.,** Effects of intraluminal epidermal growth factor on mucosal proliferation in the small intestine of adult rats, *Gastroenterology*, 91, 1134, 1986.
62. **May, R.J., Quaroni, A., Kirsch, K., and Isselbacher, K. J.,** A villous cell derived inhibitor of intestinal cell proliferation, *Am. J. Physiol.*, 241, G520, 1981.
63. **Biller, J. A., Montgomery, R.K., Grand, R. J., Klagsbrun, M., and Rosenthal, A.,** Use of 3T3 cell growth factor assay for the delineation and characterization of humoral trophic factors involved in intestinal adaptation in the rat, *Gastroenterology*, 91, 448, 1986.
64. **Jacobowitz, D.,** Histochemical studies of the autonomic innervation of the gut, *J. Pharmacol. Exp. Ther.*, 149, 358, 1965.
65. **Tutton, P. J. M.,** Neural and endocrine control systems acting on the population kinetics of the intestinal epithelium, *Med. Biol.*, 55, 201, 1977.
66. **Levine, G. M., Kotler, D. P., and Yezdimir, E. A.,** Luminal nutrition obviates sympathectomy-induced atrophy in the rat, in *Mechanisms of Intestinal Adaptation,* Robinson, J. W. L., Dowling, R. H., and Riecken, E.-O., Eds., MTP Press, Lancaster, England, 1982, 311.
67. **Gallavan, R. H. and Chou, C. C.,** Possible mechanisms for the initiation and maintenance of postprandial intestinal hyperemia, *Am. J. Physiol.*, 249, G301, 1985.
68. **Shepherd, A. P. and Riedel, G. L.,** Laser-doppler blood flowmetry of intestinal mucosal hyperemia induced by glucose and bile, *Am. J. Physiol.*, 248, G393, 1985.
69. **Dowling, R. H.,** Compensatory changes in intestinal absorption, *Br. Med. Bull.*, 23, 275, 1967.
70. **Clarke, R. M.,** Luminal nutrition versus function work-load as controllers of mucosal morphology and epithelial replacement in rat small intestine, *Digestion*, 15, 411, 1977.
71. **Richter, G. C., Levine, G. M., and Shiau, Y.-F.,** Effects of luminal glucose versus nonnutritive infusates on jejunal mass and absorption in the rat, *Gastroenterology*, 85, 1105, 1983.
72. **Altmann, G. G.,** Influence of bile and pancreatic secretions on the size of the intestinal villi in the rat, *Am. J. Anat.*, 132, 167, 1971.
73. **Roy, C. C., Laurendeau, G., Doyon, G., Chartrand, L., and Rivest, M. R.,** The effect of bile and of sodium taurocholate on the epithelial cell dynamics of the rat small intestine, *Proc. Soc. Exp. Biol. Med.*, 194, 1000, 1975.
74. **Fine, H., Levine, G. M., and Shiau, Y.-F.,** Effects of cholecystokinin and secretin on intestinal structure and function, *Am. J., Physiol.*, 245, G358, 1982.
75. **Weser, E., Babbitt, J., Hoban, M., and Vanderventer, A.,** Intestinal adaptation. Different growth response to disaccharides compared with monosaccharides in rat small bowel, *Gastroenterology*, 91, 1521, 1986.
76. **Karasov, W. H., Solberg, D. H., Chang, S. D., Hughes, M., Stein, E. D., and Diamond, J. M.,** Is intestinal transport of sugars and amino acids subject to critical period programming?, *Am. J. Physiol.*, 249, G770, 1985.
77. **Karasov, W. H. and Diamond, J. M.,** Adaptive regulation of sugar and amino acid transport by vertebrate intestine, *Am. J. Physiol.*, 8, G443, 1983.
78. **Solberg, D. H. and Diamond, J. M.,** Comparison of different dietary sugars as inducers of intestinal sugar transporters, *Am. J. Physiol.*, 252, G574, 1987.
79. **Spector, M. H., Traylor, J., Young, E. A., and Weser, E.,** Stimulation of mucosal growth by gastric and ileal infusion of single amino acids in parenterally nourished rats, *Digestion*, 21, 33, 1981.
80. **Stein, E. D., Chang, S. D., and Diamond, J. M.,** Comparison of different amino acids as inducers of intestinal amino acid transport, *Am. J. Physiol.*, 252, G626, 1987.
81. **Ferraris, R. P., Kwan, W. W., and Diamond, J.,** Regulatory signals for intestinal amino acid transporters and peptidases, *Am. J. Physiol.*, 255, G151, 1988.
82. **Jain, R., Eikenburg, B. E., and Johnson, L. R.,** Stimulation of ornithine decarboxylase activity in digestive tract mucosa, *Am. J. Physiol.*, 253, G303, 1987.
83. **Dembinski, A. B., Yamaguchi, T., and Johnson, L. R.,** Stimulation of mucosal growth by a dietary amine, *Am. J. Physiol.*, 247, G352, 1984.
84. **Seidel, E. R., Haddox, M. I., and Johnson, L. R.,** Ileal mucosal growth during intraluminal infusion of ethylamine or putrescine, *Am. J. Physiol.*, 249, G434, 1985.

85. **Tabata, K. and Johnson, L. R.,** Mechanism of induction of mucosal ornithine decarboxylase by food, *Am. J. Physiol.*, 251, G370, 1986.
86. **Morin, C. L., Ling, V., and Bourassa, D.,** Small intestinal and colonic changes induced by a chemically defined diet, *Dig. Dis. Sci.*, 25, 123, 1980.
87. **Balint, J. A., Fried, M. D., and Imai, C.,** Ileal uptake of oleic acid: evidence for adaptive response to high fat feeding, *Am. J. Clin. Nutr.*, 33, 2276, 1980.
88. **Jacobs, L. R.,** Effect of short-term dietary fat on cell growth in rat gastrointestinal mucosa and pancreas, *Am. J. Clin. Nutr.*, 37, 361, 1983.
89. **Thomson, A. B. R., Keelan, M., Clandinin, M. T., and Walker, K.,** Dietary fat selectively alters transport properties of rat jejunum, *J. Clin. Invest.*, 77, 279, 1986.
90. **Sircar, B., Johnson, L. R., and Lichtenberger, L. M.,** Effect of synthetic diets on gastrointestinal mucosal DNA synthesis in rats, *Am. J. Physiol.*, 244, G327, 1983.
91. **Tabata, K. and Johnson, L. R.,** Ornithine decarboxylase and mucosal growth in response to feeding, *Am. J. Physiol.*, 251, G270, 1986.
92. **Dowling, R. H., Riecken, E. O., Laws, J. W., and Booth, C. C.,** The intestinal response to high bulk feeding in the rat, *Clin. Sci.*, 32, 1, 1967.
93. **Jacobs, L. R.,** Effects of dietary fiber on mucosal growth and cell proliferation in the small intestine of the rat: a comparison of oat bran, pectin and guar with total fiber deprivation, *Am. J. Clin. Nutr.*, 37, 954, 1983.
94. **Sigleo, S., Jackson, M. J., and Vahouny, G. V.,** Effects of dietary fiber constituents on intestinal morphology and nutrient transport, *Am. J. Physiol.*, 246, G34, 1984.
95. **Ecknauker, R., Sircar, B., and Johnson, L. R.,** Effect of dietary bulk on small intestinal morphology and cell renewal in the rat, *Gastroenterology*, 81, 781, 1981.
96. **Jacobs, L. R. and Lupton, J. R.,** Effect of dietary fibers on rat large bowel mucosal growth and cell proliferation, *Am. J. Physiol.*, 246, G378, 1984.
97. **Roediger, W. E. W.,** Role of anaerobic bacteria in the metabolic welfare of the colonic mucosa in man, *Gut*, 21, 793, 1980.
98. **Urban, E., Starr, P. E., and Michel, A. M.,** Morphologic and functional adaptation of large bowel after small bowel resection in the rat, *Dig. Dis. Sci.*, 28, 265, 1983.
99. **Ryan, G. P., Dudrick, S. J., Copeland, E. M., and Johnson, L. R.,** Effects of various diets on colonic growth in rats, *Gastroenterology*, 77, 658, 1979.
100. **Haentjens, P., Deluaux, G., Chayvialle, J. A., and Willems, G.,** Postprandial stimulation of epithelial cell proliferation in defunctioned colon of rats is not caused by gastrin, *Gastroenterology*, 90, 939, 1986.
101. **Delvaux, G., Caes, F., and Willems, S. G.,** Refeeding of fasting rats stimulates epithelial cell proliferation in the excluded colon, *Gastroenterology*, 86, 802, 1984.

Growth Factors

Chapter 12

INHIBITORY REGULATION OF GASTROINTESTINAL ORGAN GROWTH

Jean Morisset

TABLE OF CONTENTS

I. INTRODUCTION

In the last two decades it has become clear that the proliferation of cells *in vivo* and *in vitro* is controlled by several hormones and growth factors which are present in blood and tissue fluids.

As for other cells, proliferation of cells in the gastrointestinal tract is also controlled by multiple hormones and growth factors. The trophic action of gastrin on pancreas and gastric, duodenal, ileal, and colonic mucosa is well documented.[1,2] Similarly, the growth effects of cholecystokinin (CCK) and secretin on the exocrine pancreas are well documented,[3,4] but few studies have been undertaken to define their possible involvement on growth in other parts of the gut. CCK and secretin may prevent the intestinal mucosa hypoplasia resulting from total parenteral nutrition in the dog[5] and the rat[6] and cause small increases in rat duodenal mucosa DNA content.[7] Bombesin has been shown to have trophic effects on the pancreas, but is much less potent than CCK.[8]

Other hormones (insulin, thyroxine, hydrocortisone, and epidermal growth factor [EGF]) also exhibit specific growth effects on various gut organs. Besides being involved in the ontogeny of intestinal enzymes[9] and regulation of pancreatic amylase synthesis,[10] insulin is also associated with growth of the intestine[11] and pancreas.[12] In suckling rats, thyroxine increases cellular proliferation of the pancreas,[13] as does hydrocortisone[14] which is also involved in fetal intestinal growth.[15] Among the known growth factors, EGF can induce growth of the stomach,[16] intestine,[17] and pancreas.[18] Although somatomedin C is expected to be involved in the development of the organs of the gastrointestinal tract, the only clue so far for such an implication remains the presence of IGF-1 specific receptors on normal pancreatic acinar cells[19] and AR42J cells.[12] In the latter, IGF-1 was able to induce growth and amylase synthesis but was 1% as potent as insulin.[12]

During normal physiological development, in the process of regeneration after injury or partial destruction, or under conditions of hormonally induced growth by exogenous treatments, each organ can limit its growth or expansion by still unknown mechanisms. In this chapter, I would like to present some experimental data indicating that such a control might be accomplished by some specific cellular components, gut hormones, or growth inhibiting factors.

II. ROLE OF CHALONES

Control of cell proliferation is executed in part by means of negative feedback, utilizing cell specific messenger substances. In the early sixties, a series of theoretical models were presented which suggested that within any cell line, the functionally active end cells may signal their existence to the mitotically active precursor cells by means of a cell-specific inhibitor substance.[20] In 1962, Bullough named such a signal chalone.[21] Generally, it is assumed that only one chalone exists in each tissue, but evidence is accumulating that there may be two.[22] Furthermore, each chalone acts within its specific cell by means of its intracellular concentration.

A. DISTRIBUTION

Chalone molecules seem to be ubiquitous substances, as they have been found in all tissues examined so far. In skin, chalone was identified in epidermical cells, sebaceous glands, eccrine glands, and melanocytes. Granulocytes, erythrocytes, and lymphocytes contain the factor in blood. It has also been detected in liver, kidney, lung, intestine, uterine cervix, and testis.[23]

B. BIOCHEMICAL CHARACTERIZATION

A classic chalone must exhibit the following characteristics: it has total cell specificity without being species specific, its effects are reversible, and the molecule shows no sign of cytotoxicity.[24]

The major problem with research in this field is that the chemical nature of these chalone substances is still unknown because none of them has yet been purified to homogeneity. As of today, there are only three chalones which have been subjected to any chemical characterization and purification: the epidermal, granulocyte, and lymphocyte chalones.

It is now generally accepted that there are two epidermal chalones, G_1 and G_2. The G_1 chalone has been characterized by Marks,[25] the molecular weight of which has been initially estimated between 20,000 and 300,000, but further purification indicates that it is less than 10,000 Da. It is a stable molecule resistant to boiling, phenol, 1 *M* formic acid, trypsin, pronase, and nucleases. The G_2 chalone, although complexed with larger molecular weight substances, is a relatively small anionic glycoprotein with an estimated M_r of 20,000.[26] With regards to the lymphocyte and granulocyte chalones, very active material of M_r 500 to 2000 has been purified.[24]

In the liver, the M_r of the chalone has been in many instances estimated at 2000 for a molecule of great hydrophobicity.[27] However, a humoral inhibitor, which may play a primary role in the regulatory mechanism of cell division in the regenerating rat liver, has been characterized and identified as α_1-acid glycoprotein with a M_r of 40,000.[28] This protein is synthesized by the hepatocyte itself and possesses a suppressive effect on mitosis of this cell. Philpott has presented evidence for a chalone in the rat stomach,[29] and an aqueous extract of rabbit small intestinal mucosa was shown to inhibit thymidine incorporation into rat jejunal DNA; in this case however, organ specificity was not established since similarly prepared colon, kidney, and liver extracts also inhibited jejunal DNA synthesis.[30]

C. MECHANISM OF ACTION

Although it is difficult and speculative to describe biological effects of a substance when it has not been fully purified, some hypotheses have been presented to explain the role of chalones. Their major mode of action would be to affect cell proliferation by regulating the degree of amplification. Such a phenomenon is regulated by modulating the duration of the cell cycle by prolonging the cell cycle time without blocking cell proliferation completely. Such a prolongation of the cell cycle would lead to accelerated cell maturation, which was indeed observed with the granulocytic chalone.[31] In the mitotic cycle, the epidermal chalone seems to inhibit at two selective points, one in the late G_1 phase and the other in the late G_2 phase. The G_1 point is about 9 h before the start of DNA synthesis, while the G_2 point occurs only minutes before mitosis; the probability of a cell successfully passing these two critical points is inversely proportional to the chalone concentration. Passage through late G_2 seems to be under the control of the G_2 chalone, as it is effective after the S phase prior to mitosis.[24]

In the liver, mitotic inhibition by the α_1-acid glycoprotein depends on its intracellular concentration, which in turn appears to be under the influence of the extracellular concentration of α_1-antitrypsin synthesized by hepatocytes. The following mechanism seems to operate: after partial hepatectomy for example, the plasma concentration of α_1-antitrypsin decreases rapidly, leading to the activation of endogenous proteases, which may alter the glycoprotein structure of the cell membrane of the hepatocyte, facilitating the extrusion of α_1-acid glycoprotein. Indeed, the plasma concentration of α_1-acid glycoprotein increases fourfold over that of nonhepatectomized animals at the time of maximal mitosis.[32] Along with the regenerative proliferation of the hepatocyte, the concentration of α_1-antitrypsin increases in plasma, which may suppress the activity of endogenous proteases and hinder the hepatic cells from excreting α_1-acid glycoprotein and thus inhibiting cell division. Therefore, maximal DNA synthesis and mitosis occur when the intracellular concentration of α_1-acid glycoprotein is at its lowest level. This hepatic cell-cycle control under the dual influence of α_1-antitrypsin and α_1-acid glycoprotein has been extensively studied. Unfortunately, a similar mechanism has not been investigated in other organs and it is therefore premature to propose this as the general mechanism of action of chalones.

III. ROLE OF INHIBITORY HORMONES AND GROWTH FACTORS

It would be unwise to believe that control of organ growth is ultimately restricted to the influence of stimulatory trophic hormones and growth factors. Indeed, the antitrophic action of some gastrointestinal hormones and growth factors has been reported, but the mechanism by which these actions occur is far from being understood.

Among the gastrointestinal hormones, secretin has been suggested as a potential antitrophic factor. Although it does not have any inhibitory effect when given alone, secretin inhibited the trophic effect of gastrin on the stomach, duodenum,[33] and colon.[34] Unfortunately, these studies present secretin as an inhibitor of gastrin-stimulated DNA synthesis with the assumption that such an effect will lead to an eventual decrease in cell number, which has however never been reported. A similar inhibitory effect of vasoactive intestinal peptide (VIP) on pentagastrin-stimulated colonic DNA synthesis has also been reported.[35]

In the late seventies, it was shown that the introduction of oleic acid into the colon or distal part of the small intestine inhibited pancreatic secretion in cats[36] and rats.[37] Following extraction and partial purification, the name pancreatone was given to this factor by Harper,[36] while Sarles' group proposed anti-cholecystokinin peptide (ACP).[38] The purified fraction of this factor was free of neurotensin, somatostatin, enteroglucagon, and VIP, and it was not a pancreatic polypeptide.[36,38] From a recent study, there has been some indication that pancreatone or ACP might be peptide YY.[39] Although first described as an antisecretory factor, pancreatone was also shown to reduce pancreatic growth, causing acinar cell atrophy rather than a reduction in cell number.[40]

If we can get a general consensus that an antitrophic action or effect of a hormone or factor implicates decreases in cell numbers from cells in culture or in total DNA content in an organ, then the above antitrophic effects ascribed to secretin, VIP, and pancreatone on the basis of reduced DNA synthesis or cell size will have to be confirmed with studies on total tissue DNA content.

In addition to its well known mitogenic effects,[41] EGF inhibits the proliferation of a variety of human cancer cells lines,[42-44] cultured rat pituitary carcinoma cells,[45] and cultured RIE-1 rat intestinal cells.[46] In a recent study,[47] it was shown that the addition of EGF to cultured human fetal jejunum decreased thymidine incorporation into total DNA within 24 h. A marked decrease in epithelial labeling index indicated that the decreased DNA synthesis is attributable to the epithelial proliferative cells and not only to the developing mesenchyme. Furthermore, EGF counteracted the stimulatory effect of hydrocortisone on DNA synthesis. In the adult rat pancreas, *in vivo* EGF treatment inhibited thymidine incorporation into DNA which resulted in a reduction of total DNA content.[48] Even though the exact mechanism whereby EGF exerts its inhibition on all these cells is not known, such an inhibitory effect on cell proliferation points out the increasing evidence for a bifunctional role of this growth factor in the regulation of cell growth.

IV. ROLE OF SOMATOSTATIN

Besides the potential role of tissue specific chalones for inhibition of cell proliferation, such controls can also be accomplished by growth inhibiting hormones, and one of the most plausible candidates remains somatostatin.

This hormone was originally discovered in extracts from the hypothalamus of sheep[49] and was later found in many organs of the digestive tract and pancreas. The molecular form first described consisted of 14 amino acids in a cyclic arrangement and was called somatostatin-14. A few years later, larger molecular sizes were purified from the intestine and these include somatostatin 28, 25, and 20.[50,51]

Somatostatin-like immunoreactive material has been identified as early as 8 weeks in the pancreas, duodenum, and intestine of human fetuses, at 12 weeks in antrum and colon, and at 14 weeks in fundus.[52-54] In the rat, immunoreactive somatostatin-containing D cells were first observed in the pancreatic islets on day 17 of gestation and in the duodenum on day 18.[55] The early appearance of the hormone and its specific distribution in the pancreas and gastrointestinal tract may argue in favor of an inhibitory role for this peptide in the development of these organs. This possibility is further strengthened by the observations that exogenous somatostatin-14 and its analog (L-5-Br-Trp^8-somatostatin) can reduce normal and caerulein-stimulated pancreatic growth[56,57] as well as the development of transplanted pancreatic ductal and acinar tumor cells.[58] In adult rats,[59] infusion of somatostatin reduced gastrin-stimulated DNA synthesis and cell division in both fundic and antral progenitor cells. In cases of chronic hypergastrinemia, induced either by transposition of the antrum to the colon[60] or omeprazole treatment,[61] long-acting somatostatin appears to antagonize the growth-promoting effect of endogenous gastrin since it decreased mucosal heights of total glandular stomach and antrum, and parietal and gastrin cell labeling indices. Somatostatin is also very effective in reducing jejunal cell renewal[62] and the proliferative activity in ileal crypts cells following jejunoileal resection.[63] Such an inhibition of growth and differentiation was also observed in cultured intestinal mucosa.[64] In a small intestinal crypt cell line (IEC-6),[65] gastrin and EGF increased DNA synthesis and caused a significant stimulation in cell replication. Addition of somatostatin to the cell culture did not modify basal rate of DNA synthesis or cell replication but it selectively and completely inhibited EGF-induced DNA synthesis and cell replication, without interfering with the gastrin effect. It has been proposed that the trophic effect of EGF is associated with its receptor-tyrosine kinase activity resulting in phosphorylation of tyrosine residues of the EGF receptor protein. Somatostatin, by stimulating a phosphoprotein phosphatase activity, would counteract the EGF-stimulated phosphorylation process and thus inhibit its trophic effects.

In the whole organism, somatostatin may function as a neurohumoral regulator, a neurotransmitter, an endocrine or paracrine hormone, or as a lumone.[66] Indeed, it is released into the blood, synaptic clefts, the intercellular space, and the lumen of the digestive tract in response to a variety of stimuli.[67,68] Since somatostatin is a potent inhibitor of secretion from many different endocrine[49,69,70] and exocrine cells,[71,72] a local paracrine action represents an attractive mode of action.[73] Indeed, it has been observed that some somatostatin cells are equipped with long processes which make contact with neighboring G cells in the antropyloric mucosa and with parietal cells in the oxyntic mucosa. In the endocrine pancreas, similar structures are present in the vicinity of the endocrine α and β cells.[74]

As of today, somatostatin can be presented as the ultimate molecule which can inhibit (1) various types of hormone secretion, (2) multiple physiological secretory processes, and (3) stimulated cell proliferation in different organs. In the next few pages, I will summarize what is known about the mechanisms of action of this hormone, its receptor present on cells of the digestive tract, their interaction with guanine nucleotide binding proteins, their relation with second messenger production, and finally their modulation.

A. SOMATOSTATIN RECEPTOR BIOCHEMICAL CHARACTERIZATION

As with any other hormone, somatostatin interacts on its target cells via specific receptors. These receptors have been characterized on isolated cells,[75] cell membranes,[76] and cytosol[77] and secretion vesicles[78] using various iodinated analogs of somatostatin 14 and 28 as radioligands (Table 1). For endocrine[78] and exocrine pancreatic cells[79] and gastric mucosa cells,[80] the apparent receptor affinities, as estimated by Scatchard analysis (K_D), vary between 0.05 and 4.5 n*M*. However, the somatostatin receptor characterized from the intestinal mucosa cytosol seems unique since it has two classes of binding site both of relatively low affinities, K_H: 70 n*M*, K_L: 1050 n*M*,[81] whereas a single class of binding sites has been found in all the other cells (Table 1). The presence of somatostatin receptors on insulinoma cells with comparable affinity and

TABLE 1
Biochemical Characteristics of the Somatostatin Receptor

Species	Ligand	Tissue	K_D (n*M*)	B_{max}	Mol wt (kDa)	Ref.
Rat	^{125}I-Tyr^{1}-SS-14	Pancreatic acinar cell membrane	0.32	600 fmol/mg prt	90	85
	^{125}I-Tyr^{1}-SS-14	Pancreatic acinar cell membrane	0.33	142 fmol/mg prt	90	86
	^{125}I-Tyr^{11}-SS-14	Pancreatic acinar cell membrane	0.48	266 fmol/mg prt	200	
					80	87
					70	
Guinea pig	^{125}I-Tyr^{11}-SS-14	Pancreatic acinar cell membrane	0.15	178 fmol/mg prt	—	105
	^{125}I-Tyr^{11}-SS-14	Pancreatic acinar cell	0.28	72 fmol/mg prt	—	84
Rat	^{125}I-Tyr^{1}-SS-14	Pancreatic acinar cell cytosol	0.1	30 fmol/mg prt	—	77
		Gastric parietal cell cytosol	0.03	14 fmol/mg prt	—	77
Guinea pig	[Leu^{8},D-Trp^{22}I^{125}Tyr^{25}]SS28	Pancreatic acinar cell membrane	0.05	157 fmol/mg prt	—	83
Hamster	[Leu^{8},D-Trp^{22}I^{125}Tyr^{25}]SS28	Insulinoma cell	0.25	68 fmol/mg prt	—	82
Rat	^{125}I-Tyr^{1}-SS-14	Islet cell	0.15	4.8 fmol/mg prt	—	78
	^{125}I-Tyr^{11}-SS-14	Intestinal mucosa cytosol	70	4.6 pmol/mg prt	—	81

Note: (K_D) Dissociation constant; (B_{max}) receptor concentration; (mol wt) estimated molecular weight of the receptor.

concentration to those of normal cells confirms that these receptors are retained in tumoral tissue.[82] Receptor densities vary from one cell type to the other; the islet cells have the lowest concentration at 4.8 fmol/mg protein.[78] Densities evaluated in pancreatic acinar cells or membranes vary from 72 to 600 fmol/mg protein. The highest concentration was found in intestinal mucosa cytosol at 277 pmol/mg protein for the lower affinity sites.[81] The different receptor subtypes do not correspond to the different forms of somatostatin (14 or 28) for pancreatic acinar cells or for islet B cells. Indeed, receptors from normal islet B cells[78] and insulinoma cells[82] exhibit similar affinities of 0.15 and 0.25 n*M*, for the analogs used, somatostatin-14 or somatostatin-28, respectively. Similarly, membrane receptors from guinea pig pancreatic acinar cells bind somatostatin 14 and 28 in the same range of concentrations, although competition experiments with natural somatostatin 14 and 28 indicate a slightly higher affinity for somatostatin 14 than for somatostatin 28.[83,84]

To evaluate the nature of the somatostatin receptor and determine its molecular size, the following biochemical studies were performed. Ligands were first allowed to bind to the receptor. The complex was then covalently cross-linked and characterized by electrophoresis in SDS-polyacrylamide gels. Unfortunately to date, all such studies were performed using pancreatic acinar cell membranes from rat or guinea pig. Therefore the only information available pertained to pancreatic tissue. The first report[85] presents the somatostatin receptor as a 90 kDa protein. This was later confirmed by a different group[86] (Table 1). Using a different cross-linking agent, disuccinimidyl suberate (DSS) instead of n-hydroxysuccinimidyl-4-azidobenzoate (HSAB), Srikant and Patel[87] identified three proteins of apparent M_r of 200,000, 80,000, and 70,000 under reducing conditions or not. In a more detailed experiment, Zeggari et al.[88] evaluated the somatostatin receptor using different cross-linkers on membranes solubilized with Zwittergent 3-14. Of the four cross-linkers used, *N*-5-azido-2-nitrobenzoyloxysuccinimide (ANB NOS), *N* succinimidyl 6-(azido-2′-nitrophenylamine) hexanoate (SANAH), HSAB, and DSS, ANB-NOS was the most efficient and irreversible cross-linker while DSS was virtually ineffective; this last cross-linker was used to identify the three previously described somatostatin receptor proteins.[87] Furthermore, for equivalent amounts of initially membrane-bound radioactivity, about twice as much radioactivity was eluted from a gel filtration column after the membranes were prelabeled with ^{125}I-[leu^8, dTrp^{22}, Tyr^{25}]S28 than with ^{125}I-Tyr^{11}-S14. This suggests that the radioligand S28 is more efficiently cross-linked to the receptor by ANB-NOS. However, with both S28 and S14 radioligands, an apparent molecular mass of 93 and 94 kDa was determined, confirming the 90 kDa size of the receptor.[88] The glycoprotein nature of the somatostatin receptor was established by its elution from a WGA-agarose column by *N*-acetylglucosamine or *N,N′,N″*-triacetylchitotriose which has a higher affinity for WGA.

Although somatostatin is recognized as an inhibitor of many physiological functions carried out by organs of the digestive tract, there is, however, little information available with respect to the mechanism by which it affects intracellular signal transduction in any given cell.

B. INTERACTION WITH GUANINE NUCLEOTIDE BINDING PROTEINS

The coupling of a guanine nucleotide binding protein to the pancreatic somatostatin receptor has been postulated from the observation that somatostatin binding to its receptor was significantly decreased by guanine nucleotides.[87,89] Further characterization of this guanine nucleotide regulation of the somatostatin receptor has indicated that the receptor interacts with a guanine nucleotide inhibitory protein (Ni).[90] This conclusion was drawn from the observations that the islet-activating protein (IAP): (1) inhibited somatostatin binding to its receptor, (2) induced ADP-ribosylation of a 41 kDa protein corresponding to the alpha subunit of Ni, and (3) that such ribosylation by IAP prevented the inhibition by somatostatin of VIP-stimulated cAMP formation in pancreatic acinar cells.[90]

C. INHIBITION OF SECOND MESSENGER AND OTHER INTRACELLULAR ACTIONS

It is now well established that hormones act on their target cells by stimulating at least one of two distinct biochemical pathways to generate their physiological response. These two pathways involve phosphatidylinositol breakdown by phospholipase C or activation of adenylate cyclase and production of cAMP.

Evidence from studies performed on intestine, gastric parietal cells, pancreatic acinar, and ductal cells indicate that somatostatin acts by inhibiting stimulated adenylate cyclase activity. Indeed, somatostatin inhibits VIP-stimulated colonic secretion through inhibition of cAMP production.[91] Similarly, somatostatin inhibits histamine-stimulated aminopyrine accumulation in purified parietal cells[92] via a selective inhibitory effect on cAMP production and cAMP-dependent protein kinase activation.[93] The *in vivo* inhibition of CCK- and acetylcholine-stimulated pancreatic enzyme secretion by somatostatin could not be reproduced *in vitro* on dispersed pancreatic acini,[94] probably because these two agonists induce enzyme secretion via the phosphatidylinositol breakdown pathway[95] which has never yet been observed to be affected by somatostatin.[75] However, in these dispersed pancreatic acini, a very good correlation has been found between somatostatin concentrations which inhibit 50% of secretin-stimulated cAMP production and receptor binding affinities.[96] We do not know presently the physiological meaning of this observation since *in vitro*, physiological concentrations of secretin do not induce enzyme release. On the contrary, the somatostatin inhibition of secretin-stimulated cAMP formation in isolated pancreatic duct cells is interesting because these cells control secretion of water and electrolytes from the pancreas.[97]

Besides the interaction of somatostatin with its plasma membrane receptor, cytosolic receptors have also been characterized in gastric mucosal cells[98] and pancreatic acini.[77] These receptors have been postulated to be phosphoprotein phosphatase (PPPase) regulatory subunits that would dissociate from the enzyme complex upon somatostatin binding, thus releasing an active PPPase. Activation of the phosphatase would dephosphorylate substrates previously phosphorylated by the action of stimulatory hormones and this mechanism could thus be partially responsible for the inhibitory effects of somatostatin.

In anterior pituitary cells and pancreatic islets,[99] peculiar vesicles appear to translocate somatostatin receptor from the cell interior to the plasma membrane; in these vesicles, receptors have a cytoplasmic orientation as visualized from gold-conjugated somatostatin binding to these vesicles.[99] In islet cells,[100] since glucose stimulates insulin release, it would concomitantly initiate translocation of somatostatin receptors from the cytosol to the plasma membrane. Such a process would sensitize insulin-B cells to somatostatin and provide a specific anatomical locus for somatostatin to inhibit hormone release integrated to the exocytosis process at the point of fusion of the secretion vesicle with the surface membrane. In pituitary cells,[101] somatostatin inhibits fusion of calmodulin-stimulated secretion vesicles with the plasma membranes. This mechanism has been described only in secretion vesicles involved in hormone secretion and its existence in the exocytosis process of exocrine glands remains to be established.

In summary, somatostatin can inhibit hormone-stimulated physiological responses through three possible mechanisms: (1) inhibition of stimulated adenylate cyclase via coupling of a guanine nucleotide inhibitory protein, (2) activation of a phosphoprotein phosphatase, or (3) binding to specific receptors on secretion vesicles.

D. MODULATION

Somatostatin has the ability to modulate the density of its receptor on the membrane. Somatostatin depletion in gastric corporeal, antral, and duodenal mucosae by cysteamine correlates with an increase of somatostatin binding sites within 4 h.[102] On the contrary, in streptozotocin diabetic rats, pancreatic acinar cell somatostatin receptors were down-regulated by hypersomatostatinemia that persists up to a week after the induction of diabetes.[87] In this last situation, reduction of somatostatin receptors may have occurred via internalization and subsequent degradation. Internalization of the intact somatostatin radioligand has recently been reported[103] via vesicles that are transported to the nuclear compartment but not to lysosomes. In this case, internalization may be a means of delivering the hormone or the hormone-receptor complex to intracellular specific sites, the nucleus being one of them, by which the hormone would play a post membrane receptor role. The presence of cytosolic somatostatin receptor in the pancreas would be consistent with this mode of action.[77]

Somatostatin receptor activity may not only be regulated by its specific hormone since exposure to CCK and analogs in a pancreatic plasma membrane preparation directly reduced somatostatin binding to its receptor without a significant change in its affinity. The observation that such a modification was not induced by carbachol suggests that mobilization of membrane-bound Ca^{2+} may not be involved in the regulation of somatostatin binding to its receptor.[85] However, since CCK has been shown to induce phosphatidylinositol breakdown in the pancreas,[95] the possibility exists that activation of a protein kinase may affect somatostatin binding to its receptor.

In rat pancreatic acinar carcinoma AR42J cells, Viguerie et al.[104] observed a reduction in somatostatin receptor concentration 24 h after the addition of dexamethasone without any change in the affinity of the receptor. Such a reduction in somatostatin receptor concentration was accompanied by an increase in the sensitivity of these cells to the inhibitory effect of somatostatin on cAMP formation stimulated by VIP. The mechanism involved in this hypersensitivity to somatostatin remains unknown. It is not however related to a dexamethasone effect on the somatostatin receptor itself since its affinity was not altered. One possibility is that the glucocorticoid may have increased the effective coupling of Ni to the somatostatin receptor.

Simultaneous exposure of pancreatic acini to a tumor promoting phorbol ester, 12-O-tetradecanoyl phorbol 13-acetate (TPA), and somatostatin resulted in a 2.5-fold decrease in receptor affinity without any change in binding capacity.[75] However, pancreatic acinar cell[86] or plasma membrane[76] pretreatment with either TPA or a synthetic diacylglycerol analog, 1-oleyl-2-acetylglycerol (OAG), resulted in a subsequent decrease in somatostatin binding to its receptor due to a reduction in maximum binding capacity without change of affinity. A simultaneous pretreatment with TPA and the calcium ionophore A23187 caused a synergistic inhibition of somatostatin binding although the ionophore alone had no effect. These data suggest that activation of protein kinase C can modulate somatostatin binding to its receptor in association with mobilized intracellular calcium and such a mechanism may explain the previously postulated effects of CCK.[85] Furthermore, the observation that the combined treatment of TPA and somatostatin caused a loss in somatostatin receptor affinity without change of capacity while a pretreatment with TPA, followed by exposure to somatostatin, resulted in reduced receptor capacity without change of affinity indicates that the presence or absence of somatostatin during TPA exposure modulates the receptor differently. It remains to be determined whether modifications occurred at the receptor level or at postreceptor loci such as guanine nucleotide regulatory proteins.

To our knowledge, there has been no study yet performed on the development of the somatostatin receptor in any tissue of the body. Such information would be needed to correlate the inhibitory functions of the hormone to the occupation of its receptor during development.

V. CONCLUSION

Over the years, emphasis has been focused in the search for hormones and growth factors associated with proliferation and differentiation of specific organs or cells in culture. Such studies have been so far very successful. Some attempts have also been made to identify inhibitory hormones or factors that could counteract the proliferative effects of the already known trophic hormones or growth factors; however, breakthroughs in this line of research have not been astonishing.

The chalone story seemed promising for a while but problems with purification of these molecules seem to have slowed down progress in this field as evidenced by the very few publications on this topic over the last 10 years. The antitrophic effects of secretin and pancreatone remain to be confirmed by others, giving special attention to the demonstration that these two substances can reduce proliferation of cells in tissue culture or can reduce total DNA content of specific target organs *in vivo*. Studies on the effects of EGF are becoming very stimulating with the discovery of its dual effects on cell growth either as a stimulatory or inhibitory factor. Specific conditions which delineate and isolate these opposite effects will have to be clarified to avoid misuse of this factor. Finally, of all the antitrophic factors examined, somatostatin now seems to be the most reliable and the best understood. This hormone will probably be the most frequently used to suppress secretory functions of many digestive tract organs as well as to control tumor cell growth.

REFERENCES

1. **Johnson, L. R.,** Regulation of gastrointestinal growth, in *Physiology of the Gastrointestinal Tract*, Johnson, L. R., Ed., Raven Press, New York, 1981, 169.
2. **Johnson, L. R. and Guthrie, P. D.,** Proglumide inhibition of trophic action of pentagastrin, *Am. J. Physiol.*, 246, G62, 1984.
3. **Dembinski, A. B. and Johnson, L. R.,** Stimulation of pancreatic growth by secretin, caerulein, and pentagastrin, *Endocrinology*, 106, 323, 1980.
4. **Solomon, T. E., Vanier, M., and Morisset, J.,** Cell site and time course of DNA synthesis in pancreas after caerulein and secretin, *Am. J. Physiol.*, 245, G99, 1983.
5. **Hugues, C. A., Bates, T., and Dowling, R. H.,** Cholecystokinin and secretin prevent the intestinal mucosal hypoplasia of total parenteral nutrition in the dog, *Gastroenterology*, 75, 34, 1978.
6. **Fine, H., Levine, G. M., and Shian, Y. F.,** Effects of cholecystokinin and secretin on intestinal structure and function, *Am. J. Physiol.*, 245, G358, 1983.
7. **Morisset, J. and Genik, P.,** Effect of acute and chronic administration of secretin and caerulein on rat duodenal and gastric growth, *Regul. Pept.*, 5, 111, 1983.
8. **Lhoste, E., Aprahamian, M., Pousse, A., Hoeltzel, A., and Stock-Damge, C.,** Trophic effect of bombesin on the rat pancreas: is it mediated by the release of gastrin or cholecystokinin, *Peptides*, Suppl. 3, 89, 1985.
9. **Ménard, D. and Malo, C.,** Insulin-evoked precocious appearance of intestinal sucrase activity in suckling mice, *Dev. Biol.*, 69, 661, 1979.
10. **Soling, H. D. and Unger, K. O.,** The role of insulin in the regulation of α-amylase synthesis in the rat pancreas, *Eur. J. Clin. Invest.*, 2, 199, 1972.
11. **Cezard, J. P., Forgue-Lafitte, M. E., Chamblier, M. C., and Rosselin, G. E.,** Growth-promoting effect, biological activity, and binding of insulin in human intestinal cancer cells in culture, *Cancer Res.*, 41, 1148, 1981.
12. **Mossner, J., Logsdon, C. D., Williams, J. A., and Goldfine, I. D.,** Insulin, via its own receptor, regulates growth and amylase synthesis in pancreatic acinar AR42J cells, *Diabetes*, 34, 891, 1985.
13. **Sesso, M. A.,** Effet de la thyroxine et de la cortisone sur les grains de sécrétion, les acides nucléiques et les activités amylolytique et protéolytique du pancréas chez le jeune rat, *C.R. Acad. Sci. Paris*, 254, 569, 1962.
14. **Morisset, J. and Jolicoeur, L.,** Effect of hydrocortisone on pancreatic growth in rat, *Am. J. Physiol.*, 239, G95, 1980.
15. **Arsenault, P. and Ménard, D.,** Influence of hydrocortisone on human fetal small intestine in organ culture, *J. Pediatr. Gastroenterol. Nutr.*, 4, 893, 1985.
16. **Johnson, L. R. and Guthrie, P. D.,** Stimulation of rat oxyntic gland mucosal growth by epidermal growth factor, *Am. J. Physiol.*, 238, G45, 1980.
17. **Arsenault, P. and Ménard, D.,** Stimulatory effects of epidermal growth factor on deoxyribonucleic acid synthesis in the gastrointestinal tract of the suckling mouse, *Comp. Biochem. Physiol.*, 86B, 123, 1987.
18. **Dembinski, A., Gregory, H., Konturek, S. J., and Polanski, M.,** Trophic action of epidermal growth factor on the pancreas and gastroduodenal mucosa in rats, *J. Physiol.*, 325, 35, 1982.
19. **Williams, J. A., Bailey, A., Humbel, R., and Goldfine, I. D.,** Insulin like growth factors bind to specific receptors in isolated pancreatic acini, *Am. J. Physiol.*, 246, G96, 1984.

20. **Osgood, E. E.,** A unifying concept of the etiology of the leukemias, lymphomas, and cancers, *J. Natl. Cancer Inst.*, 18, 155, 1957.
21. **Bullough, W. S.,** The control of mitotic activity in adult mammalian tissues, *Biol. Rev.*, 37, 307, 1962.
22. **Elgjo, K.,** Epidermal chalone: cell cycle specificity of two epidermal growth inhibitors, *Natl. Cancer Inst. Monogr.*, 38, 71, 1973.
23. **Bullough, W. S.,** Mitotic control in adult mammalian tissues, *Biol. Rev.*, 50, 99, 1975.
24. **Patt, L. M. and Houck, J. C.,** The incredible shrinking chalone, *FEBS Lett.*, 120, 163, 1980.
25. **Marks, F.,** The epidermal chalones, in *Chalones*, Houck, J. C., Ed. Elsevier, New York, 1976, 173.
26. **Isaksson-Forser, G., Burton, D., Korsgaard, R., Elgjo, K., and Iversen, O.,** Partial purification of epidermal G_2 chalone, *Virchows Arch. (B) Cell Pathol.*, 26, 97, 1977.
27. **Verly, W. G., Deschamps, Y., Pushpatadam, J., and Desrosiers, M.,** The hepatic chalone. I. Assay method for the hormone and purification of the rabbit liver chalone, *Can. J. Biochem.*, 49, 1376, 1971.
28. **Onda, H. and Yoshikawa, J.,** α_1-acid glycoprotein as a hepatocyte-specific mitosis-inhibiting protein in regenerating rat liver, *Gann*, 66, 227, 1975.
29. **Philpott, G. W.,** Tissue specific inhibition of cell proliferation in embryonic stomach epithelium in vitro, *Gastroenterology*, 61, 25, 1971.
30. **Sassier, P. and Bergeron, M.,** A method for the study of the inhibition of DNA synthesis by rabbit tissue extracts, *Can. J. Physiol. Pharmacol.*, 54, 367, 1976.
31. **Benestad, H. B. and Rytomaa, T.,** Regulation of maturation rate of mouse granulocytes, *Cell Tissue Kinet.*, 10, 461, 1977.
32. **Onda, H.,** α_1-acid glycoprotein and α_1-antitrypsin as mitotic inhibitors in regenerating rat liver, *Gann*, 68, 301, 1977.
33. **Johnson, L. R. and Guthrie, P. D.,** Secretin inhibition of gastrin-stimulated deoxyribonucleic acid synthesis, *Gastroenterology*, 67, 601, 1974.
34. **Johnson, L. R. and Guthrie, P. D.,** Effect of secretin on colonic DNA synthesis, *Proc. Soc. Exp. Biol. Med.*, 158, 521, 1978.
35. **Johnson, L. R.,** New aspects of the trophic action of gastrointestinal hormones, *Gastroenterology*, 72, 788, 1977.
36. **Harper, A. A., Hood, A. J. C., Mushens, J., and Smy, J. R.,** Inhibition of external pancreatic secretion by extracts of ileal and colonic mucosa, *Gut*, 15, 825, 1974.
37. **Laugier, R. and Sarles, H.,** Action of oleic acid on the exocrine pancreatic secretion of the conscious rat: evidence for an anticholecystokinin-pancreozymin factor, *J. Physiol.*, 271, 81, 1977.
38. **Sarles, H., Hage, G., Laugier, R., Demol, P., and Bataille, D.,** Present status of the anticholecystokinin hormone, *Digestion*, 19, 73, 1979.
39. **Tatemoto, K.,** Isolation and characterization of peptide YY(PYY), a candidate gut hormone that inhibits pancreatic exocrine secretion, *Proc. Natl. Acad. Sci. U.S.A.*, 79, 2514, 1982.
40. **Hally, A. D., Harper, A. A., Hood, A. J. C., Mushens, J., and Smy, J. R.,** Effect of pancreatone on pancreatic growth in rats, *J. Physiol.*, 285, 34P, 1978.
41. **Morisset, J.,** Receptors of the gastrointestinal hormones and their development, chap. 3, this volume.
42. **Barnes, D. W.,** Epidermal growth factor inhibits growth of A431 human epidermoid carcinoma in serum-free cells culture, *J. Cell Biol.*, 93, 1,1982.
43. **Imai, Y., Leung, C. K. H., Friesen, G. H., and Shin, R. P. C.,** Epidermal growth factor receptors and effect of epidermal growth factor on growth of human breast cancer cells in long-term tissue culture, *Cancer Res.*, 42, 4394, 1982.
44. **Korc, M., Padilla, J., and Grosso, D.,** Epidermal growth factor inhibits the proliferation of a human endometrial carcinoma cell line, *J. Clin. Endocrinol. Metab.*, 62, 874, 1986.
45. **Schonburnn, A., Krasnoff, M., Westendorf, J. W., and Tashjian, A. H.,** Epidermal growth factor and thyrotropin-releasing hormone act similarly on a clonal pituitary cell strain, *J. Cell Biol.*, 85, 786, 1980.
46. **Blay, J. and Brown, K. D.,** Contradistinctive growth responses of cultured rat intestinal epithelial cells to epidermal growth factor depending on cell population density, *J. Cell Physiol.*, 192, 343, 1986.
47. **Ménard, D., Arsenault, P., and Pothier, P.,** Biologic effects of epidermal growth factor in human fetal jejunum, *Gastroenterology*, 94, 656, 1988.
48. **Morisset, J., Korc, M., and Larose, L.,** Antagonistic and additive trophic effects of caerulein and EGF in rat pancreas, *Gastroenterology*, 90, 1557, 1986.
49. **Brazeau, P., Vale, W., Burgus, R., Ling, N., Butcher, M., Rivier, J., and Guillemin, R.,** Hypothalamic polypeptide that inhibits the secretion of immunoreactive pituitary growth hormone, *Science*, 179, 77, 1973.
50. **Pradayrol, L., Jornvall, H., Mutt, V., and Ribet, A.,** N-terminally extended somatostatin: the primary structure of somatostatin-28, *FEBS Lett.*, 109, 55, 1980.
51. **McIntosh, C. H. S.,** Gastrointestinal somatostatin: distribution, secretion and physiological significance, *Life Sci.*, 37, 2043, 1985.
52. **Arimura, A., Sato, H., Dupont, A., Nishi, N., and Schally, A. V.,** Somatcstatin: abundance of immunoreactive hormone in rat stomach and pancreas, *Science*, 189, 1007, 1975.

53. **Polak, J. M., Pearse, A. G. E., Grimelins, L., Bloom, S. R., and Arimura, A.,** Growth-hormone release inhibiting hormone (GH-RH) in gastrointestinal and pancreatic D cells, *Lancet*, 1, 1220, 1975.
54. **Chayvialle, J. A., Paulin, G., Dubois, P. M., Descos, F., and Dubois, M. P.,** Ontogeny of somatostatin in the human gastro-intestinal tract, endocrine pancreas and hypothalamus, *Acta Endocrinol.*, 94, 1, 1980.
55. **Dupouy, J. P., Chatelain, A., and Dubois, M. P.,** Normal development of cells with somatostatin immunoreactivity in the pancreas and duodenum of the rat fetus and newborn, *Cell Tissue Res.*, 231, 463, 1983.
56. **Morisset, J., Genik, P., Lord, A., and Solomon, T. E.,** Effects of chronic administration of somatostatin on rat exocrine pancreas, *Regul. Peptides*, 4, 49, 1982.
57. **Morisset, J.,** Somatostatin: a potential antigrowth factor for the exocrine pancreas, *Regul. Peptides*, 10, 11, 1984.
58. **Redding, T. W. and Schally, A. V.,** Inhibition of growth of pancreatic carcinomas in animal models by analogs of hypothalamic hormones, *Proc. Natl. Acad. Sci. U.S.A.*, 81, 248, 1984.
59. **Lehy, T., Dubrasquet, M., and Bonfils, S.,** Effect of somatostatin on normal and gastrin-stimulated cell proliferation in the gastric and intestinal mucosae of the rat, *Digestion*, 19, 99, 1979.
60. **Lehy, T., Dubrasquet, M., Brazeau, P., and Bonfils, S.,** Inhibitory effect of prolonged administration of long-acting somatostatin on gastrin-stimulated fundic epithelial cell growth in the rat, *Digestion*, 24, 246, 1982.
61. **Cadiot, G., Lehy, T., Bonfils, S., and Lewin, M. J. M.,** Long-acting somatostatin analog (SMS201-995) partly antagonizes the trophic effect due to omeprazole in rat gastric mucosa, *Gastroenterology*, 92, 1336, 1987.
62. **Bosshard, A., Pansu, D., Chayvialle, J. A., and Reinberg, A.,** Effect of fasting and somatostatin administration on circadian rhythms of jejunal cell renewal and plasma gastrin level in rats, *Digestion*, 23, 245, 1982.
63. **Holmes, S. J. K., Jaspan, J. B., and Moossa, A. R.,** The effect of somatostatin on postresectional ileal hyperplasia, *Endocrinology*, 111, 1397, 1982.
64. **Stange, E. F., Schneider, A., Schusdziarra, V., and Ditschuneit, H.,** Inhibitory effects of somatostatin on growth and differentiation in cultured intestinal mucosa, *Horm. Metabol. Res.*, 16, 74, 1984.
65. **Conteas, C. N. and Nandi Majumdar, A. P.,** The effects of gastrin, epidermal growth factor, and somatostatin on DNA synthesis in a small intestinal crypt cell line (EIC-6), *Proc. Soc. Exp. Biol. Med.*, 184, 307, 1987.
66. **Lucey, M. R.,** Endogenous somatostatin and the gut, *Gut*, 27, 457, 1986.
67. **Arimura, A. and Fishback, J. B.,** Somatostain: regulation of secretion, *Neuroendocrinology*, 33, 246, 1981.
68. **Koop, H., Simson, G., McIntosh, C., Bedell, R. A., Arnold, R., and Creutzfeldt, W.,** Somatostatin and gastrin release into the gastric lumen in rats, *Can. J. Physiol. Pharmacol.*, 59, 574, 1981.
69. **Bloom, S. R., Mortimer, C. H., Thorner, M. O., Besser, G. M., Hall, R., Gomez-Pan, A., Roy, V. M., Russell, R. C. G., Coy, D. H., Hastin, A. J., and Schally, A. V.,** Inhibition of gastrin and gastric acid secretion by growth hormone release-inhibiting hormone, *Lancet*, 2, 1106, 1974.
70. **Schlegel, W., Raptis, S., Harvey, K. F., Oliver, J. M., and Pfeiffer, E. F.,** Inhibition of cholecystokinin pancreozymin release by somatostatin, *Lancet*, 2, 166, 1977.
71. **Boden, G., Sivitz, M. C., Owen, O. T., Ersa-Koumar, N., and Landor, J. H.,** Somatostatin suppresses secretin and pancreatic exocrine secretion, *Science*, 190, 163, 1975.
72. **Gomez-Pan, A., Reed, J. D., Albinus, M., Schaw, B., and Hall, R.,** Direct inhibition of gastric acid and pepsin secretion by growth hormone release-inhibiting hormone in cats, *Lancet*, 1, 888, 1975.
73. **Schubert, M. L., Edwards, N. F., Arimura, A., and Makhlouf, G. M.,** Paracrine regulation of gastric acid secretion by fundic somatostatin, *Am. J. Physiol.*, 252, G485, 1987.
74. **Larsson, L. I., Goltermann, N., de Magistris, L., Rehfeld, J. F., and Schwartz, T. W.,** Somatostatin cell processes as pathways for paracrine secretion, *Science*, 205, 1393, 1979.
75. **Zeggari, M., Susini, C., Viquerie, N., Estève, J. P., Vaysse, N., and Ribet, A.,** Tumor promoter inhibition of cellular binding of somatostatin, *Biochem. Biophys. Res. Commun.*, 128, 850, 1985.
76. **Matozaki, T., Sakamoto, C., Nagao, M., and Baba, S.,** Protein kinase C regulates somatostatin binding to its receptors on rat pancreatic acinar cell membranes, *Biomed. Res.*, 8, 39, 1987.
77. **Reyl-Desmars, F. and Lewin, M. J. M.,** Evidence for an intracellular somatostatin receptor in pancreas: a comparative study with reference to gastric mucosa, *Biochem. Biophys. Res. Commun.*, 109, 1324, 1982.
78. **Mehler, P. S., Sussman, A. L., Maman, A., Leitner, J. W., and Sussman, K. E.,** Role of insulin secretagogues in the regulation of somatostatin binding by isolated rat islets, *J. Clin. Invest.*, 66, 1334, 1980.
79. **Tran, V. T., Flint Beal, M., and Martin, J. B.,** Two types of somatostatin receptors differentiated by cyclic somatostatin analogs, *Science*, 228, 492, 1985.
80. **Reyl, F., Silve, C., and Lewin, M. J. M.,** Somatostatin receptors on isolated gastric cells, in *Hormone Receptors in Digestion and Nutrition*, Rosselin, G., Fromageot, P., and Bonfils, S., Eds., Elsevier/North-Holland Biochemical Press, 1979, 391.
81. **Arilla, E., Pilar Lopez-Ruiz, M., Guijarro, L. G., Prieto, J. C., Gomez-Pan, A., and Hirst, B.,** Characterization of somatostatin binding sites in cytosolic fraction of rat intestinal mucosa, *Biochim. Biophys. Acta*, 802, 203, 1984.
82. **Reubi, J. C., Rivier, J., Perrin, M., Brown, M., and Vale, W.,** Specific high affinity binding sites for somatostatin-28 on pancreatic B cells: differences with brain somatostatin receptors, *Endocrinology*, 110, 1049, 1982.

83. **Zeggari, M., Viguerie, N., Susini, C., Estève, J. P., Vaysse, N., Rivier, J., Wunsch, E., and Ribet, A.,** Characterization of pancreatic somatostatin binding sites with a ^{125}I-somatostatin 28 analog, *Peptides*, 7, 953, 1986.
84. **Estève, J. P., Susini, C., Vaysse, N., Antoniotti, H., Wunsch, E., Berthon, G., and Ribet. A.,** Binding of somatostatin to pancreatic acinar cells, *Am. J. Physiol.*, 247, G62, 1984.
85. **Sakamoto, C., Goldfine, I. D., and Williams, J. A.,** The somatostatin receptor on isolated pancreatic acinar cell plasma membranes. Identification of subunit structure and direct regulation by cholecystokinin., *J. Biol. Chem.*, 259, 9623, 1984.
86. **Matozaki, T., Sakamoto, C., Nagoa, M., and Baba, S.,** Phorbol ester or diacylglycerol modulates somatostatin binding to its receptors on rat pancreatic acinar cell membranes, *J. Biol. Chem.* 261, 1414, 1986.
87. **Srikant, C. B. and Patel, Y. C.,** Somatostatin on rat pancreatic acinar cells. Pharmacological and structural characterization and demonstration of down-regulation in streptozotocin diabetes, *J. Biol. Chem.*, 261, 7690, 1986.
88. **Zeggari, M., Viguerie, N., Susini, C., Garnier, M., Estève, J. P., and Ribet, A.,** Solubilization and characterization of guinea-pig pancreatic somatostatin receptors., *Eur. J. Biochem.*, 164, 667, 1987.
89. **Susini, C., Bailey, A., Szecowka, J., and Williams, J. A.,** Characterization of covalently cross-linked pancreatic somatostatin receptors, *J. Biol. Chem.*, 261, 16738, 1986.
90. **Sakamoto, C., Matozaki, T., Nagoa, M., and Baba, S.,** Coupling of guanine nucleotide inhibitory protein to somatostatin receptors on pancreatic acinar membranes, *Am. J. Physiol.*, 253, G308, 1987.
91. **Carter, R. F., Bitar, K. N., Nfass, A. M., and Makhlouf, G. M.,** Inhibition of VIP-stimulated intestinal secretion and cyclic AMP production by somatostatin in the rat, *Gastroenterology*, 74, 726, 1978.
92. **Chew, C. S.,** Inhibitory action of somatostatin on isolated gastric glands and parietal cells, *Am. J. Physiol.*, 245, G221, 1983.
93. **Gespach, C., Mangeat, P., Marchis-Mouren, G., and Rosselin, G.,** Selective inhibition by somatostatin-14 and 28 of cAMP-production and cAMP-dependent protein kinase activation induced by histamine, vasoactive intestinal peptide (VIP) and PGE_2 in guinea pig gastric glands, Fourth International Symposium on Gastrointestinal Hormones, Stockholm, Sweden, 1982, 13.
94. **Singh, M.,** Effect of somatostatin on amylase secretion from in vivo and in vitro rat pancreas, *Dig. Dis. Sci.*, 31, 506, 1986.
95. **Putney, J. W., Burgess, G. M., Halenda, S. P., McKinney, J. S., and Rubin, R. P.,** Effects of secretagogues on [^{32}P] phosphatidyl-inositol 4,5-biphosphate metabolism in the exocrine pancreas, *Biochem. J.*, 212, 483, 1983.
96. **Taparel, D., Susini, C., Estève, J. P., Diaz, J., Cazaubon, C., Vaysse, N., and Ribet, A.,** Somatostatin analogs: correlation of receptors affinity with inhibition of cyclic AMP formation in pancreatic acinar cells, *Peptides*, 6, 109, 1985.
97. **Vaysse, N., Susini, C., Estève, J. P., Pradayrol, L., Balas, D., and Ribet, A.,** Action of somatostatin, big somatostatin and secretin on rat pancreatic duct cells, in *Hormone Receptors in Digestion and Nutrition*, Rosselin, G., Fromageot, P., and Bonfils, S., Eds., Elsevier/North-Holland Biochemical Press, 1979, 215.
98. **Reyl, F. J. and Lewin, M. J. M.,** Intracellular receptor for somatostatin in gastric mucosal cells: decomposition and reconstitution of somatostatin-stimulated phosphoprotein phosphatases. *Proc. Natl. Acad. Sci. U.S.A.*, 79, 978, 1982.
99. **Draznin, B., Mehler, P. S., Leitner, J. W., Sussman, K. E., Dahl, R., Vatter, A., and Melmed, S.,** Localization of somatostatin receptors in secretion vesicles in anterior pituitary cells and pancreatic islets, *J. Receptor Res.*, 5, 83, 1985.
100. **Draznin, B., Leitner, J. W., and Sussman, K. E.,** A unique control mechanism in the regulation of insulin secretion. Secretagogue-induced somatostatin receptor recruitment, *J. Clin. Invest.*, 75, 1510, 1985.
101. **Draznin, B., Steiner, C., Sherman, N., and Sussman, K.,** Somatostatin inhibits fusion of pituitary secretion vesicles with the plasma membranes, *Biochem. Biophys. Res. Commun.*, 139, 673, 1986.
102. **Rossowski, W. J., Ozden, A., Ertan, A., and Arimura, A.,** Regulation of somatostatin-14 and gastrin I binding sites in rat gastrointestinal mucosa by ulcerogenic dose of cysteamine, *Life Sci.*, 40, 1783, 1987.
103. **Viguerie, N., Estève, J. P., Susini, C., Vaysse, N., and Ribet, A.,** Processing of receptor-bound somatostatin: internalization and degradation by pancreatic acini, *Am. J. Physiol.*, 252, G535, 1987.
104. **Viguerie, N., Estève, J. P., Susini, C., Logsdon, C. D., Vaysse, N., and Ribet, A.,** Dexamethasone effects on somatostatin receptors in pancreatic acinar AR4-2J cells, *Biochem. Biophys. Res. Commun.*, 147, 942, 1987.
105. **Taparel, D., Estève, J. P., Susini, C., Vaysse, N., Balas, D., Berthon, G., Wunsch, E., and Ribet, A.,** Binding of somatostatin to guinea pig pancreatic membranes; regulation by ions, *Biochem. Biophys. Res. Commun.*, 115, 827, 1983.

Chapter 13

HORMONAL EFFECTS ON GASTROINTESTINAL CANCER GROWTH

Courtney M. Townsend, Jr., Pomila Singh, and James C. Thompson

TABLE OF CONTENTS

I. INTRODUCTION

Carcinomas of the large bowel, stomach, and pancreas account for more than 150,000 new cases of cancer and cause the death of 75,000 patients yearly.[1] Treatment of patients with these cancers relies heavily on adequate surgical resection. This is effective only when localized tumor is present. There is no effective systemic treatment for metastatic colon, stomach, or pancreatic cancer.

Endocrine control of cancer growth was first demonstrated by the Scottish surgeon Beatson,[2] who performed bilateral oophorectomy for treatment of breast cancer. In 1952, Huggins and Bergenstal[3] reported that bilateral adrenalectomy caused regression of breast and prostate cancer in humans. The evolution of the concept that cancers which arise from classical hormone target tissues are either hormone-dependent or -independent (terminology first used in 1945 by Huggins and Scott)[4] has allowed the development of successful although temporary strategies to arrest growth of hormone-responsive cancers. Hormone-responsive cancers may be inhibited either by hormone deprivation or by the administration of antagonistic hormones or by excessive amounts of trophic hormones.

Gastrointestinal hormones have been found to stimulate growth of the normal mucosa of the gut and growth of the pancreas. These findings raise the question of whether these agents also act to regulate growth of some cancers of the gut and pancreas.

II. EFFECTS OF GASTROINTESTINAL HORMONES ON CARCINOGENESIS IN THE GUT AND PANCREAS

Chemical carcinogenesis is currently viewed as a process comprised of two steps, initiation and promotion. Exposure to specific chemical agents initiates cell mutations; cell division allows the malignant phenotype to be expressed and to be transferred to progeny. Maskens[5] has confirmed the two-step model in rat colon carcinogenesis. Any agent which begins the process of carcinogenesis is an *initiator*; an agent which acts only after the carcinogenic process has been initiated is a *promoter*. Cocarcinogens do not cause cancer but act before, during, or after carcinogenesis to increase the carcinogenic effect. Although all factors involved in tumor development due to carcinogens are not known, cell proliferation is a central requirement in all models examined to date.[6,7] Agents which stimulate cell proliferation produce more S-phase cells (sensitive to DNA damage), increase the possibility of malignant transformation, reduce the latent period of tumor appearance, and stimulate tumor growth.

Chemical carcinogens alter the normal pattern of cell proliferation. In mice exposed to dimethylhydrazine (DMH), abnormal colonic crypts predate the appearance of adenoma or carcinoma.[8-11] In patients with preneoplastic syndromes (familial polyposis or Gardner's syndrome) and colon cancer, similar alterations of cell proliferation of grossly normal-appearing colonic mucosa are noted.[8,10-13]

The role that gut hormones (that are known to stimulate cellular proliferation) may play in the process of carcinogenesis in the gut and pancreas has been studied. Some studies have found that hormones increased carcinogenesis, while others found that gut hormones reduced carcinogenesis in the gut and pancreas.

A. GASTRIC CARCINOGENESIS

Studies on the effect of gastrin on carcinogen-induced stomach cancers provide conflicting results. Some investigators have found that gastrin increased the incidence of carcinogen-induced gastric cancer in rats,[14] but others have found that gastrin inhibited the incidence of carcinogen-induced stomach cancers in rats.[15] Some investigators believe that the trophic effect of gastrin enhances the effects of carcinogens, while others believe that the trophic effect of gastrin allows healing of carcinogen-induced injury and thus prevents tumor development.

Atrophic gastritis and endogenous hypergastrinemia significantly increased the incidence of *N*-methyl-*N'*-nitro-*N*-nitrosoguanidine (MNNG)-induced gastric cancers in rats.[16] In this model, atrophic gastritis was induced either by chronic taurocholate administration or by anti-gastric mucosal antibodies. Tetragastrin administration to rats was found to significantly decrease the severity of MNNG-induced gastric mucosal atrophy.[17] Gastric mucosal atrophy is associated with achlorhydria, hypergastrinemia, and an increased incidence of carcinoma in man; however, the concomitant increase in gastrin-stimulated acid secretion in some experimental models may explain, in part, the differences between rats and man.

The timing of gastrin administration during MNNG carcinogenesis is an important determinant of enhanced tumor development.[14,18] Gastrin given during exposure to MNNG increased gastric cancer development,[14] whereas gastrin administration later did not enhance tumor development.[15,18]

Acid may inhibit the carcinogenic actions of MNNG on the stomach. Increased acid secretion, stimulated by either gastrin or histamine administration, has been reported to reduce the incidence of MNNG-induced gastric carcinomas.[19] Tetragastrin combined with propranolol (begun after completion of MNNG treatment) stimulates acid secretion and inhibits gastric carcinogenesis in rats.[20]

Bilateral vagotomy (without pyloroplasty) performed in rats 1 week after 15 weeks of MNNG treatment significantly increased both atypical glandular hyperplasia and cancer when compared to rats treated with MNNG alone;[21] however, levels of gastrin and acid secretion were not reported. Deveney et al.[22] performed antrectomy or antral transposition 2 weeks before MNNG administration was begun. They failed to find an increased incidence of gastric cancer in rats made hypergastrinemic by antral transposition; however, antrectomy increased the incidence of gastric cancers. They conclude that acid, not endogenous gastrin, affected carcinogenesis.

Discrepancies among these studies of gastric carcinogenesis may be explained by differences in experimental design.

Long-term administration of high doses of omeprazol produces endogenous hypergastrinemia, hyperplasia of enterochromaffin like cells, and gastric carcinoid tumors in rats.[23] Development of carcinoids can be prevented by antrectomy.[24] Three patients have been reported in whom antrectomy caused disappearance of gastric carcinoids in the remaining stomach.[25,26]

B. COLONIC CARCINOGENESIS

Endogenous hypergastrinemia, produced by antral exclusion performed after carcinogen treatment was begun, did not alter tumor incidence but increased DNA synthesis in carcinogen-induced colon tumors.[27] On the other hand, when antral exclusion was performed 2 weeks before carcinogen treatment, tumor incidence was increased.[28] However, Oscarson et al.[29] reported that endogenous hypergastrinemia (induced by fundusectomy at the onset of carcinogen treatment) did not potentiate DMH-induced colon carcinogenesis.

Secretin inhibited ^{3}H-thymidine incorporation[30] and pentagastrin increased the labeling index and metaphase index (but not the mitotic index)[31] in carcinogen-induced tumors in rats.

The timing of hormone administration in relation to MNNG administration has been evaluated.[32-35] When tetragastrin and intrarectal MNNG were given simultaneously, colon carcinogenesis was significantly inhibited; when tetragastrin was begun after MNNG was stopped, no protective effect was noted.[32,33] It is thought that gastrin reduced MNNG-induced erosions, ulcerations, and the number of foci of atypical regenerative glandular hyperplasia. These authors conclude that gastrin-induced cell proliferation promoted mucosal healing and inhibited carcinogenesis.[34] Iishi et al.[35] found that vasoactive intestinal peptide (VIP), begun 3 weeks before and continued throughout administration of azoxymethane (AOM), significantly enhanced colonic carcinogenesis in rats.

The effects of dietary manipulations (which decrease or increase colonic mucosal cell proliferation) on colon carcinogenesis have been studied.[36,37] Tatsuta et al.[36] found that a defined liquid diet (given to rats treated with AOM) caused colonic mucosal atrophy, decreased serum gastrin levels, and increased the incidence of colon tumors. They found that tetragastrin treatment of rats fed a fiber-free diet (and treated with AOM) significantly reduced colon carcinogenesis;[37] however, tetragastrin did not influence carcinogenesis in rats fed 20% wheat bran. These authors suggest that the hyperplastic effect of endogenous gastrin on colonic mucosa prevents colon carcinogenesis; however, no group of AOM-treated rats given a liquid diet and exogenous gastrin was studied. Since bran produced colonic mucosal hyperplasia, the authors state that although tetragastrin is trophic for colonic mucosa, gastrin must have some other effect as well on carcinogenesis.

The role of other agents associated with gut mucosal cell proliferation has been examined. Removal of the major source of epidermal growth factor (EGF) in mice (the submandibular salivary gland) protected them against carcinogen-induced colon cancers.[38] Exogenous administration of EGF, on the other hand, did not alter DMH-induced colonic carcinogenesis;[39] however, the number of squamous cell carcinomas of the anus was increased. These findings appear to reflect tissue-specific difference in carcinogen-promoting effects of EGF. Two groups of investigators have failed to demonstrate an effect of endogenous hyperenteroglucagonemia on colonic carcinogenesis.[40,41]

The role that gut hormones play in colonic carcinogenesis cannot be determined from studies published to date; these studies are largely descriptive. The differences between the studies of Tatsuta and colleagues (who propose that the trophic effect of gastrin is anti-carcinogenic) and other investigators may be due to differences in methods used to induce hypergastrinemia (exogenous versus endogenous), the dose of hormone given, the time of administration of hormone during carcinogenesis, or the experimental systems employed. None of the studies reported effects of the doses of tetragastrin employed on growth of normal gastric and colonic mucosa. These variables must be more completely studied. Mechanisms by which hormones stimulate or inhibit carcinogenesis have not been fully evaluated. Gastrin stimulates cell proliferation in the gut. However, the effects of gastrin on carcinogenesis may be entirely indirect; for example, acid secretion or hormones (secretin or somatostatin) released by gastrin may directly affect carcinogenesis. Future studies should systematically examine these possibilities by using agents that block acid secretion, block gastrin release, or block the action of gastrin on cellular receptors.

C. PANCREATIC CARCINOGENESIS

There are two animal models of pancreatic carcinoma: the nitrosamine-treated hamster and the azaserine-treated rat. Histologically, human pancreatic adenocarcinoma is most often (90%) ductal adenocarcinoma. Nitrosamine-treated hamsters develop ductal carcinomas which are identical to the human cancers, whereas azaserine-treated rats develop acinar carcinomas. The role that gut hormones that affect normal pancreatic cellular proliferation may play in pancreatic carcinogenesis has been evaluated in both animal models. Studies published to date have examined either the effect of dietary factors (which may release endogenous hormones) or the effect of exogenously administered hormones on pancreatic carcinogenesis.

Administration of azaserine to rats fed raw soya flour[42] or high dietary fat (20% corn oil)[43] enhances the development of pancreatic cancers. The combination of di(2-hydroxypropyl)nitrosamine and raw soya flour produces acinar cell tumors in the rat while nitrosamine alone does not.[44] These findings suggest that the mechanism of nitrosamine carcinogenesis is fundamentally different between rats and hamsters. Cholecystokinin (CCK) stimulates pancreatic hyperplasia; it is thought that the trypsin-inhibiting potency of raw soya interferes with the postulated luminal trypsin/plasma CCK feedback mechanism which results in increased endogenous release of CCK. Bombesin and cerulein (a CCK analog) stimulate the early stages of azaserine carcinogenesis in the rat pancreas.[45]

Administration of several related nitrosamines induces ductal carcinoma in hamsters. Results of studies on the effects of gut hormones on pancreatic carcinogenesis in hamsters are conflicting. Johnson et al.[46] reported that the octapeptide fragment of CCK (CCK-8) inhibited development of nitrosamine-induced pancreatic cancer in hamsters. Andrén-Sandberg et al.[47] failed to demonstrate any effect of cerulein on development of pancreatic cancer produced by *N*-nitrosobis(2-hydroxypropyl)amine (BHP) in hamsters. The authors concluded that cerulein acts on a site different than BHP. This conclusion is not founded by a single experiment in which only a single dose of cerulein was used. It is quite likely that the dose of cerulein used was insufficient to stimulate hyperplasia, which is required for cocarcinogen effect. In contrast, Satake et al.[48] found that weekly administration of cerulein enhanced the carcinogenic effect of BHP in hamsters. Howatson and Carter found that both CCK[49] and secretin[50] enhanced pancreatic carcinogenesis induced by *N*-nitrosobis-(2-oxypropyl)amine (BOP) in hamsters. EGF given during BOP administration promotes pancreatic carcinogenesis in hamsters.[51]

Both the somatostatin analog (RC-160) and luteinizing hormone-releasing hormone (LHRH) significantly decreased BOP pancreatic carcinogenesis and prolonged survival in hamsters.[52]

Agents which stimulate hyperplasia of the normal pancreas also appear to stimulate carcinogenesis of the pancreas. Future studies should systematically examine the timing of hormone administration and should examine the effects of hormone antagonists.

III. EFFECTS OF GASTROINTESTINAL HORMONES ON GROWTH OF GUT AND PANCREATIC CANCERS

The trophic effect of gut hormones on a number of established gastrointestinal cancers is well-documented. Takahashi and colleagues[53] found that gastrin-enhanced growth of human stomach and colon cancers transplanted into nude mice could be inhibited by secretin. We have reported[54] that the combination of cerulein and secretin, but neither agent alone, produced significant stimulation of growth of a transplantable hamster pancreatic ductal cancer (H2T). We have reported that chronic administration of pentagastrin stimulated growth of mouse colon cancer (MC-26) *in vivo* and significantly decreased survival due to accelerated tumor growth in mice treated with pentagastrin.[55] Sumiyoshi et al.[56] found that pentagastrin stimulated the growth and stimulated cellular levels of cyclic adenosine 3′,5′monophosphate (cAMP) and cAMP-dependent protein kinase activity of human gastric adenocarcinoma xenografts in athymic mice.

In some studies, both hormone receptor antagonists and gut hormones have been found to inhibit cancer growth. Beauchamp et al.[57] reported that proglumide, an antagonist of gastrin and CCK receptors,[58] inhibited growth of MC-26 mouse colon cancers and prolonged survival of tumor-bearing mice. In this study, no exogenous gastrin treatment was given. Alexander et al.[59] found that asperlicin, a competitive, nonpeptide CCK antagonist, significantly inhibited the growth of a human pancreatic adenocarcinoma (SKI) that has CCK receptors (CCK-R). The findings from our two studies suggest that endogenous gastrin and CCK play a role in growth of hormone-responsive gut and pancreatic cancers.

Gut hormones which are trophic for normal tissue may inhibit rather than stimulate growth of some cancers. CCK has been found to inhibit growth of cholangiocarcinoma and a gastric carcinoma. Bombesin inhibited growth of a pancreatic cancer. CCK-8 inhibited growth of the human cholangiocarcinoma (SLU-132) implanted in nude mice.[60] It was also found that the inhibitory effects of cerulein are direct and receptor-mediated.[61] SLU-132 contains a significant number (30 to 60 fmol/mg protein) of specific, high-affinity ($K_D = 0.029$ n*M*) CCK-R. Cerulein significantly inhibited tumor growth. CR1409 (a CCK-R antagonist) alone did not affect tumor growth, but CR1409 prevented cerulein-mediated inhibition of tumor growth. Some fragments of CCK (CCK-8 and Glt-CCK-8) inhibited the growth of a poorly differentiated human gastric cancer (SC-6-JCK) growing in athymic nude mice;[62] other fragments (pGlu-CCK-8, CCK-12) did not. CCK-8 inhibited pentagastrin-stimulated increased cAMP content and cAMP-dependent protein kinase activity of this tumor. These studies demonstrate inhibitory effects of CCK on growth of some cancers; the mechanisms have not been defined.

We found that chronic bombesin treatment inhibited growth of a human ductal pancreatic adenocarcinoma (SKI) xenografted into nude mice and stimulated the growth of the normal nude mouse pancreas.[63] This demonstration of both trophic and inhibitory responses by the same hormone suggests that the actions of bombesin (or possibly the hormones that it releases) are site-specific and interact differently in normal and neoplastic pancreatic tissues.

There are several possible explanations for the apparent discrepancies in the effect of gut hormones on cancers: different doses of the same hormone may act differently; multiple forms of hormone receptors may exist and serve different functions; hormones may act directly on target cells through hormone-receptor regulation (homologous or heterologous); or hormones may act indirectly by releasing other agents (autocrine, paracrine, or endocrine) which act on the target cells. Further studies are required to precisely define the mechanisms of hormone modulation of cancer growth.

Studies of the effects of gut hormones *in vitro* have not demonstrated profound stimulation of cancer cell proliferation; the reasons are not clear. Pentagastrin increases DNA synthesis in cultured rat gastric and duodenal mucosal cells.[64] Growth of the human colon carcinoma cell line (HC84S) has been found to be mildly stimulated by the combination of gastrin and EGF *in vitro*.[65] Tetragastrin and cerulein cause proliferation of BV-9 rat gastric cells *in vitro*.[66] Pentagastrin added to the primary cultures of normal human colonic epithelial cells and colon cancer cells increased cell numbers 72 h after treatment.[67] Moyer et al.[68] demonstrated that glucagon increased cell numbers of human colorectal cancer cells after 6 d of incubation, suggesting a direct stimulatory role for this hormone. Ochiai et al.[69] found that gastrin increased cAMP-dependent protein kinase activity and growth of human gastric adenocarcinoma cells in culture. Kusyk et al.[70] have shown that the maximum stimulatory effect of gastrin on growth of human colon cancer cells *in vitro* occurs during the synthesis phase (S-phase) of the cell cycle.

Some human and some rat colon cancer cell lines are sensitive to mouse EGF and human urogastrone when tested in clonogenic assay or when examined for ^{3}H-thymidine uptake.[71] EGF stimulated a 20 to 30% increase in MIA PaCa-2A cell number at 18 h after a single dose of EGF in serum-free medium.[72] Somatostatin alone did not affect cell proliferation but did inhibit the effect of EGF.

Gastric inhibitory peptide (GIP), glucagon, and VIP stimulate cAMP production in the human stomach carcinoma cell line HGT-1.[73] This cell line appears to have separate binding sites for each of the three hormones, since combinations of the hormones were additive in stimulating cAMP production.

The majority of *in vitro* studies have demonstrated only limited activity of gut hormones. This may be a reflection of tumor cell heterogeneity; only a small proportion of cells in an uncloned cell line may be capable of responding to gut hormones. Alternatively, gut hormones may require other factors which are not present in culture medium to exert their maximum effect.

IV. PEPTIDE HORMONE RECEPTORS AND GASTROINTESTINAL CANCER

Receptors for steroid and peptide hormones have been detected in a wide variety of human tumors. However, not all cancers of endocrine-responsive tissue possess specific hormone receptors. The reasons for this are unknown.

There are specific receptors for VIP in human colon,[74] human pancreatic,[75] and human gastric[76] carcinoma cell lines. Secretin[75,77] and cerulein[77] receptors have been described in a human pancreatic cell line.

We have developed a highly sensitive assay for the measurement of gastrin receptors (GR) in normal and neoplastic gut tissues.[78] We have found high-affinity GR on gut cancer cell lines *in vitro*[78] and in freshly resected human stomach and colon cancers.[79] We have so far identified two human stomach cancer cell lines, AGS and SII (as well as four of the seven different clones derived from the AGS cell line), three human colon cancer cell lines (LoVo, LS-180, and WiDr), and one mouse colon cancer cell line (MC-26) that possess GR. The affinity of these gastrin-binding sites on the cancer cell lines (K_D = 0.5 n*M*) appears to be similar to that observed for normal cells from the fundic mucosa of rats.

We have found that the trophic effects of gastrin on MC-26 mouse colon cancers are mediated by the interaction of gastrin with its GR[80] and that GR are sensitive to circulating levels of gastrin.[80,81] Both proglumide and somatostatin analog 201-995 (Sandoz, East Hanover, NJ) inhibit the trophic effect of pentagastrin on MC-26 mouse colon adenocarcinoma.[81] This inhibition may be related to the decrease in high-affinity receptors for gastrin on MC-26 colon cancers that we observed in mice treated with either agent.

The relative binding affinity (RBA) of four analogs of gastrin, for inhibition of binding of ^{125}I-gastrin to GR on MC-26 cells, was investigated.[82] Removal of even one *N*-terminal amino acid resulted in a significant loss (to 15%) in the binding affinity of the analog (gastrin 2-17) for binding to GR on MC-26 cells. A further reduction of *N*-terminal amino acids (beyond position 8 — pentagastrin) resulted in a significant loss of RBA to only 0.3%. Proglumide (a gastrin and CCK-receptor antagonist) had a low RBA (0.0003%). No specific binding sites for CCK could be measured on MC-26 cells, and CCK did not inhibit gastrin binding to MC-26 cells. Asperlicin, a specific CCK-receptor antagonist, did not inhibit gastrin binding. From these studies we conclude that GR on normal and neoplastic gut tissue in mice are different. We have recently found by use of cross-linking techniques that there are different molecular species of GR on normal mouse fundic mucosa and MC-26 cells.[83]

We have found that estradiol (E_2) treatment significantly up-regulates GR levels on fundic mucosa membranes of ovariectomized guinea pigs and up-regulates GR levels and stimulates growth of MC-26 cancers in E_2-treated female mice. We have found that chronic E_2 treatment of ovariectomized guinea pigs and mice elevates serum gastrin levels.[84,85] These studies indicate the steroid regulation of GR levels and growth of normal and neoplastic gut tissues may be indirectly mediated via an increase in gastrin levels.

Weinstock and Baldwin[86] reported the presence of low-affinity gastrin binding sites (K_D varied between 0.1 and 1.3 μ*M*) on five human gastric carcinoma cell lines. These K_D values are significantly higher than those reported for gastrin binding to parietal cells or to the MC-26 mouse colon carcinoma cell line (K_D = 0.5 n*M*). These authors have been unable to demonstrate any effect of G-17 occupation of low-affinity binding sites on cell proliferation *in vitro*; the physiologic significance of these binding sites is unclear.

We have found that the presence of CCK receptors (CCK-R) can predict responsiveness of human pancreatic cancers to cerulein treatment.[87] CCK-R were measured, by use of a specific binding assay which we have developed, on two moderately differentiated human pancreatic adenocarcinomas, SKI, a tumor with CCK-R, and CAV, a tumor without CCK-R, growing in nude mice. Cerulein produced significant stimulation of SKI tumor growth; this effect was inhibited by proglumide. Growth of CAV was not stimulated by cerulein treatment. Furthermore, cerulein treatment maintained the presence of high-affinity CCK-R (K_D = <1.0 n*M*) in SKI tumors while tumors in the control and cerulein plus proglumide groups had only nonsaturable (K_D = >10.0 n*M*) CCK binding sites. Cerulein regulation of high-affinity CCK-R is similar to our findings of gastrin regulation of high-affinity GR in MC-26 colon cancers.[80,81] These findings indicate that growth-stimulatory effects of cerulein were mediated via CCK-R. The presence or absence of CCK-R on human pancreatic adenocarcinomas may then predict responsiveness of the tumor to hormone treatment. However, the type of response may not be predicted by the presence of hormone receptors; in pancreatic cancer with CCK-R, cerulein stimulated growth, whereas in cholangiocarcinoma with CCK-R, cerulein inhibited growth.

We found that chronic treatment with SMS 201-995, a long-acting somatostatin analog, produced similar inhibition of the growth of these two human pancreatic adenocarcinomas growing in nude mice. This inhibitory effect of SMS 201-995 is not due to interference with endogenous CCK, since SKI has CCK-R and CAV does not.[88]

We have shown by use of cross-linking techniques that hamster H2T pancreatic ductal cancer has a receptor for VIP which is not present on a human pancreatic ductal cancer cell line (MIA).[89] Chronic administration of VIP significantly inhibited growth of H2T cancers growing in hamsters; VIP did not affect growth of MIA cancers growing in nude mice.[90] The differential response of hamster and human pancreatic cancer to VIP treatment may be due to the presence or absence of VIP receptors.

Smith et al.[91] found elevated fasting plasma gastrin levels in 8 of 24 patients with adenomatous polyps and 8 of 20 patients with colon cancer. Five cancer patients had fasting gastrin values over 100 p*M*. The cause of elevated gastrin levels and whether the elevated gastrin levels precede or follow the development of polyps and colon cancer are unknown.

We have analyzed 67 primary human colon cancers for GR.[92] We found a spectrum of GR expression. Some cancers had no detectable GR. Some cancers had high-affinity (K_D = <1.0 n*M*) GR, while other cancers had only low-affinity (K_D = >1.0 n*M*) GR. Twenty patients (29.9%) had cancers with GR content >10 fmol/mg protein. Mean GR content was significantly greater (11.8 ± 2.9 fmol/mg protein) in Dukes' stage A and B cancers compared to stage C and D cancers (6.2 ± 1.6 fmol/mg protein). A significantly greater percentage (52.4%) of patients in the early stages (A and B) had tumors with GR >10 fmol/mg protein compared to patients with more advanced (C and D) cancers (19.6%). GR content of normal colon mucosa correlated with the GR receptor content of the cancers. Early follow-up data (40 to 972 d, mean = 292 d) revealed that no patient with GR content >10 fmol/mg protein had developed recurrence or had died, regardless of Dukes' stage (11B; 6C; 2D). In contrast, 5 of 19 Dukes' C patients and 8 of 16 Dukes' D patients with GR content <10 fmol/mg protein had developed recurrence or died. Although a longer follow-up period is required, GR content of colon cancers may have prognostic significance. We find it interesting that 40% of the patients reported by Smith et al.[91] had elevated gastrin levels and that 30% of our patients had increased levels of GR. Furthermore, our finding that GR levels were low in both the normal mucosa and in cancers in some patients and high in other patients may be related to serum gastrin levels. At this point this is only an interesting speculation; we have not studied gastrin levels in our patients.

Polyamines (putrescine, spermine, and spermidine) are ubiquitous low molecular weight organic cations which are required by mammalian cells for growth and differentiation.[93]

A marker of cell growth in rapidly proliferating and regenerating normal tissue[94-96] is increased activity of the enzyme ornithine decarboxylase (ODC) (the rate-limiting enzyme in the biosynthesis of polyamines) and increased synthesis of polyamines. Abnormalities of ODC activity and polyamine biosynthesis are found during experimental carcinogenesis and in both benign and malignant colonic neoplasms.[97,98] Upp et al.[97] and Kingsnorth et al.[98] have found that levels of polyamines are elevated in colorectal cancer compared to normal, uninvolved mucosa.

We have examined the relationship between polyamine levels and GR in samples of colon cancer and mucosa from 40 patients and in mucosa from 11 patients without cancer.[97] We found that polyamine levels in colon cancers were significantly higher than in normal colon mucosa from the same patients. Spermidine and spermine were significantly higher in colon mucosa from patients with cancer compared to patients without cancer. Spermidine and the spermidine/spermine ratio, an index of cell proliferation, were increased in colon cancers with GR compared to cancers without GR. Some patients with colon cancers have elevated fasting gastrin levels.[91] Our findings that patients with GR in colon cancers have elevated polyamine levels is evidence that endogenous gastrin may play a trophic role in some human colon cancers.

We believe that there is compelling evidence to indicate that gut hormones can indeed be trophic for both normal and neoplastic gut and pancreatic cells and that the effect is most likely mediated by the interaction of gut hormones with their receptors. These studies provide further evidence that the determination of gut hormone receptors may provide the ability to predict prognosis and may predict response of cancers to hormone manipulation in a manner similar to the ability of determinations of estrogen and progesterone receptors to predict prognosis and response to hormone manipulation in breast cancer.

V. GROWTH FACTORS AND POSSIBLE AUTOCRINE CONTROL OF GUT AND PANCREATIC CANCER GROWTH

Growth factors and gut hormones may function in an autocrine manner to control growth of gut and pancreatic cancers. Coffey and associates[99] found that several different colonic carcinoma cell lines produced both α- and β-transforming growth factor-like substances (TGF-α, TGF-β). Receptors for both EGF and TGF-β were detected in the colon cancer cell lines. Roberts et al.[100] reported that TGF-β had no effect on the growth of the HT29 colonic carcinoma cell line, but it inhibited a variety of other noncolonic transformed epithelial cell lines. We find it interesting that none of the colon carcinoma cell lines tested responded to exogenously administered TGF-β or EGF. A human colon cancer cell line has been found to produce both inhibitory (tumor-inhibiting factor) and stimulatory (transforming growth factor) substances.[101] Transforming growth factors enhanced the growth of a slow-growing colon cancer cell line but had no effect on growth of the cells that produced it. Increased levels of insulin-like growth factor II messenger RNA (mRNA) were found in rectosigmoid and rectal cancers.[102] Thus, growth factors may either stimulate or inhibit cell proliferation. We believe that it is possible that gut hormones may act through modulation of synthesis or release of growth factors which possess different properties.

EGF receptors have been found in colon, gastric, and pancreatic cancers[103-110] and in normal gastric and colonic mucosa.[104,105] The relationships between EGF receptors and biologic behavior of these cancers has not been established.

EGF mildly increased the number of PANC-I cells after 6 d in cultures in which the serum concentration was 0.1%.[106] Pancreatic carcinoma cell lines (ASPC-I, T_3M_4, PANC-I, COLO 357, and MIA PaCa-2) produce transforming growth factor alpha (TGF-α) and express TGF-α mRNA.[108] The amount of TGF-α protein found in the supernatant of these cells did not correlate with the amount of TGF-α mRNA. Differential growth effects of TGF-α and EGF on colony formation of PANC-I cells in soft agar were noted. The role of TGF-α as an autocrine growth factor for pancreatic cancer has not been conclusively demonstrated.

Production of human EGF-like immunoreactivity by five human gastric cancer cell lines (MKN-1, MKN-28, MKN-45, MKN-74, and KATO-III) has been described.[109] The presence of both EGF and EGF receptors has been studied in gastric cancers.[110] No EGF immunoreactivity was found in early cancers, but EGF-positive tumor cells were detected in 38 of 130 (29.2%) of advanced cancers. EGF receptor was found in 33.8% of advanced carcinomas but in only one early carcinoma; 17 advanced carcinomas had synchronous expression of EGF and EGF receptors.

These findings that some cancers produce EGF and TGF-α and have receptors for these growth factors raise the possibility that EGF and TGF-α may possibly function as autocrine growth factors for some human gut and pancreatic cancers.

Bombesin has been implicated as a mitogen or as an autocrine growth factor for normal and malignant cells.[111-118] Bombesin is produced by some human small-cell lung cancers; a monoclonal antibody to bombesin inhibited growth[111] and administration of bombesin stimulated the growth[118] of xenografts of human small-cell lung cancer in nude mice. Our study demonstrates that bombesin can, indeed, act as an autocrine growth factor for human small-cell lung cancer.

We find it interesting that transforming growth factor-α, EGF, and bombesin (agents which affect gastrointestinal function) are produced by some gut and pancreatic cancers and that these agents have growth-modulating effects on some cancers. In the future, gut hormones may be found to modulate gut and pancreatic cancer growth through release of cellular growth factors which act in an autocrine fashion or, alternatively, tumor-produced gut hormones may be found to be autocrine growth factors for gut and pancreatic cancers. As specific autocrine growth factors are identified, specific antagonists may be synthesized to combat the process of malignant transformation and growth of cancers.

The growth of some, but not all, breast cancers can be inhibited by hormone manipulation. Identification of patients who have hormone-dependent breast cancers can be accomplished by analysis of breast cancers for the presence of estrogen and progesterone receptors. Breast cancers with abundant estrogen and progesterone receptors respond, in general, to endocrine manipulation while most others with low concentrations of receptors (and almost all of those without receptors) do not respond to hormone treatment.[119] Not all breast cancers with hormone receptors respond to hormone manipulation; we expect that not all gut and pancreatic cancers with hormone receptors will respond to hormones. In fact, we expect to find that not all cancers with hormone receptors will respond in the same way to the same hormone.

The trophic effect of gut hormones may have important clinical applications for treatment of gut and pancreatic cancers. We have now developed methods by which we can quantitate gut hormone receptors in normal and neoplastic gut mucosa and pancreas. We expect to find a spectrum of gut hormone receptors; some cancers will possess specific hormone receptors and some will not. In tumors with gut hormone receptors, growth may be manipulated (stimulated or inhibited) by hormones. By assaying tumors for hormone receptors and studying the effects of hormones, both on tumors that do and those that do not possess hormone receptors, we should be able to more clearly define the role of gut hormones in growth control of gut and pancreatic cancers. Analysis of gut and pancreatic tumors for gut hormone receptors may thus allow us to select those patients with these cancers who would respond to treatment with antihormones or hormone ablation. It may be possible in the future to develop therapeutic strategies for patients with gut and pancreatic cancers that are based upon manipulation of the serum concentrations or effects of gut hormones in a manner similar to current strategies that are successfully employed in the treatment of patients with breast cancer.

REFERENCES

1. **Silverberg, E. and Lubera, J. A.,** Cancer statistics, 1988, *Ca-A,* 38, 5, 1988.
2. **Beatson, G. T.,** On the treatment of inoperable cases of carcinoma of the mammal. Suggestions for a new method of treatment, with illustrative cases, *Lancet,* 2, 104, 1896.
3. **Huggins, C. and Bergenstal, D. M.,** Inhibition of human mammary and prostatic cancers by adrenalectomy, *Cancer Res.,* 12, 134, 1952.
4. **Huggins, C. and Scott, W. W.,** Bilateral adrenalectomy in prostatic cancer. Clinical features and urinary excretion of 17-ketosteroids and estrogen, *Ann. Surg.,* 122, 1031, 1945.
5. **Maskens, A. P.,** Confirmation of the two-step nature of chemical carcinogenesis in the rat colon adenocarcinoma model, *Cancer Res.,* 41, 1240, 1981.
6. **Ryser, H. J.-P.,** Chemical carcinogenesis, *N. Engl. J. Med.,* 285, 721, 1971.
7. **Terpstra, O. T., Dahl, E. P., Williamson, R. C. N., Ross, J. S., and Malt, R. A.,** Colostomy closure promotes cell proliferation and dimethylhydrazine-induced carcinogenesis in rat distal colon, *Gastroenterology,* 81, 475, 1981.
8. **Lipkin, M. and Deschner, E.,** Early proliferative changes in intestinal cells, *Cancer Res.,* 36, 2665, 1976.
9. **Richards, T. C.,** Early changes in the dynamics of crypt cell populations in mouse colon following administration of 1,2-dimethylhydrazine, *Cancer Res.,* 37, 1680, 1977.
10. **Lipkin, M.,** Phase 1 and phase 2 proliferative lesions of colonic epithelial cells in diseases leading to colonic cancer, *Cancer,* 34, 878, 1974.
11. **Deschner, E. and Maskens, A. P.,** Significance of the labeling index and labeling distribution as kinetic parameters in colorectal mucosa of cancer patients and DMH treated animals, *Cancer,* 50, 1136, 1982.

12. **Lipkin, M., Blattner, W. E., Fraumeni, J. F., Jr., Lynch, H. T., Deschner, E., and Winawer, S.,** Tritiated thymidine (ϕ_p,ϕ_h) labeling distribution as a marker for hereditary predisposition to colon cancer, *Cancer Res.*, 43, 1899, 1983.
13. **Lipkin, M.,** Method for binary classification and risk assessment of individuals with familial polyposis based on [^{3}H]TdR labelling of epithelial cells in colonic crypts, *Cell Tissue Kinet.*, 17, 209, 1984.
14. **Tahara, E. and Haizuka, S.,** Effect of gastro-entero-pancreatic endocrine hormones on the histogenesis of gastric cancer in rats induced by *N*-methyl-*N'*-nitro-*N*-nitrosoguanidine; with special reference to development of scirrhous gastric cancer, *Gann*, 66, 421, 1975.
15. **Tatsuta, M., Ito, H. T., Okuda, S., Taniguchi, H., and Tamuria, H.,** Effect of prolonged administration of gastrin on experimental carcinogenesis in rat stomach induced by *N*-methyl-*N'*-nitro-*N*-nitrosoguanidine, *Cancer Res.*, 37, 1808, 1977.
16. **Kishimoto, S., Kunita, S., Shimizu, S., Koh, H., Yamamoto, M., Kajiyamam, G., and Miyoshi, A.,** Effects of atrophic changes in mucosa and endogenous gastrin (hypergastrinemia) on development of experimental gastric cancer and its growth in rats: a pilot study, *Hiroshima J. Med. Sci.*, 32, 213, 1983.
17. **Tatsuta, M., Yamamura, H., Taniguchi, H., and Tamura, H.,** Gastrin protection against chemically induced gastric adenocarcinomas in Wistar rats: histopathology of the glandular stomach and incidence of gastric adenocarcinoma, *J. Natl. Cancer Inst.*, 69, 59, 1982.
18. **Tahara, E., Shimamoto, F., Taniyama, K., Ito, H., Kosako, Y., and Sumiyoshi, H.,** Enhanced effect of gastrin on rat stomach carcinogenesis induced by *N*-methyl-*N'*-nitro-*N*-nitrosoguanidine, *Cancer Res.*, 42, 1781, 1982.
19. **Tatsuta, M., Itoh, T., Okuda, S., Wada, A., Taniguchi, H., Tamura, H., and Yamamura, H.,** Effects of gastrin and histamine on gastric carcinogenesis induced in rats by *N*-methyl-*N'*-nitro-*N*-nitrosoguanidine, *Eur. J. Cancer*, 16, 631, 1980.
20. **Tatsuta, M., Iishi, H., Yamamura, H., and Taniguchi, H.,** Enhancement by propranolol of the inhibitory effect of tetragastrin on gastric carcinogenesis induced by *N*-methyl-*N'*-nitro-*N*-nitrosoguanidine in Wistar rats, *Cancer Res.*, 47, 111, 1987.
21. **Tatsuta, M., Yamamura, H., Iishi, H., Ichii, M., Noguchi, S., Baba, M., and Taniguchi, H.,** Promotion by vagotomy of gastric carcinogenesis induced by *N*-methyl-*N'*-nitro-*N*-nitrosoguanidine in Wistar rats, *Cancer Res.*, 45, 194, 1985.
22. **Deveney, C. W., Freeman, H., and Way, L. W.,** Experimental gastric carcinogenesis in the rat. Effects of hypergastrinemia and acid secretion, *Am. J. Surg.*, 139, 49, 1980.
23. **Carlsson, E., Larsson, H., Mattsson, H., Ryberg, B., and Sundell, G.,** Pharmacology and toxicology of omeprazole — with special reference to the effects on the gastric mucosa, *Scand. J. Gastroenterol.*, 21(118), 31, 1986.
24. **Sundler F., Carlsson, E., Hakanson, R., Larsson, H., and Mattsson, H.,** Inhibition of gastric acid secretion by omeprazole and ranitidine. Effects on plasma gastrin and gastric histamine, histidine decarboxylase activity and ECL cell density in normal and antrectomized rats, *Scand. J. Gastroenterol.*, 21(118), 39, 1986.
25. **Richards, A. T., Hinder, R. A., and Harrison, A. C.,** Gastric carcinoid tumours associated with hypergastrinaemia and pernicious anaemia — regression of tumours by antrectomy. A case report, *S. Afr. Med. J.*, 72, 51, 1987.
26. **Eckhauser, F. E., Lloyd, R. V., Thompson, N. W., Raper, S. E., and Vinik, A. I.,** Antrectomy for multicentric argyrophyl gastric carcinoids: a preliminary report, *Surgery*, 104, 1046, 1988.
27. **McGregor, D. B., Jones, R. D., Karlin, D. A., and Romsdahl, M. M.,** Trophic effects of gastrin on colorectal neoplasms in the rat, *Ann Surg.*, 195, 219, 1982.
28. **Karlin, D. A., McBath, M., Jones, R. D., Elwyn, K. E., and Romsdahl, M. M.,** Hypergastrinemia and colorectal carcinogenesis in the rat, *Cancer Lett.*, 29, 73, 1985..
29. **Oscarson, J. E. A., Veen, H. F., Ross, J. S, and Malt, R. A.,** Dimethylhydrazine-induced colonic neoplasia: dissociation from endogenous gastrin levels, *Surgery*, 91, 525, 1982.
30. **Elwyn, K. E., Jones, R. D., and Romsdahl, M. M.,** Inhibitory effects of secretin on gastrin-stimulated rat colon neoplasms, *Cancer*, 55, 1186, 1985.
31. **Lamote, J. and Willems, G.,** Stimulating effect of pentagastrin on cancer cell proliferation kinetics in chemically induced colon cancer in rats, *Regul. Pept.*, 20, 1, 1988.
32. **Tatsuta, M., Yamamura, H., Ichii, M., and Taniguchi, H.,** Effect of prolonged administration of gastrin on experimental carcinogenesis in rat colon induced by intrarectal instillation of *N*-methyl-*N'*-nitro-*N*-nitrosoguanidine, *Cancer Res.*, 43, 2258, 1983.
33. **Tatsuta, M., Yamamura, H., Iishi, H., Noguchi, S., Ichii, M., and Taniguchi, H.,** Gastrin has no promoting effect on chemically induced colonic tumors in Wistar rats, *Eur. J. Cancer Clin. Oncol.*, 21, 741, 1985.
34. **Tatsuta, M., Iishi, H., Ichii, M., Noguchi, S., Yamamura, H., and Taniguchi, H.,** Effect of tetragastrin on the colonic mucosa of rats during intrarectal administration of *N*-methyl-*N'*-nitro-*N*-nitrosoguanidine, *Cancer Res.*, 46, 4539, 1986.
35. **Iishi, H., Tatsuta, M., Baba, M., Okuda, S., and Taniguchi, H.,** Enhancement by vasoactive intestinal peptide of experimental carcinogenesis induced by azoxymethane in rat colon, *Cancer Res.*, 47, 4890, 1987.

36. **Tatsuta, M., Yamamura, H., Iishi, H., Ichii, M., Noguchi, S., Baba, M., and Taniguchi, H.,** Effect of a chemically defined diet in liquid form on colon carcinogenesis in rats, *J. Natl. Cancer Inst.*, 75, 911, 1985.
37. **Tatsuta, M., Iishi, H., Yamamura, H., and Taniguchi, H.,** Inhibition by tetragastrin of experimental carcinogenesis in rat colon: effect of wheat bran consumption, *Int. J. Cancer*, 41, 239, 1988.
38. **Li, A. K. C., Schattenkerk, M. E., de Vries, J. E., Ford, W. D. A., and Malt, R. A.,** Submandibular siladenectomy retards dimethylhydrazine-induced colonic carcinogenesis, *Gastroenterology*, 78 (Abstr.), 1207, 1980.
39. **Kingsnorth, A. N., Abu-Khalaf, M., Ross, J. S., and Malt, R. A.,** Potentiation of 1,2-dimethylhydrazine-induced anal carcinoma by epidermal growth factor in mice, *Surgery*, 97, 696, 1985.
40. **Bristol, J. B., Ghatei, M. A., Smith J. H. F., Bloom, S. R., and Williamson, R. C. N.,** Elevated plasma enteroglucagon alone fails to alter distal colonic carcinogenesis in rats, *Gastroenterology*, 92, 617, 1987.
41. **Savage, A. P., Matthews, J. L., Ghatei, M. A., Cooke, T., and Bloom, S. R.,** Enteroglucagon and experimental intestinal carcinogenesis in the rat, *Gut*, 28, 33, 1987.
42. **McGuinness, E. E., Morgan, R. G. H., Levison, D. A., Hopwood, D., and Wormsley, K. G.,** Interaction of azaserine and raw soya flour on the rat pancreas, *Scand. J. Gastroenterol.*, 16, 49, 1981.
43. **Roebuck, B. D., Kaplita, P. V., Edwards, B. R., and Praissman, M.,** Effects of dietary fats and soybean protein on azaserine-induced pancreatic carcinogenesis and plasma cholecystokinin in the rat, *Cancer Res.*, 47, 1333, 1987.
44. **Levison, D. A., Morgan, R. G., Brimacombe, J. S., Hopwood, D., Coghill, G., and Wormsley, K. G.,** Carcinogenic effects of di(2-hydroxypropyl)nitrosamine (DHPN) in male Wistar rats: promotion of pancreatic cancer by a raw soya flour diet, *Scand. J. Gastroenterol.*, 14, 217, 1979.
45. **Lhoste, E. F. and Longnecker, D. S.,** Effect of bombesin and caerulein on early stages of carcinogenesis induced by azaserine in the rat pancreas, *Cancer Res.*, 47, 3273, 1987.
46. **Johnson, F. E., LaRegina, M. C., Martin, S. A., and Bashiti, H. M.,** Cholecystokinin inhibits pancreatic and hepatic carcinogenesis, *Cancer Detection Prevention*, 6, 389, 1983.
47. **Andrén-Sandberg, A., Dawiskiba, S., and Ihse, I.,** Studies of the effect of cerulein administration on experimental pancreatic carcinogenesis, *Scand. J. Gastroenterol.*, 19, 122, 1984.
48. **Satake, K., Mukai, R., Kato, Y., and Umeyama, K.,** Effects of cerulein on the normal pancreas and on experimental pancreatic carcinoma in the Syrian golden hamster, *Pancreas*, 1, 246, 1986.
49. **Howatson, A. G. and Carter, D. C.,** Pancreatic carcinogenesis-enhancement by cholecystokinin in the hamster-nitrosamine model, *Br. J. Cancer*, 51, 107, 1985.
50. **Howatson, A. G. and Carter, D. C.,** Pancreatic carcinogenesis: effect of secretin in the hamster-nitrosamine model, *J. Natl. Cancer Inst.*, 78, 101, 1987.
51. **Chester, J. F., Gaissert, H. A., Ross, J. S., and Malt, R. A.,** Pancreatic cancer in the Syrian hamster induced by *N*-nitrosobis(2-oxopropyl)-amine: cocarcinogenic effect of epidermal growth factor, *Cancer Res.*, 46, 2954, 1986.
52. **Paz-Bouza, J. I., Redding, T. W., and Schally, A. V.,** Treatment of nitrosamine-induced pancreatic tumors in hamsters with analogs of somatostatin and luteinizing hormone-releasing hormone, *Proc. Natl. Acad. Sci. U.S.A.*, 84, 1112, 1987.
53. **Takahashi, T., Yamaguchi, T., Bando, K., Tanaka, J., Ogata, B. N., and Kohno, K.,** Effects of gastrointestinal hormones on protein synthesis and growth of gastrointestinal cancers (abstract booklet), *International Cancer Congress*, 97, 1982.
54. **Townsend, C. M., Jr., Franklin, R. B., Watson, L. C., Glass, E. J., and Thompson, J. C.,** Stimulation of pancreatic cancer growth by caerulein and secretin, *Surg. Forum*, 32, 228, 1981.
55. **Winsett, O. E., Townsend, C. M., Jr., Glass, E. J., and Thompson, J. C.,** Gastrin stimulates growth of colon cancer, *Surgery*, 99, 302, 1986.
56. **Sumiyoshi, H., Yasui, W., Ochiai, A., and Tahara, E.,** Effects of gastrin on tumor growth and cyclic nucleotide metabolism in xenotransplantable human gastric and colonic carcinomas in nude mice, *Cancer Res.*, 44, 4276, 1984.
57. **Beauchamp, R. D., Townsend, C. M., Jr., Singh, P., Glass, E. J., and Thompson, J. C.,** Proglumide, a gastrin receptor antagonist, inhibits growth of colon cancer and enhances survival in mice, *Ann. Surg.*, 202, 303, 1985.
58. **Cifarelli, A., Setnikar, I., and Vidal y Plana, R. R.,** Antagonism of proglumide with human gastrin at the receptor site, *Hepatogastroenterology*, 27 (Suppl.), 124, 1980.
59. **Alexander, R. W., Upp, J. R., Jr., Singh, P., Hugh, T., Poston, G. J., Townsend, C. M., Jr., and Thompson, J. C.,** Asperlicin inhibits the growth of a xenografted human pancreatic carcinoma, *Gastroenterology*, 92 (Abstr.), 1293, 1987.
60. **Hudd, C., Euhus, D. M., LaRegina, M. C., Herbold, D. R., Palmer, D. C., and Johnson, F. E.,** Effect of cholecystokinin on human cholangiocarcinoma xenografted into nude mice, *Cancer Res.*, 45, 1372, 1985.
61. **Hugh, T., Alexander, R. W., Upp, J. R., Jr., Poston, G. J., Schauweker, T., Singh, P., Townsend, C. M., Jr., and Thompson, J. C.,** Caerulein inhibits growth of a human cholangiocarcinoma with cholecystokinin receptors, *Gastroenterology*, 92, 1801, 1987

62. **Yasui, W., Sumiyoshi, H., Ochiai, A., and Tahara, E.,** Cholecystokinin inhibition of tumor growth and gastrin-stimulated cyclic adenosine 3′:5′-monophosphate metabolism in human gastric carcinoma in nude mice, *Cancer Res.*, 46, 740, 1986.
63. **Alexander, R. W., Upp, J. R., Jr., Poston, G. J., Townsend, C. M., Jr., Singh, P., and Thompson, J. C.,** Bombesin inhibits growth of human pancreatic adenocarcinoma in nude mice, *Pancreas*, 3: 297, 1988.
64. **Lichtenberger, L., Miller, L. R., Erwin, D. N., and Johnson, L. R.,** Effect of pentagastrin on adult rat duodenal cells in culture, *Gastroenterology*, 65, 242, 1973.
65. **Murakami, H. and Masui, H.,** Hormonal control of human colon carcinoma cell growth in serum-free medium, *Proc. Natl. Acad. Sci. U.S.A.*, 77, 3464, 1982.
66. **Kobori, O., Vuillot, M.-T., and Martin, F.,** Growth responses of rat stomach cancer cells to gastro-entero-pancreatic hormones, *Int. J. Cancer*, 30, 65, 1982.
67. **Sirinek, K. R., Levine, B. A., and Moyer, M. P.,** Pentagastrin stimulates in vitro growth of normal and malignant human colon epithelial cells, *Am. J. Surg.*, 149, 35, 1985.
68. **Moyer, M. P., Aust, J. B., Dixon, P. S., Levine, B. A., and Sirinek, K. R.,** Glucagon enhances growth of cultured human colorectal cancer cells in vitro, *Am. J. Surg.*, 150, 676, 1985.
69. **Ochiai, A., Yasui, W., and Tahara, E.,** Growth-promoting effect of gastrin on human gastric carcinoma cell line TMK-1, *Jpn. J. Cancer Res.*, 76, 1064, 1985.
70. **Kusyk, C. J., McNiel, N. O., and Johnson, L. R.,** Stimulation of growth of a colon cancer cell line by gastrin, *Am. J. Physiol.*, 251, G597, 1986.
71. **Maley, M. A. L., Agrez, M. V., and House, A. K.,** Growth factor-induced proliferative responses of human and DMH-induced rat colorectal tumour cell lines, *Aust. J. Exp. Biol. Med. Sci.*, 64, 445, 1986.
72. **Liebow, C., Hierowski, M., and duSapin, K.,** Hormonal control of pancreatic cancer growth, *Pancreas*, 1, 44, 1986.
73. **Gespach, C., Emami, S., and Rosselin, G.,** Gastric inhibitory peptide (GIP), pancreatic glucagon and vasoactive intestinal peptide (VIP) are cAMP-inducing hormones in the human gastric cancer cell line HGT-1. Homologous desensitization of VIP receptor activity, *Biochem. Biophys. Res. Commun.*, 120, 641, 1984.
74. **Laburthe, M., Rousset, M., Chevalier, G., Boissard, C., Dupont, C., Zweibaum, A., and Rosselin, G.,** Vasoactive intestinal peptide control of cyclic adenosine 3′:5′-monophosphate levels in seven human colorectal adenocarcinoma cell lines in culture, *Cancer Res.*, 40, 2529, 1980.
75. **Estival, A., Mouniclou, P., Trocheris, V., Scemama, J. L., Clemente, F., Hollande, E., and Ribet, A.,** Presence of VIP receptors in human pancreatic adenocarcinoma cell line. Modulation of the cAMP response during cell proliferation, *Biochem. Biophys. Res. Commun.*, 111, 958, 1983.
76. **Emami, S., Gespach, C., Forgue-LaFitte, M.-F., Broer, Y., and Rosselin, G.,** Histamine and VIP interactions with receptor-cyclic AMP systems in the human gastric cancer cell line HGT-1, *Life Sci.*, 33, 415, 1983.
77. **Estival, A., Clemente, F., and Ribet, A.,** Adenocarcinoma of the human exocrine pancreas: presence of secretin and caerulein receptors, *Biochem. Biophys. Res. Commun.*, 102, 1336, 1981.
78. **Singh, P., Rae-Venter, B., Townsend, C. M., Jr., Khalil, T., and Thompson, J. C.,** Gastrin receptors in normal and malignant gastrointestinal mucosa: age-associated changes, *Am. J. Physiol.*, 249, G761, 1985.
79. **Rae-Venter, B., Townsend, C. M., Jr., Thompson, J. C., and Simon, P. M.,** Gastrin receptors in human colon carcinoma, *Gastroenterology*, 80 (Abstr.), 1256, 1981.
80. **Singh, P., Walker, J. P., Townsend, C. M., Jr., and Thompson, J. C.,** Role of gastrin and gastrin receptors on growth of a transplantable mouse colon cancer (MC-26) in Balb/c mice, *Cancer Res.*, 46, 1612, 1986.
81. **Singh, P., Le, S., Beauchamp, R. D., Townsend, C. M., Jr., and Thompson, J. C.,** Inhibition of pentagastrin-stimulated up-regulation of gastrin receptors and growth of mouse colon tumor *in vivo* by proglumide, a gastrin receptor antagonist, *Cancer Res.*, 47, 5000, 1987.
82. **Singh, P., Kumar, S., Townsend, C. M., Jr., and Thompson, J. C.,** Structural requirements for the binding of gastrin to mouse colon cancer cells, *Surg. Forum*, 38, 171, 1987.
83. **Chicoine, L., Guo, Y.-S., Thompson, J. C., Townsend, C. M., Jr., and Singh, P.,** Differences in the molecular size of gastrin receptors (GR) on normal and cancerous gut tissues from mice, *Gastroenterology*, 94, A67, 1988.
84. **Singh, P., Townsend, C. M., Jr., Singh, H., Rubin, N., and Thompson, J. C.,** Gastrin receptors in normal and neoplastic gastrointestinal mucosa, in *Gastrin and Cholecystokinin. Chemistry, Physiology and Pharmacology*, Bali, J.-P. and Martinez, J., Eds., Elsevier Science Publishers, Amsterdam, 1987, 99.
85. **Chicoine, L., Guo, Y.-S., Thompson, J. C., Townsend, C. M., Jr., and Singh, P.,** Regulation of gastrin receptor and gastrin levels by gonadal steroids in vivo in rodents, *Gastroenterology*, 94, A67, 1988.
86. **Weinstock, J. and Baldwin, G. S.,** Binding of $gastrin_{17}$ to human gastric carcinoma cell lines, *Cancer Res.*, 48, 932, 1988.
87. **Upp, J. R., Jr., Singh, P., Townsend, C. M., Jr., and Thompson, J. C.,** Predicting response to endocrine therapy in human pancreatic cancer with cholecystokinin receptors, *Gastroenterology*, 92 (Abstr.), 1677, 1987.
88. **Upp, J. R., Jr., Olson, D., Poston, G. J., Alexander, R. W., Townsend, C. M., Jr., and Thompson, J. C.,** Inhibition of growth of two human pancreatic adenocarcinomas in vivo by somatostatin analog SMS 201-995, *Am. J. Surg.*, 155, 29, 1988.

89. **Yao, C.-Z., Poston, G. J., Townsend, C. M., Jr., and Thompson, J. C.,** Covalent cross linking identification of vasoactive intestinal peptide receptors in human colon and pancreatic cancer and hamster pancreatic cancer, *Gastroenterology,* 92 (Abstr.), 1702, 1987.
90. **Poston, G. J., Yao, C. Z., Upp, J. R., Jr., Alexander, R. W., Townsend, C. M., Jr., and Thompson, J. C.,** Vasoactive intestinal peptide inhibits the growth of hamster pancreatic cancer but not human pancreatic cancer in vivo, *Pancreas,* 3, 439, 1988.
91. **Smith, J. P., Wood, J. G., and Solomon, T. E.,** Elevated gastrin levels in patients with colorectal cancer and adenomatous polyps, *Gastroenterology,* 92 (Abstr.), 1646, 1987.
92. **Upp, J. R., Jr., Singh, P., Townsend, C. M., Jr., and Thompson, J. C.,** The clinical significance of gastrin receptors in human colon cancers, *Gastroenterology,* 94 (Abstr.), A472, 1988.
93. **Pegg, A. E. and McCann, P. P.,** Polyamine metabolism and function, *Am. J. Physiol.,* 243, C212, 1982.
94. **Janne, J., Poso, H., and Raina, A.,** Polyamines in rapid growth and cancer, *Biochim. Biophys. Acta,* 473, 241, 1978.
95. **Luk, G. D. and Baylin, S. B. L.,** Polyamines and intestinal growth — increased polyamine biosynthesis after jejunectomy, *Am. J. Physiol.,* 245, G656, 1983.
96. **Luk, G. D. and Baylin, S. B.,** Ornithine decarboxylase as a biologic marker in familial colonic polyposis, *N. Engl. J. Med.,* 311, 80, 1984.
97. **Upp, J. R., Jr., Saydjari, R., Townsend, C. M., Jr., Singh, P., Barranco, S. C., and Thompson, J. C.,** Polyamine levels and gastrin receptors in colon cancers, *Ann. Surg.,* 207, 0022, 1988.
98. **Kingsnorth, A. N., Lumsden, A. B., and Wallace, H. M.,** Polyamines in colorectal cancer, *Br. J. Surg.,* 71, 791, 1984.
99. **Coffey, R. J., Jr., Shipley, G. D., and Moses, H. L.,** Production of transforming growth factors by human colon cancer lines, *Cancer Res.,* 46, 1164, 1986.
100. **Roberts, A. B., Anzano, M. A., Wakefield, L. M., Roche, N. S., Stern, D. F., and Sporn, M. B.,** Type β transforming growth factor: a bifunctional regulator of cellular growth, *Proc. Natl. Acad. Sci. U.S.A.,* 82, 119, 1985.
101. **Levine, A. E., McRae, L. J., Hamilton, D. A., Brattain, D. E., Yeoman, L. C., and Brattain, M. G.,** Identification of endogenous inhibitory growth factors from a human colon carcinoma cell line, *Cancer Res.,* 45, 2248, 1985.
102. **Tricoli, J. V., Rall, L. B., Karakousis, C. P., Herrera, L., Petrelli, N. J., Bell, G. I., and Shows, T. B.,** Enhanced levels of insulin-like growth factor messenger RNA in human colon carcinomas and liposarcomas, *Cancer Res.,* 46, 6169, 1986.
103. **Bradley, S. J., Garfinkle, G., Walker, E., Salem, R., Chen, L. B., and Steele, G., Jr.,** Increased expression of the epidermal growth factor receptor on human colon carcinoma cells, *Arch. Surg.,* 121, 1242, 1986.
104. **Sakai, K., Mori, S., Kawamoto, T., Taniguchi, H., Kobori, O., Morioka, Y., Kuroki, T., and Kano, K.,** Expression of epidermal growth factor receptors on normal human gastric epithelia and gastric carcinomas, *J. Natl. Cancer Inst.,* 77, 1047, 1986.
105. **Yasui, W., Sumiyoshi, H., Hata, J., Kameda, T., Ochiai, A., Ito, H., and Tahara, E.,** Expression of epidermal growth factor receptor in human gastric and colonic carcinomas, *Cancer Res.,* 48, 137, 1988.
106. **Korc, M. and Magun, B. E.,** Binding and processing of epidermal growth factor in PANC-I human pancreatic carcinoma cells, *Life Sci.,* 36, 1849, 1985.
107. **Korc, M., Meltzer, P., and Trent, J.,** Enhanced expression of epidermal growth factor receptor correlates with alterations of chromosome 7 in human pancreatic cancer, *Proc. Natl. Acad. Sci. U.S.A.,* 83, 5241, 1986.
108. **Smith, J. J., Derynck, R., and Korc, M.,** Production of transforming growth factor α in human pancreatic cancer cells: evidence for a superagonist autocrine cycle, *Proc. Natl. Acad. Sci. U.S.A.,* 84, 7567, 1987.
109. **Mori, K., Ibaragi, S., Kurobe, M., Furukawa, S., and Hayashi, K.,** Production of an hEGF-like immunoreactive factor by human gastric cancer cells depends on differentiational state of the cells, *Biochem. Biophys. Res. Commun.,* 145, 1019, 1987.
110. **Yasui, W., Hata, J., Yokozaki, H., Nakatani, H., Ochiai A., Ito H., and Tahara E.,** Interaction between epidermal growth factor and its receptor in progression of human gastric carcinoma, *Int. J. Cancer,* 41, 211, 1988.
111. **Cuttitta, F., Carney, D. N., Mulshine, J., Moody, T. W., Fedorko, J., Fischler, A., and Minna, J. D.,** Bombesin-like peptides can function as autocrine growth factors in human small-cell lung cancer, *Nature,* 316, 823, 1985.
112. **Rozengurt, E. and Sinnett-Smith, J.,** Bombesin stimulation of DNA synthesis and cell division in cultures of Swiss 3T3 cells, *Proc. Natl. Acad. Sci. U.S.A.,* 80, 2936, 1983.
113. **Willey, J. C., Lechner, J. F., and Harris, C. C.,** Bombesin and the C-terminal tetradecapeptide of gastrin-releasing peptide are growth factors for normal human bronchial epithelial cells, *Exp. Cell Res.,* 153, 245, 1984.
114. **Weber, S., Zuckerman, J. E., Bostwich, D. G., Bensch, K. G., Sikic, B. I., and Raffin, T. A.,** Gastrin releasing peptide is a selective mitogen for small cell lung carcinoma *in vitro*, *J. Clin. Invest.,* 75, 306, 1985.

115. **Carney, D. N., Cuttitta F, Moody, T. W., and Minna, J. D.,** Selective stimulation of small cell lung cancer clonal growth by bombesin and gastrin-releasing peptide, *Cancer Res.*, 47, 821, 1987.
116. **Carney, D. N., Cuttitta, F. C., Gazadar, A. F., and Minna, J. D.,** Autocrine clonogenic factor(s) (CF) are produced by cell lines (CL) of small cell lung cancer (SCLC), *Proc. Am. Soc. Clin. Oncol.*, 2 (Abstr.), 14, 1983.
117. **Zachary, I. and Rozengurt, E.,** High-affinity receptors for peptides of the bombesin family in Swiss 3T3 cells, *Proc. Natl. Acad. Sci. U.S.A.*, 82, 7616, 1985.
118. **Alexander, R. W., Upp, J. R., Jr., Poston, G. J., Gupta, V., Townsend, C. M., Jr., and Thompson, J. C.,** Effects of bombesin on growth of human small cell lung carcinoma in vivo, *Cancer Res.*, 48, 1439, 1988.
119. **Horowitz, K. G., McGuire, W. L., Pearson, O. H., and Segaloff, A.,** Predicting response to endocrine therapy in human breast cancer; a hypothesis, *Science*, 189, 726, 1975.

Index

INDEX

F

G

H

I

O

P

R

S

T

U

V

W